The
History of Bacteriology

UNIVERSITY OF LONDON HEATH CLARK LECTURES, 1936
*delivered at The London School of Hygiene
and Tropical Medicine*

By

WILLIAM BULLOCH, M.D., F.R.S.

*Emeritus Professor of Bacteriology in the
University of London*

DOVER PUBLICATIONS, INC.
NEW YORK

In memory of my former teachers

D. J. HAMILTON, *Aberdeen*

F. V. BIRCH-HIRSCHFELD, *Leipzig*

ÉMILE ROUX, *Paris*

C. J. SALOMONSEN, *Copenhagen*

Published in Canada by General Publishing Company, Ltd., 30 Lesmill Road, Don Mills, Toronto, Ontario.
Published in the United Kingdom by Constable and Company, Ltd., 10 Orange Street, London WC2H 7EG.

This Dover edition, first published in 1979, is an unabridged republication of the work originally published by Oxford University Press in 1938. The present edition is reprinted by special arrangement with Oxford University Press.

International Standard Book Number: 0-486-23761-3
Library of Congress Catalog Card Number: 78-73065

Manufactured in the United States of America
Dover Publications, Inc.
180 Varick Street
New York, N.Y. 10014

PREFACE

IT is a curious fact that, in spite of the vast and growing proportions of almost every branch of bacteriology, so little has been written on the history of the science. Indeed, the only attempt at presenting the historical development of the subject was the book by Friedrich Löffler, published in 1887, and entitled, *Vorlesungen über die geschichtliche Entwickelung der Lehre von den Bacterien . . . Erster Theil bis zum Jahre 1878.* In his preface, Löffler announced that the work would be completed by the publication of the second part at the end of the year 1887. That is more than half a century ago, but the complete book has not yet appeared, and will not appear because its author has been dead nearly a quarter of a century.

In 1930 I wrote the chapter on the 'History of Bacteriology' for the Medical Research Council's *System of Bacteriology*, and this article met with a considerable amount of approval both at home and abroad, and apparently supplied a want. In 1936 I was asked to deliver the Heath Clark Lectures in the University of London, and it was suggested to me that I should deal with the 'Development of Bacteriology'. I delivered the Heath Clark Lectures in the London School of Hygiene and Tropical Medicine in January and February 1937. As part of these lectures traversed the same ground as my article in the *System of Bacteriology*, I applied, through Sir Edward Mellanby, F.R.S., Secretary of the Medical Research Council, for the permission of the Controller of H.M. Stationery Office to utilize the said article as a basis for the present book. This request was very generously granted by H.M. Stationery Office, whose property it was, and permission was also given to utilize the illustrations which had appeared in the *System of Bacteriology*. For this generosity I desire to thank the Controller of H.M. Stationery Office. I have added a good deal to what I already wrote in the *System of Bacteriology*, making the subject more complete and up to date. I have also supplied an extensive bibliography and biographical notices of some of the early workers in bacteriology, a subject which apparently is very little known among bacteriologists. The collection of material for a history of bacteriology is a work of years and is not always easy or possible unless one has access to the largest libraries. Had it been easy it would not lave been left so long undone. I have gone carefully through the literature contained in this book. At the same time I should like to express my thanks to Clifford Dobell, F.R.S., for help and advice generously given over many years, and for the loan of valuable papers and rare books which I could not have obtained otherwise without a great deal of trouble. A word of thanks also remains to the Oxford University Press for their share in the production of this book in its final form.

WILLIAM BULLOCH.

LONDON HOSPITAL, LONDON, E.

June 1938

CONTENTS

LIST OF PLATES

ILLUSTRATIONS

I

ANCIENT DOCTRINES ON THE NATURE OF CONTAGION

I

ANCIENT DOCTRINES ON THE NATURE OF CONTAGION

THE results of investigations of the last century showed that a very large number of diseases of man and animals result from the entry into the body of certain living agents invisible to the naked eye but capable of being seen by the aid of the microscope. These diseases are spoken of as infective, and they occur either sporadically or infecting numbers of individuals in epidemics or epizootics. Many epidemics are of national or international importance, and they may be in the form of world-wide pandemics. Such occurrences have been witnessed at all periods of written history, but it should be realized that an exact diagnosis of many outbreaks incompletely described hundreds or thousands of years ago is not possible to-day. The words λοιμός, λοιμώδης, νόσος, *pestis*, *pestilentia*, were applied by the ancients to any epidemic disease irrespective of its symptoms or nature. Duclaux has truly said that 'even in science homonyms are not synonyms thirty years apart; the same tinsel covers very different small models', and Greenwood has correctly emphasized the same fact that popular medical history 'catches at words or phrases in ancient literature which seem to suggest that the author had an inkling of what we now suppose to be the truth'.

The idea of communication by contact—in other words 'contagion'—is exceedingly old and did not originate in connexion with disease at all. Physical properties such as heat or cold are communicable, and this must have been an observation of primitive man thousands of years ago. When we come to written medical history it has been doubted by many scholars whether the classical writers of Greece and Rome had any clear conception of what we now call 'infection' or 'contagion'. This is the view of writers like C. G. Gruner, K. Sprengel, Choulant, Daremberg, Hirsch, and Haeser, and with special reference to the theory of contagion by Marx, Yeats, Omodei, Francis Adams, Puschmann, and Bernheim. C. F. H. Marx (1796–1877) in his *Origines contagii* (1824) has dealt with this subject exhaustively, and considers that the most ancient traces of the opinion that disease may be communicated by touch is found among the customs of the Egyptians and Jews.

In the Bible there are several references to contagious diseases. Thus the Mosaic regulations in regard to leprosy were minutely described in Leviticus (chaps. xiii and xiv), where an interesting account of the diagnosis of the disease is given. When leprosy is probable the priest 'shall look on him and pronounce him unclean', and he shall be shut up and inspected at the end of seven days. If the disease has not progressed the patient shall be shut up for another seven days and examined again, and 'if the plague be dim and the plague be not spread in the skin the priest shall pronounce him clean, and he shall wash his clothes and be clean' (Lev. xiii. 6). If the disease has spread the patient is pronounced to be leprous, 'and the leper in whom the plague is, his clothes shall be rent and the hair of his head shall go loose, and he shall cover his upper lip and shall cry, Unclean, unclean. All the days wherein the plague is in him he shall be unclean; he is unclean: he shall dwell alone: without the camp shall his dwelling be' (Lev. xiii. 45). The disease was also believed to be spread by the garments which were unclean, and the order was given that the garments should be destroyed in the fire (Lev. xiii. 52). The curious law of the leper is fully described in Leviticus (chap. xiv) and is mixed up with sacrifices, burnt offerings and inunctions. Personal hygiene and the hygiene of the house is also dealt with. When a house was infected the order was that it should 'be scraped within round about, and they shall pour out the mortar that they scrape off without the city into an unclean place' (Lev. xiv. 41). After being reconditioned the house may still be infected, and if this be so the house shall be broken down and all its parts shall be carried into an unclean place, and those who take part in the demolition must wash their clothes. Very strict rules were also laid down in Leviticus (chap. xv) in connexion with discharges from various parts of the bodies of men and women and the necessity of purifying clothing that had become soiled.

The origin of disease was also referred to in the Bible and was closely connected with punishment for rousing the wrath of the Lord. Miriam, the sister of Moses and Aaron, took a prominent part in the exodus, but at Hazeroth, after leaving Sinai, Miriam and Aaron spoke against Moses 'because of the Cushite woman he had married' (Num. xii. 1), and the Lord spake unto all three of them and came down in a pillar of cloud and called Aaron and Miriam out of the tent and showed he was on Moses' side, and the anger of the Lord was kindled against them (Num. xii. 9); and the cloud removed from over the tent, and 'behold, Miriam was leprous, as white as snow'; and Aaron was horrified and said 'Oh my lord, lay not, I pray thee, sin upon us, for that we have done foolishly, and for that we have sinned: let her not, I pray, be as one dead, of whom the flesh is half consumed'; and Moses interceded for Miriam his sister, and the Lord said 'Let her be shut up without the camp seven days, and after that she shall be brought in again'. And apparently she was healed and the people journeyed on from Hazeroth.

Azariah, son and successor of Amaziah, King of Judah, was smitten by the Lord with leprosy (2 Kings xv. 5), but he was not so fortunate as Miriam, for we are told that 'he was a leper unto the day of his death, and dwelt in a several house' (i.e. a lazar house).

These quotations clearly indicate that the Jews had sound beliefs in regard to the possible spread of disease by contact, but their ideas of causation were connected with supernatural things. Much of the idea of contagious transmission of disease was lost in the Greek and Roman period, and most scholars agree that in the Hippocratic treatises there is no certain reference to contagion as we understand it to-day. It is indeed strange, as Marx has pointed out, that apparently lay writers like the historians, philosophers, and even the poets, came to understand the propagation of pestilence by touching the diseased and by fomites before the medical profession adopted this idea.

Thucydides appears to have been one of the first to make any positive allusion to the contagious nature of certain plagues. In his history of the great Athenian pestilence of 430 B.C. he relates that the disease first attacked (ἤψατω) the men of the Piraeus, and it was thought that the Peloponnesians had poisoned the wells. At the siege of Potidaea, Thucydides also described a disease which destroyed the Athenian army. It was believed to have been brought from Athens.

References of a similar kind are to be found in the writings of Dionysius of Halicarnassus, who tells us that in a great epidemic those who wished to alleviate the sufferings of others by touching, or had communication with the sick, caught the same distemper, so that many houses became desolate for want of attendants.

Diodorus Siculus, Dion Cassius of Nicaea, Appian and Livy have also referred to epidemics of contagion, and even poets like Lucretius and Virgil referred to epidemics and epizootics. Of a particular pestilence in the reign of Commodus, Dion Cassius relates that many persons not only in the city but in the whole Roman Empire were killed by wicked wretches, who for a stipulated reward dipped small needles into the pestilent poison and thus communicated the disease to others.

The causes of epidemic diseases continued to be a great mystery in ancient times. For centuries these diseases were regarded as supernatural and mainly as divine judgements to punish the wickedness of mankind. Sin was the work of a spirit or demon who possessed the power of evil over men. The divine punishments were to be avoided by sacrifices and lustrations to appease the anger of incensed heaven. Thus, in the *Iliad*, Apollo was the deity who with his darts inflicted epidemic sickness on the army before Troy, and the disease could only be allayed by the supplication of Chryses. The Israelites also had a theurgical theory of pestilence, for we are told that 'the Lord sent a pestilence upon Israel from the morning even to the time appointed:

and there died of the people from Dan even to Beer-sheba seventy thousand men' (2 Sam. xxiv. 15).

This supernatural theory of disease lasted for many centuries but was gradually displaced by the idea that pestilence is due to natural, especially cosmo-telluric, phenomena such as eclipses, comets, earthquakes, inundations, and particular changes in the air which was believed to be polluted or defiled by 'miasms' ($\mu\iota\alpha\sigma\mu\alpha$, stain). The modifications of the atmosphere as a result of climate or season was a favourite doctrine of Hippocrates. Just as heat and cold, moisture and dryness, succeed each other throughout the year, he believed that the body underwent analogous changes which influence the diseases of the period, and from this was elaborated the doctrine of pathological 'constitutions'—a doctrine held for centuries and not yet extinct. In the Hippocratic treatise 'on winds' air is regarded as the cause of disease, and it is stated that when air is infected with miasms which are inimical to mankind people become ill. In fact, bad air was the main aetiological agent in the Hippocratic pathology. The miasmatic theory, as has been pointed out by Greenwood, differs, however, from the modern idea of *contagium animatum* in that the latter assigns an active role to the successive recipients. This was not included in the medical doctrines of the ancients.

After the fall of Rome and the decline of civilization there was a cessation in the acquisition of knowledge. At the most the writers of successive centuries contented themselves with verbose commentaries on the works of Hippocrates, Galen, and other early medical writers. Even as late as the sixth century of the present era the pandemic of plague which devasted the earth in the reign of Justinian was regarded as caused essentially by the vitiation of the atmosphere engendered from the putrefaction of animal substances. This was supposed to have begun in the neighbourhood of Pelusium between the Serbonian bog and the eastern channel of the Nile.

While the Arabian physicians of the East and West added a good deal to our knowledge of different contagious diseases it cannot be said that they advanced beyond the Hippocratic and Galenical doctrines of aetiology.

During the Middle Ages an unbroken series of great epidemics and epizootics gave much opportunity for observation and reflection, and by degrees the doctrines of infection and contagion began to emerge and again it was a lay writer who emphasized the belief in contact or contagion. This was the great Giovanni Boccaccio (1313–75), who, in his famous *Decameron*, completed in 1358, referred to the plague of 1348 when, as he tells us, there came to Florence the death-dealing pestilence 'which through the operation of the heavenly bodies or of our own iniquitous dealings being sent down upon mankind for our correction by the just wrath of God had some years before appeared in the parts of the East'. As every one knows, the *Decameron* is a series of tales told by seven ladies and three gentlemen who left Florence and betook themselves to a country villa to escape the plague of the town and they whiled away ten days with their stories. Boccaccio gives an account of the swellings in the groin and armpits, and speaks of the plague as a contagion. He tells us that not only to talk with the patients or to deal with them in any way brought about the disease in the healthy. 'To touch their clothes or whatever other object had been used by those who had been ill caused the communication of the disease' (*Decameron*, 1st day).

According to Omodei, among the first to spread the knowledge of contagion in plague were Jacopo da Forli (Jacobus Foroliviensis) (died 1413), and at a later period Alessandro Benedetti (1460–1525), professor at Padua. In his treatise on plague Benedetti not only maintained that the disease may be contracted by touch, but that the morbific principle is imbibed and retained in articles used by the sick. From these observations he inculcated the necessity of purifying the clothes.

Studies on epidemics of small-pox, typhus, measles, the English sweats from 1485 to 1551, the great pandemic of syphilis in the fifteenth and sixteenth centuries, resulted in the advancement of our knowledge of infection and contagion. The merit of placing this knowledge on a surer basis we owe to Girolamo Fracastoro, commonly known by the latinized form of his name as Hieronymus Fracastorius. He was a scholar, poet, and thinker, and was born in Verona about 1478. He died in 1553, being then about 75 years of age. Like other Veronese youths Fracastorius went to the University of Padua, where he met students from all parts of Europe including the famous astronomer Nicholaus Koppernigk (Copernicus), Marcantonio della Torre, professor of anatomy, and Pietro Pomponazzi, astronomer and physician. In 1501 Fracastorius was appointed lecturer on logic at Padua and continued in the post for about six years, when he left as a result of the political disturbances which culminated in the League of Cambrai (1508), between Louis XII of France and the Emperor Maximilian, for the dismemberment of Venice. Pope Julius II and Ferdinand the Catholic were mixed up in the turmoil which ensued. The league included Verona among the cities which were to be the spoil of Maximilian. Bloodshed followed, Padua was sacked and the university closed its doors. Fracastorius betook himself to Pordenone (in the Udine) with other refugees from Padua under the Venetian general Alviano, whom he accompanied to Verona in 1509. Alviano was defeated and Verona was handed over to Maximilian. Fracastorius lived some years in Verona and witnessed a great epidemic of plague there. He then settled with his family in a villa at Incaffi on the shores of Lake Garda, between the river Adige and the lake. Here in his villa he spent a life of study and reflection and wrote his famous Latin poem *Syphilis sive morbus Gallicus* (published at Verona 1530). From time to time he made short stays in Verona when things were quiet politically. In 1528 Giberti, Bishop of Verona, became the patron of Fracastorius, and in 1534 gave him a villa at Malcesine at the upper end of Lake Garda, and here he lived some years. In 1538 he published at Venice his *Homocentrica sive de stellis*. In 1547 he took part in the famous Council of Trent, being appointed medical adviser to the Council, but he left after an outbreak of typhus which caused the Council to transfer its activities to Bologna. On 6 August 1553 Fracastorius had a stroke and died the same evening. The place of his burial is unknown.

The chief work of Fracastorius which concerns us here is a small quarto book of 77 pages of text and was published at Venice in 1546 under the title 'De sympathia et antipathia rerum / liber unus / De contagione et contagiosis / morbis et curatione / libri III.'

This work is rare and is written in Latin in a condensed style not always easy to translate or to interpret. There is an extensive literature upon the book and French, German, and English translations, the last being the most accurate and complete by Wilmer Cave Wright, Ph.D., Professor of Greek at Bryn Mawr College, Philadelphia, and published in 1930.

Fracastorius' work on contagion is in three books of which the first and most important deals with the theory of contagion, the second contains accounts of different contagious diseases, while the third book is concerned with their cure. In the first chapter contagion is. defined as an infection which passes from one individual to another. It is something different from the corruption which occurs in milk or meat. Contagion presents three different types: (1) contagion by contact alone; (2) contagion by fomites; and (3) contagion at a distance. In the second type the contagion leaves a *foyer* which may disseminate the contagion, as e.g. itch, phthisis, area, or elephantiasis. Fracastorius uses in a special sense the word 'fomes', which was used by classical writers to denote 'touchstone' or 'tinder'. He says: 'I call "fomites" clothes, wooden things and other things of that sort which in themselves are not corrupted but are able to preserve the original germs of the contagion and to give rise to its transference to others.' In a third category he includes disease, which may be contracted at a distance, such as pestilent fevers, phthisis, lippitudes (blear eyes), and exanthemata like small-pox. Not all contagions are transmissible at a distance but all are communicable by contact. He thinks (chap. iii) that the infection which appears among fruits, e.g. grape to grape, apple to apple, operates in the same kind of fashion as that produced by contact. Fracastorius tried hard to probe the essence of contagion and spoke of the *seminaria* of disease, which may be translated the 'seeds' or 'germs' of disease, remembering, however, that he knew nothing about germs as we understand them to-day. An example of the kind of language which Fracastorius used may be given here. 'Contagion', he says, 'is a precisely similar putrefaction which passes from one thing to another: its germs (*seminaria*) have great activity, they are made up of a strong and viscous combination, and they have not only a material but also a spiritual antipathy to the animal organism.' It is not clear from this what he really meant and how we are to interpret his views according to our present theory of infection. Some have believed that Fracastorius considered the 'seeds' of disease to be living, but we must remember that he regarded them as analogous to exhalation, and he compared contagion to the exhalation of an onion in causing lachrymation.

Fracastorius (chap. xii) draws attention to the fact that all contagions do not behave alike. Some attack one organ, others another organ. He illustrates this in an admirable description of a contagious disease which appeared in 1514 in cattle. There is no doubt that he had before him a typical outbreak of foot-and-mouth disease and he believed it was caused by the air, for, as he says, 'air is very apt to transmit the contagion because it easily takes the infection up and is necessary for life'. He could not altogether emancipate himself from the views of his predecessors. Thus he says (chap. xiii), while speaking of the signs of contagion: 'premonitory signs occur in the heavens, the air, water or earth, conjunctions of planets, great rains with subsequent drought lead to putrefaction.'

In his second book of *De Contagione* Fracastorius clearly describes the history of many contagions and makes valuable observations. With regard to variola and measles he pointed out that they attack children by preference. Every one is attacked once, but 'it is rare for people who have had these diseases to have them again' (lib. ii, cap. ii). He gives (lib. ii, cap. vi) an excellent, indeed a classical, account of typhus fever ('De febre quam lenticulas vel punticulas aut peticulas vocant'). This chapter is quoted verbatim in Haeser's great *Lehrbuch der Geschichte der Medicin* (1882, Bd. III. 370–3). Indeed Haeser (loc. cit., p. 358) describes Fracastorius as the 'Begründer der wissenschaftlichen Epidemiographie'.

Fracastorius was a believer in the infectivity of consumption (lib. ii, cap. ix) and thought that the virus was very tenacious and persisted in the clothing for as long as two years. He considered that the 'germs' are infective only for the lungs but not for the other organs. 'It is extraordinary', he tells us, 'to see in families up to the fifth and sixth generation all the members die of phthisis at the same age.' Hydrophobia (cap. x) is propagated only by the bite of the rabid dog. The incubation period, which is mostly 30 days, may be as long as 8 months, as occurred in a case which came under his own notice. Fracastorius also gives his mature views on syphilis and its cause (loc. cit., cap. xi and xii). His description of the disease is lucid and accurate, and he noted the occurrence of the disease in children from the milk of infected mothers. He also gave a description of gummata or 'gummositates', as he calls them. He was, however, baffled in trying to find the cause of syphilis, but thought the contagion came from the air which had become putrid. He made the interesting observation that the disease changed its character with time. The third book of *De Contagione* deals with the cure of contagious diseases in eleven chapters, but need not concern us here.

Fracastorius' work on contagion must be regarded as a great, if for a time isolated, landmark in the doctrine of infective diseases. It was the result of the wide and practical study of epidemics of plague, typhus, syphilis, and foot-and-mouth disease occurring in northern Italy in his time, and of a still more intensive contemplation and reasoning on what he had seen and how he could interpret it. During his lifetime and for years afterwards Fracastorius' works were widely read and studied, but by degrees their great importance was lost sight of by the end of the sixteenth ecntury, and much that he knew and taught had to be rediscovered in the eighteenth and nineteenth centuries.

It is interesting to note that exactly a century after the death of Fracastorius the great William Harvey, in a letter dated 1653 to the most excellent and learned John Nardi of Florence, stated that he found it very difficult to conceive how pestilence or leprosy should be communicated to a distance by contagion by a zymotic element contained in woollen or linen things, household furniture, even the walls of a house, cement, rubbish, &c., as we find it stated in the book of Leviticus (chap. xiv). 'How, I ask,' says Harvey, 'can contagion long lurking in such things leave them in fine and after a long lapse of time produce its like in another body? Nor in one or two only but in many without respect of strength, sex, age, temperament or mode of life, and with such violence that the evil can by no art be stayed or mitigated.'

A clear idea of the opinions on the cause of plague may also be gained in the *Loimographia* (1666) of the London apothecary William Boghurst, one of the heroes of the Great Plague of London (1665), one who scorned flight from the doomed city and stayed behind to give his services to the miserable people afflicted with the disease. Although Boghurst was a man of learning, an original observer and an acute critic, and notwithstanding a large experience of plague in the great epidemic, the best explanation he could offer of its cause was that 'plague or pestilence is a most subtle, peculiar, insinuating, venomous, deleterious exhalation arising from the maturation of the ferment of the faeces of the earth extracted in the aire by the heat of the sun and difflated from place to place by the winds and most tymes gradually but sometymes immediately aggressing apt bodyes'—a view not very far removed from the ancients but very different from that urged by Fracastorius in his great *De Contagione* 126 years previously.

II
CONTAGIUM ANIMATUM

II

CONTAGIUM ANIMATUM

THE theoretical speculations of Fracastorius and his immediate successors suggesting the existence of seeds or germs
(*seminaria*) of disease led to the revival of the doctrine of
contagium animatum, a doctrine that was later to receive
an objective basis which came from the discovery of the
microscope at the beginning of the seventeenth century.
By means of this instrument an undreamt-of world of
microscopic creatures was revealed. It has been stated by
more than one author that the first to suggest that there
were creatures of microscopic dimensions was Athanasius
Kircher (1602–80), who wrote a large number of treatises
covering such subjects as optics, acoustics, mathematics,
the earth, the heavens, magnetism, plague, and miracles.
He was born at Geisa (Eisenach) and became a Jesuit.
After studying classical and oriental languages he taught
for a time at Münster, Cologne, Coblenz, and Mainz,
and in 1630 was Professor in Würzburg. During the Thirty
Years War he went to Avignon, and later was in Rome
and Vienna. He died at the age of 78 in 1680. The
works usually cited as important in the history of bacteriology are his *Ars magna lucis et umbrae* (1646), in which
(p. 834) he gives an illustration of a magnifying glass, and
his *Scrutinium physico-medicum pestis* (1658), where he
refers to effluvium as being alive and constituted of very
small imperceptible living bodies. The latter 'since they
are very fine and subtle and very light are driven about not
otherwise than the atoms by the very movement of the air.
But since they have a certain sluggishness and are of glutinous tenacity they find it easy to insinuate themselves into
the innermost fibres of clothes, ropes and linen sheets. Of
a truth whatever is full of pores, like wood, bones, cork,
and even metals, they penetrate by their fineness and found
new hot-beds of contagion.' I quote his words to show how
vaguely he wrote, and it is not certain that his meaning can

be accurately determined in the light of present knowledge. Indeed much of his language is incomprehensible, or at any rate ambiguous. Kircher's biographer in the *Allgemeine Deutsche Biographie* asserts that none of his works constitute a scientific advance, and there is much boasting of his scientific discoveries, rambling mysteries about numbers, and other nonsense. With regard to Kircher's *Scrutinium physico-medicum pestis* Dobell (1932) writes that to him it 'appears as a farrago of nonsensical speculation by a man possessed of neither scientific acumen nor medical instinct'.

During Kircher's lifetime he received ardent support from Christian Lange (1619–62), who studied at Leipzig and after extensive travels settled in that town as professor of physiology, later of anatomy and surgery, and finally of pathology. Lange was a mystic and a firm believer in *pathologia animata*, the doctrine of the production of disease by the entry of minute living agents into the body. After his death a large work of Lange's entitled *Pathologia animata* was published at Frankfort in 1688 and ran to 698 pages. Lange also brought out (1659) an edition of Kircher's *Scrutinium*, and it was in fact largely through this that Kircher's views became widely known.

Of great importance for the advancement of knowledge was the invention of the microscope. The history of this wonderful instrument has been traced exhaustively by the Dutchman Pieter Harting in his work *Het Mikroskoop* (1848–54) in four volumes, and I have drawn on this work for the following statements. The art of making convex or concave lenses is old, but the invention of the microscope took place at the end of the sixteenth or beginning of the seventeenth century. The double or compound microscope was invented before the simple or single instrument which was really the outcome of the experience gained in the making of small lenses for the compound microscope. It is not known with certainty who invented the compound microscope and an extensive literature exists on the subject. The rival claims of Italians and Dutchmen have been carefully dealt with by Harting, who

EMIL VON BEHRING
(1854–1917)

FERDINAND COHN
(1828–1898)

finally came to the conclusion that the credit lay between Cornelius Drebbel of Alkmaar (N. Holland) and the two Janssens, Hans and Zaccharias, of Middelburg. In the opinion of Harting the invention took place in Middelburg (near Flushing) before 1610 and possibly about 1590. For a number of years little heed was paid to the invention of the Janssens, but by degrees the observations of men like Robert Hooke, Marcello Malpighi, and Nehemiah Grew showed that the microscope could be of great practical use. At first the double microscopes were very cumbersome and imperfect, and the grinding of lenses was so difficult that an early substitute was found in melted globules of glass. Hooke's microscope was of this type.

The earliest English work on microscopic observations was Henry Power's *Experimental philosophy in three books; containing new Experiments, microscopical, mercurial, magnetical* (1664). He described a number of magnified objects from the vegetable and animal kingdoms, but gave only one rude figure illustrating a piece of ribbon magnified. In 1665 appeared the famous *Micrographia* of the great inventor and mechanician Robert Hooke (1635–1703), who, among many microscopic illustrations drawn by himself, figured one of a flea in gigantic proportions. Early but very good illustrations of microscopes are to be found in the work of Louis Joblot (1718). The solar microscope invented by Leeuwenhoek was perfected by J. N. Lieberkühn in 1739. The invention of achromatic lenses for the correction of chromatic aberration was made by Chester Moor Hall (1703–71), a barrister of the Inner Temple, in 1733, and independently by the optician John Dollond (1706–61). This constituted a great advance. In 1824 Chevalier of Paris used combinations of lenses, and the principle of the modern microscope was worked out by Joseph Jackson Lister (1786–1869) in 1830. He was a wine merchant in the City of London and is well known as the father of Joseph, Lord Lister. By degrees the microscope passed into its present form, and fundamental advances were made by the mathematician Ernst Abbe (1840–1905) in conjunction with Carl Zeiss in the eighties of last century. To-day the instrument has reached a wonderful state of perfection and is familiar to all.

In the seventeenth century the principal objects studied microscopically were small animals, and to this period belongs the discovery of the itch acarus by August Hauptmann (1657), Michael Ettmüller (1682), and G. C. Bonomo (1687). The real discoverer of an invisible world of microscopic living creatures was the great Dutch microscopist Antony van Leeuwenhoek (1632–1723). Until recently this wonderful man was more or less of an enigma, but thanks to the studies of Clifford Dobell prolonged over twenty years we now have a living picture of this great observer. Dobell's *Antony van Leeuwenhoek and his 'little animals'* (1932) is a beautiful piece of literary research augmented by a wide acquaintance with the subject-matter, and it will no doubt remain as the most authoritative work on Leeuwenhoek. I am greatly indebted to Dobell's work for the notes I here present. Leeuwenhoek was born in Delft in Holland 24 October 1632, and died in Delft 26 August 1723, nearly 91 years of age. His father was a basket-maker, and Antony was sent to school near Leyden. In 1648, being then 16, he went to Amsterdam and was placed in a draper's shop to learn the business. In this shop he rose to be book-keeper or cashier, and after some years returned to his native Delft and remained there from 1654 to 1723, a period of 70 years. In Delft he bought a house and shop and set up as a draper and haberdasher. He also held the office of Chamberlain of the Council Chamber of the worshipful Sheriffs of Delft. According to the terms of his appointment, which was made in 1660, his duties were

'to open and to shut the foresaid Chamber at both ordinary and extraordinary assemblies of the foresaid gentlemen . . . to show towards these Gentlemen all respect, honour and reverence . . . and to keep to himself whatever he may overhear in the Chamber; to clean the foresaid chamber properly and to keep it neat and tidy; to lay the fire at such times as it may be required and at his own convenience, and carefully to preserve for his own profit what coals may remain unconsumed'.

His salary began at 314 florins per annum but rose to 400 florins. Dobell considers that Leeuwenhoek's municipal offices were really sinecures and the menial part of them carried out by proxy. He is strongly of opinion that the common idea that Leeuwenhoek was a kind of beadle is without basis. Leeuwenhoek was also a qualified surveyor and had passed a mathematical examination for this post. He was also official wine gauger of the town of Delft, a post in which he had to assay all the wines and spirits entering the town and to calibrate the vessels containing them. From 1660 nothing was heard of Leeuwenhoek outside Delft, but in 1673 he suddenly became known as an amateur scientist offering a paper to the Royal Society for publication in the *Philosophical Transactions*, whose first editor was Henry Oldenburg (?1615–77), Secretary of the Royal Society. Oldenburg carried on a large correspondence with students of science all over Europe. Indeed his correspondence was so extensive as to arouse suspicion and he was committed to the Tower for a couple of months and was visited there by John Evelyn the diarist, who was one of the original fellows of the Royal Society. Oldenburg, to his credit, was one of the few who stayed in London during the Great Plague in 1665, and he was also there during the Great Fire of 1666. One of Oldenburg's correspondents was the famous Reinier de Graaf (1643–73), the discoverer of the Graafian follicle. He was a friend of Leeuwenhoek and practised in Delft. In 1673 he introduced Leeuwenhoek to Oldenburg's notice in a letter in which he says: 'I am writing to tell you that a certain most ingenious person here (i.e. Delft) named Leewenhoeck has devised microscopes which far surpass those which we have hitherto seen manufactured by Eustachio Divini and others.'

Apparently in the period from 1660 to 1673 Leeuwenhoek, the draper of Delft, had in his spare time acquired the art of making lenses and mounting them to form microscopes, and he had begun to turn his lenses on to all manner of objects. De Graaf had himself examined many of Leeuwenhoek's preparations.

The first letter of Leeuwenhoek to the Royal Society

(*Phil. Trans.* (1673), vol. viii, no. 94, p. 6037) dealt with mould, the mouth parts and eye of the bee, and on the louse. From this time for the next fifty years he sent letters to the Royal Society covering an enormous field of observations. He was not much of a scholar of languages and knew only Dutch, but many of his letters were translated into Latin and English and no doubt suffered thereby. All his observations were communicated in letters only, all written by himself in a beautiful hand. In the Library of the Royal Society are still preserved about 200 of these letters. Leeuwenhoek was elected F.R.S. on 29 January 1679–80 (o.s.) and his election was unanimous. He was greatly gratified by this recognition among fellow scientists, and on the receipt of his fellowship diploma wrote to say that he held himself 'most straitly pledged . . . to strive with all my might and main, all my life long, to make myself more worthy of this honour and privilege'. Leeuwenhoek never came to London to sign the register nor did he ever attend a meeting of the Royal Society, although on one occasion before his election he paid a visit to England (1668).

In Delft in his later years Leeuwenhoek became an object of curiosity to visitors, many of whom called to see his microscopes. Among well-known visitors were people of title, including the Tsar Peter the Great, who at that time (1698) was staying in Holland to learn shipbuilding.

Leeuwenhoek led a laborious life of study and lived long past the allotted span. He died apparently of a broncho-pneumonia on 26 August 1723 being nursed by his daughter Maria, whose fine and tender devotion to her old father makes her name revered. He was buried in the old church in Delft on 31 August 1723, and his monument can still be seen there. To the Royal Society he bequeathed a small cabinet, lacquered black, and gilded, containing 26 of his microscopes, 'every one of them ground by myself and mounted in silver and furthermore set in silver . . . that I extracted from the ore . . . and therewithal is writ down what object standeth before each little glass'. The cabinet was duly received by the Royal Society and treasured for a century but it is now lost. Leeuwenhoek's microscopes vanished from the Society's collection after they had been borrowed for inspection.

All the microscopes made by Leeuwenhoek were of the single or simple type. They were in fact magnifying glasses but he himself left no description of them or how he made them, and he had some secret method for the examination of the smallest creatures which he studied. He left behind him 247 finished microscopes and 172 lenses merely mounted between small plates, an amazing total of 419 lenses which he had made with his own hands. After his daughter's death (1745) they were put up for auction and realized the sum of 737 florins and 3 stivers (about £61 10s. in English money).

The microscopes bequeathed by Leeuwenhoek to the Royal Society were examined by Henry Baker (1740) and the greatest magnifier had its focus at $\frac{1}{20}$-inch distance from its centre and magnified 160 linear diameters—200 diameters according to present reckoning (Dobell (1932), p. 320). Each of the 'microscopes' consisted of a single biconvex lens, and in operation the lens was the fixed element while the object examined was movable towards the lens by a system of screws. The construction of Leeuwenhoek's microscopes may be understood from the following description and figure (Fig. 1) taken from Dobell (loc. cit., p. 328). The figure is a composite drawing based on examination of one of the original microscopes. The lens l (Fig. 4) is seen mounted between two thin metal plates (of brass, copper, silver, or gold), and it was held in position by four equidistant rivets in the metal. The object to be examined was held in front of the lens on the point of a short rod, the other end of which screwed into a small block of metal (figured separately in Fig. 2) which was riveted loosely on the smooth end of a long screw acting through an angle-piece (Fig. 3) attached behind the lower end of the plates by a small thumb-screw (Fig. 4). The long screw was to adjust the object in front of the lens in a vertical direction, while the pivoting of the angle-piece (Fig. 3) on its thumb-screw gave lateral motion. The object carrier could be turned on its axis by a small knob (Figs. 1 and 4). For focusing, a small thumb-screw

passed through the stage near one end (Figs. 1, 2, and 4)
and pressed vertically against the plates.

Dobell has stated his belief that Leeuwenhoek could not
have seen the details which we know he did see if he used
his lenses in the ordinary way, and he believes, and has

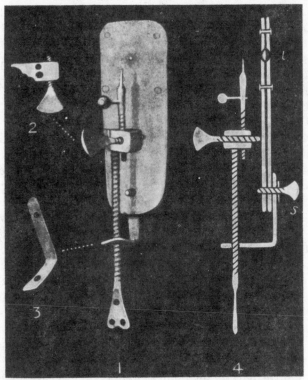

FIG. 1. Leeuwenhoek's microscope. After Dobell (1932).

brought evidence to show, that Leeuwenhoek had found
out some method of dark ground illumination. His micro-
metric estimates were based on comparison with such
things as grains of sand, hairs, human blood corpuscles,
'the diameter of a louse's eye' or 'the bigness of a hair on a
louse'. With such standards he got pretty accurate figures,
as for example the diameter of a human red blood cor-
puscle which he estimated at $\frac{1}{3000}$ inch.

I may now tell the reader something of what Leeuwenhoek found with his microscopes which entitles him to be called 'the father of protozoology and bacteriology'. In his famous 18th letter, dated 9 October 1676, he tells

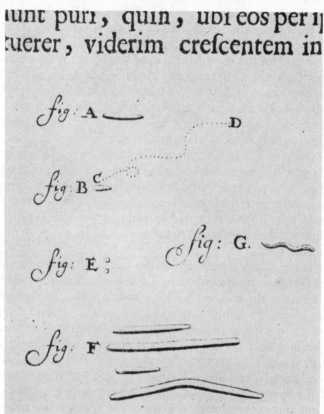

FIG. 2. Bacteria figured by Leeuwenhoek (1695).

us—I quote Dobell's translation of the original Dutch letter in the Royal Society—that in September 1675 he 'discovered living creatures in rain which had stood but a few days in a new tub that was painted blue within'. He described four sorts of animals of which one was undoubtedly Vorticella. The fourth sort 'which I saw a-moving were so small that for my part I can't assign any figure to 'em. These little animals were more than a thousand times less than the eye of a full-grown louse.' Several writers have thought that the animalcules so described were bacteria, but Dobell thinks that they were probably a species of Monas. In the same 18th letter Leeuwenhoek also described animalcules which he found in water from the river Maas, from a well in the courtyard of his house in Delft, and in sea-water which he obtained at Scheveningen when he visited that village in July 1676. Leeuwenhoek also gave a long and detailed account of 'incredibly many very little animalcules of divers sorts' which he discerned in infusions of peppercorns in water. It is not possible to identify all the forms he found in pepper water although the majority were probably protozoa. Some, however, may have been bacteria, especially those which were 'incredibly small, nay so small in my sight that even if 100 of these very wee animals lay stretched out one against another they could not reach to the length of a grain of coarse sand'.

If there is any doubt about the bacterial nature of the smallest creatures described by Leeuwenhoek in 1676 there can be none about those he described in 1683. In his 39th letter, dated 17 September 1683, he gave an unequivocal description of bacteria which he had found on his own teeth. Not only did he discover bacteria but he described and figured all the morphological types known to-day, viz. cocci, bacteria, and spiral forms (Fig. 2). The biggest sort had the shape of Fig. A and had 'a very strong and swift motion and shot through the water (or spittle) like a pike does through the water'. The second sort had the shape of Fig. B. 'These oft-times spun round like a top and every now and then took a course like that shown between C and D.' To the third sort (Fig. E) Leeuwenhoek could assign no figure, 'for at times they seemed to be oblong while anon they looked perfectly round', but 'they went ahead so nimbly and hovered so together that you might imagine them to be a big swarm of gnats or flies flying in and out among one another'. A fourth type 'consisted of a huge number of little streaks some greatly differing from others in their length but of one and the same thickness withal; one being bent crooked, another straight like Fig. F'. They appeared to be motionless.

Leeuwenhoek found the same animals on the teeth of two women and a child about 8 years old. The substance on the teeth of an old man 'who leads a sober life and never drinks brandy or tobacco' showed 'an unbelievably great company of living animalcules a-swimming more nimbly than any I had ever seen up to this time. The biggest sort . . . bent their body into curves in going forwards as in Fig. G.' Leeuwenhoek washed out his own mouth with strong wine vinegar but still found animalcules on his teeth. He estimated that 'there are more animals living in the scum on the teeth in a man's mouth than there are men in a whole kingdom'.

These 'animals' described and figured by Leeuwenhoek are clearly bacteria. The one figured at B, Dobell identifies as *Spirillum sputigenum* (*Selenomonas* Prowazek). F is *Lepothrix buccalis*, and G is a spirochaete. E represents cocci of some sort.

Leeuwenhoek returned to the subject of the animalcules on the human teeth in his 75th letter, dated 16 September 1692. In it he tells us that since 1683 he had not been able to find living creatures on his teeth, and he attributed this to the fact that he drank very hot coffee in the morning. But having carefully cleansed his mouth and taken material from his back teeth with the help of a magnifying mirror, 'I saw with as great wonderment as ever before an unconceivably great number of little animalcules. These animalcules, or most all of them, moved so nimbly among one another that the whole stuff seemed alive and a-moving.' He tried to measure the size of them by comparing them with a grain of coarse sand such as is used for scouring pewter and he found that the grain of sand was 1,000 times longer than the diameter of one of the animalcules, 'consequently then such a grain of sand was far more than a thousand millionfold bigger than one of the little creatures aforesaid'. In addition to the smallest animalcules of rounded form he saw others whose bodies were 'quite 5 or 6 times longer than they were thick and therewithal their body was of equal thickness all along so that I couldn't make out which was their head or which their tail end'. Their motion was slow. Fig. 3 (Fig. A). Others

'moved their bodies in great bends . . . and made with their bendings so swift a motion in swimming first forwards and then backwards

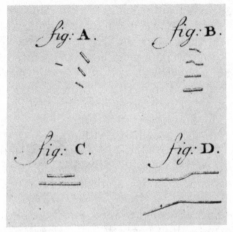

FIG. 3. Bacteria figured by Leeuwenhoek (1695)

and particularly with rolling round on their long axis that I couldn't but behold them again with great wonder and delight the more so because I hadn't been able to find them for several years, as I have already said. For I saw not alone the nimble motion of their own body; but the little animalcules too, which swam in great plenty round about these animalcules, were shoved off or driven away from them just as if you imagined you saw a butterfly or moth flitting among a swarm of gnats so that the gnats were all wafted away by the butterfly's wings'. (Fig. B.)

There were other animalcules (Fig. C)

'that were very near the same thickness but of singular length. These had so little motion that I had most times to confess they might not be living creatures at all, yet when I could keep my eye on them without getting tired I could make out that they bent their body very slow just bending it into a very faint curve so that they didn't move forward or very little'.

Animalcules of the fourth type were of the same thickness (as c) but longer. They were 'in great numbers whereof some were straight while others had a kink in them as shown in Fig. D. But the longer these animals were the less motion or life could I discern in them, yet I made sure they were living creatures or had been such when they were in the mouth and situated on the back teeth where many are generated'.

That these four types were bacteria there can be no possible doubt. Leeuwenhoek frequently expresses the feelings of pleasure he had when studying the minute creatures with his lenses. He sometimes refers to them as beasties (*beesjes*) or little creatures (*cleijne Schepsels*), and he must have spent many hours watching them. His observations are classical, and Dobell (1923) has made a just estimate of his position in the bacteriological world when he wrote that Leeuwenhoek 'was the first bacteriologist and the first protozoologist and he created bacteriology and protozoology out of nothing'.

During his lifetime Leeuwenhoek's observations became widely known and excited wonderment. Some of his discoveries were doubted, but he was so sure of himself that in a letter he wrote to Boerhaave in 1717 he said 'it does not bother me, I know I am in the right'. The distrust of some of his work must be put down in part to the imperfect optical instruments used by others and this applies particularly to the smallest animalcules which to-day can be clearly identified as bacteria. The larger animalcules were naturally much more easily seen and much of the post-Leeuwenhoek microscopy has reference to these. Indeed it was not till long after his death that bacteria were again clearly seen, and when so, all Leeuwenhoek's observations were recognized as accurate.

The first confirmation of the existence of Leeuwenhoek's larger animalcules begins with the work of Louis Joblot (1645-1723). What little is known of Joblot himself will be found in papers by Konarski (1895) and Dobell (1923). He was Professeur royal en mathématiques de l'académie royale de peinture et sculpture in Paris. His work—a quarto published in 1718 and now exceedingly scarce—is in two parts, the first dealing with the construction of microscopes, while the second, running to 96 pages, deals with the animalcules he studied. It is divided into 35 chapters and illustrated with 12 engraved plates. The second edition—the one usually met with—was published in 1754, thirty-one years after Joblot's death. Joblot's work is really the first treatise on protozoology and the animalcules of infusions, but is not of much interest from the strict bacteriological point of view except that it concerns his experiments on spontaneous generation, to be referred to later. During the eighteenth century a large number of other works appeared dealing almost entirely with the animal rather than the bacterial denizens of organic infusions. These have only indirectly a bacteriological interest apart from the origin of animalcules generally. To this group belong the works of Andry (1700), F. C. Lesser (1738), Trembley (1744), John Hill (1752), M. F. Leder-müller (1760–5), Wrisberg (1765), B. Corti (1774), J. C. Eichhorn (1775), von Gleichen (1778), and particularly in 1786 the posthumous work of the great Danish naturalist Otto Friderich Müller (1730–84), to be referred to later. The books of Henry Baker on *The microscope made easy* (1742) and *Employment for the microscope* (1753) did much to popularize microscopy in England while adding some facts on the animalcules of waters and infusions. An admirable analysis of all the early work on infusoria will be found in Bütschli (1887–9).

Early in the eighteenth century, and chiefly as the outcome of the microscopical discoveries of Leeuwenhoek, there was a recrudescence of the old doctrine which attributed certain diseases to the presence of invisible animalcules. The idea can be traced back at least to the time of Columella (*c.* 60 B.C.) and Varro (116–26 B.C.), two of the early writers on husbandry. Thus Columella, writing on the rules which should guide those who contemplate living in the country, says that

'neither a marsh nor a public highway should be near your buildings, for the former always throws up noxious and poisonous steams during the heats and breeds animals armed with poisonous stings which fly upon us in exceeding thick swarms, as also sends forth from the mud and fermented dirt envenomed pests of water snakes or serpents deprived of the moisture they enjoyed in winter, whereby hidden diseases are often contracted the causes of which even the physicians themselves cannot properly understand'.

Columella was not devoid of humour, for the reason he gives for not building a house near a highway was because persons might steal your effects and you might have to entertain them. Writing on the site on which a villa should be placed, Marcus Terentius Varro—a friend of Cicero—says you should be careful in finding out whether there are marshy lands near by, for if they become dry certain animalcules which the eye cannot discern get into the body by the mouth and nose and propagate obstinate diseases ('animalia quaedam minuta quae non possunt oculi consequi et per aëra intus in corpus per os ac nares perveniunt atque efficiunt difficilis morbos').

I have already referred to the speculations of Athanasius Kircher and Christian Lange on animate contagion, and I might add the name of C. F. Paullini (1685), who in his *Cynographia curiosa . . .*, an elaborate treatise on the dog, dealt also with contagion and miasms. On page 196 of this work he expressed himself thus: 'Omne contagion presupponit putredinem.'

In 1726 appeared in Paris an extraordinary book entitled *Systéme d'un medicin anglois sur la cause de toutes les especes de maladies . . . .* The author was stated to be a certain 'M.A.C.D.' (of which 'M' stands for monsieur). Barbier's *Dict. des ouvrages anonymes* states that the author's name was Boyle and in this respect the elision of the letter B between the A and C may not be without significance. The brochure of 34 pages of text and a table of the names of 92 of the animalcules that cause disease is illustrated by 87 minute figures of the so-called animalcules (Fig. 4). They resemble fish, shrimps, caterpillars, and animals of that kind, and most of them look as if some wet ink marks had been folded while still wet, as is done in a common amusement of children. Among the names given to the animalcules I might mention 'Apopleptiques', 'Assoupissans', 'Bubon-istes', 'Canceriques', 'Chancrifiques', 'Coursdeventristes', 'Dentaires', 'Erectifs', 'Gangrenistes', 'Gonorrhiques', 'Hepatiquans', 'Migrainistes', 'Quartains', 'Retentifs', 'Sciatiquans', 'Vapeuristes', 'Vertigiens', 'Veroliques', and so on.

In the *System of Bacteriology*, 1930, vol. i, p. 21, I took the view that the work of M.A.C.D. was a skit or lampoon on animalcules, but Dobell (1932, p. 375) has drawn attention to the existence of a letter of Vallisneri (1733) on the subject and dated Padova 15 February 1729. In this letter Vallisneri definitely asserts that M.A.C.D. was nothing but an impostor and a charlatan—an Englishman by the name of Boil. He had a trick microscope shaped like the letter M and each leg had a mirror to reflect the image. He placed a drop of urine or blood at one of the ends of the microscope and purported to demonstrate animalcules in these fluids, but by a cunning movement of the hand he contrived to show infusoria which he had previously placed in the other leg of the microscope. His fake was exposed and the swindler changed his residence without giving his new address. The object of the swindle was to sell a simple decoction which he said would kill the disease germs which he pretended to show under the microscope.

Charles Singer (1911) has drawn attention to a neglected and remarkable work of this period by one Dr. Benjamin Marten entitled *A new theory of consumptions: more especially of a phthisis or consumption of the lungs*, London, 1720; second edition 1722; pp. 154. Concerning Marten nothing is known except that he lived in Theobald's Row, London, and it has been conjectured by Singer that the work was intended for lay rather than for professional readers. Be this as it may, Marten came so near the doctrine of infective processes prevailing at the present day that his work may be said to have been a couple of centuries ahead of its time. As perhaps might have been anticipated, it left no permanent mark. The author is not mentioned by any contemporary and his book is a rarity. The most interesting part is the second chapter, 'An enquiry concerning the

Sur la cauſe des Maladies. **13**

figurez ainſi,

Si elle eſt d'une perſonne qui ait quelque *playe ulcerée*, ou qui en ſoit menacée, vous y en verrez de configurez ainſi,

Si elle d'une perſonne qui ait un *Panaris*, ou qui en ſoit menacée, vous y en verrez de configurez ainſi,

Si elle eſt d'une perſonne qui ſoit menacée d'*Apoplexie*, vous y en verrez de configurez ainſi,

Si elle eſt d'une perſonne qui ait des *Vertiges*, ou qui en ſoit menacée, vous y en verrez de configurez ainſi,

Si elle d'une perſonne qui ait le *Mal caduc*, ou qui en ſoit menacée, vous y en verrez de configurez ainſi,

Si elle eſt d'une perſonne *Folle*, ou qui ſoit me-
B iij

FIG. 4. Reproduced from M. A. C. D.'s book (1726).

prime essential and hitherto accounted inexplicable cause
of consumptions' (pp. 31–74). In this chapter Marten
thinks that van Helmont's ferments, Sylvius' salt acrimony,
Willis's sourness of the juices, Dolaeus' sharp volatile
particles, Morton's ill-natured or peculiar quality of the
humours of tubercle and such-like views leave us in the
dark as to the true and original essence of consumption.
These can only be 'secondary causes that accidentally aid
and promote some other peculiar, latent and essential
cause'. The determination of the essential cause doubtless
will be difficult but may be aided by a consideration of
microscopical observations. In Marten's opinion the
original cause of consumption

'may possibly be some certain species of animalcula or wonderfully
minute living creatures that by their peculiar shape or disagreeable
parts are inimicable to our nature but however capable of existing in
our juices and vessels and which being drove to the lungs by the
circulation of the blood or else generated there from their proper ova
or eggs with which the juices may abound or which possibly being
carried about by the air may be immediately conveyed to the lungs
by that we draw in and being there deposited as in a proper *nidus* or
nest and being produced into life coming to perfection or increasing
in bigness may by their spontaneous motion and injurious parts
stimulating and perhaps wounding or gnawing the tender vessels of
the lungs cause all the disorders mentioned viz. a more than ordinary
afflux of humours upon the part, obstruction, inflammation, exul-
ceration and all the other phenomena and deplorable symptoms of
this disease' (loc. cit., pp. 40–1).

Marten thinks that this view may seem strange to many
people, especially those who have no idea of any living
creatures 'besides what are conspicuous to the bare eye',
but to those who realized that there was an invisible world
of microscopic creatures the theory suggested might commend itself especially if one considered that there may be
myriads of other animals 'infinitely smaller and wholly
imperceptible to our eye though assisted by the best glasses
that can be made' (loc. cit., p. 43).

As there is in Marten's opinion no such thing as equivocal
generation every such minute living creature must be pro-

duced from an *ovum* or egg which must be much smaller than the adult creatures and 'consequently so light as to be capable of being carried to and fro in the air and so may be suck'd in with our breath or be lodg'd in our external pores'. Proceeding with his argument, Marten says that 'these minute creatures may deposit their ova or eggs in the fruits of the earth, and in the very food we eat'. He even supposed it possible that

'as the fluids of our parents might abound with many species of animalcula so the ova or eggs of them may have been communicated to us in our fluids with the nourishment we received through the umbilical vessels even whilst we were in the womb and may possibly lye latent in us for a longer or shorter term of years till either our vessels are become of a fit largeness to afford 'em a proper nidus or nest, or our own juices of such a property as is fit to nourish or produce them into life' (loc. cit., p. 46).

After entering the body by some route Marten supposed that the animalcules might find a nidus and pass into a state of activity and then 'by means of their spontaneous motion, ill shape or disagreeable parts may destroy the texture of our blood and juices or may stimulate and perhaps wound or gnaw the tender vessels in which they received life or into which by the circulation of the blood they may have been driven' (loc. cit., p. 48). This is an early suggestion of the process of metastasis which, considering the date of Marten's book, is very remarkable. Marten also gave due consideration to the question of specificity among animalcules and the diseases they may produce. 'Thus one species of animalcula by means of their wonderful smallness and injurious parts may instantly affect the brain and nerves and cause apoplexies and sudden death whilst other species may produce the plague, pestilential or malignant fevers, small pox, etc.' Diseases that recur in different seasons or years and maintain their type are best explained on the supposition that there are specific animalcules. Marten also discussed how certain persons can be affected with animalcules and others not so, and he concluded that if the air or food were full of animalcules all persons would get the diseases. It is reasonable to suppose, however, that only those who come into chief contact with the sick are more prone to contract the disease. With regard to consumption he thought it likely that 'by an habitual lying in the same bed with a consumptive patient, constantly eating or drinking with him or by very frequently conversing so nearly as to draw in part of the breath he emits from his lungs a consumption may be caught by a sound person'.

To account for the fact that all people do not get consumption by coming near a consumptive, Marten tells us that 'slight conversing' with a consumptive is seldom or never sufficient to catch the disease,

'there being but few if any of those minute living creatures or their eggs communicated in slender conversation and which if there are may perhaps not be produced into life or be nourished or increased in the new station they happen to be cast besides we may imagine that some persons are of such a happy constitution that if any of the ova of the minute animals that cause consumption happen to get into their bodies they may likewise be quickly forced out again'.

Marten also foreshadowed the belief that animalcules may cause some of the 'forerunning or predisposing distempers' before consumption sets in. Such distempers need not be pulmonic, but by their attack 'the juices of the body may be impoverished and the lungs impaired, that viscus also may give way to the fury of these inimicable and very minute creatures'.

I have referred in such detail to Benjamin Marten's 'New theory of Consumptions' not only because it is little known but because there is nothing in it that is not the accepted teaching on infective diseases in the twentieth century. Marten's book, however, did not alter the face of the science of the day nor indeed did it arouse any interest, and it was soon relegated to oblivion. Marten could never have dreamt that his views would turn out to be so correct. How different were they from those of his day and long after. Thus the great Linnaeus (1707–78), who was not partial to microscopic observations and doubtful of their value, did not believe in the existence of spermatozoa, for he passed the statement in Ramström's thesis (1759) that what had been taken for spermatozoa 'non esse

corporis organis praedita et animata atque adeo, necque
insecta, neque vermes sed particulas motas quarum motus
a calore dependeret liquoris'. In 1757 Nyander, another
of the pupils of Linnaeus, contended that the cheese mite
and the itch mite were identical, and that dysentery, small-
pox, plague, and syphilis were caused by mites.

Linnaeus was unable to classify the animalcules discovered by Leeuwenhoek, Joblot, and others. In his
Systema naturae, as late as the 12th edition in 1767, he
grouped them under 'Vermes' and in a class which he called
'Chaos' with six species, the last of which was *Chaos
infusorium*. At the end of this classification he placed

α Febrium exanthematicarum contagium?
β Febrium exacerbantium causa?
γ Siphilitidis virus humidum?
δ Spermatici vermiculi Leuwenh.?
ε Aethereus nimbus mense florescentiae suspensus?
ζ Fermenti putredinisque septicum Munch.?

Linnaeus regarded those as very small molecules which
lie hidden and they must be left to posterity to fathom. It
is surprising to find Linnaeus, thirty years after Leeuwenhoek's death, placing in the same class spermatozoa and 'the
ethereal clouds suspended in the sky in the month of
blossoming'. That Linnaeus's views were not universally
held even in his time is seen by reference to a work by
Marcus Antonius Plenciz published in Vienna 1762. In
this book, which is often cited as something noteworthy,
I find nothing that was not equally well dealt with by others
long before. Plenciz spoke of seeds of disease and contagion
and was of opinion that they were carried by the air and lie
dormant for a time before giving rise to countless animalcules and other animals like beetles, flies, leeches, and
especially the larvae of gnats. According to Plenciz some
of the animalcules were invisible.

III
FERMENTATION

III
FERMENTATION

THE phenomena of fermentation as seen in the production of bread, wine, and beer must have been observed in prehistoric times and throughout the ages have excited wonderment. It was, however, only in the nineteenth century that the process began to be understood, and the knowledge gained came to play a cardinal role in the founding of bacteriological science. For centuries after the revival of learning, fermentation was a favourite theme for speculation among philosophers, alchemists, and finally chemists. It was not until the appearance of the works of Cagniard-Latour, Schwann and Kützing about 1836 that fermentation was deemed to have a biological aspect.

In the times of the alchemists the word fermentation had a significance almost equivalent to what we would now call a chemical reaction, and was confounded with the processes of effervescence and ebullition, an error which arose from the bubbling and motion seen in the fermenting mass. In his writings, J. B. van Helmont constantly speaks of ferments but not in the sense of substances to excite fermentation. G. A. Spiess (1840) has pointed out that van Helmont used the word 'ferment' in a figurative sense and in general as a dynamic principle. According to van Helmont, ferments are infinite in number and kind. Thomas Willis (1621–75) says that 'fermentation hath its name from fervescency as ferment from ferviment or growing hot'. In 1659 appeared his *De fermentatione sive de motu intestino particularum in quovis corpore*, and from this small work dates the idea that fermentation is an intestine or internal motion of particles. Willis also pointed out the analogy between putrefaction and fermentation. In the eighth chapter of his book he says 'the corruption of every thing is only a separation and departure of themselves from one another into parts, of the principles before combined (the bond of the mixture being loosened), which motion by reason of the diverse disposition of their breaking forth (either with or without a stench) ends in putrefaction or rottenness' (*in putredinem desinit vel marcorum*). Although Willis dealt with the protection, against putridity, afforded by spices, brine, pickle, or sugar, he makes no kind of allusion to anything vital in the process. Very similar views were published by the German chemist and founder of the phlogiston doctrine, Georg Ernst Stahl (1660–1734), in his *Zymotechnia fundamentalis* published in 1697. This exceedingly scarce book is known chiefly through the German translation published in 1748, fourteen years after Stahl's death. Stahl developed the idea that fermentation was essentially an upheaval in the internal composition of bodies, but he also emphasized the point that the upheaval was communicable to other fermentable or putrescible substances. Stahl recognized three kinds of fermentation, viz.: (*a*) the fermentation of wine, beer, and bread; (*b*) acetic fermentation; (*c*) putrefaction and decomposition. The chief ideas of Willis and Stahl appear again and again in subsequent work on fermentation and resemble the theory of Liebig which prevailed in the nineteenth century, as was pointed out by Pasteur (1857), on the basis of the historical researches of Chevreul (1856). In his *Elementa chimiae* (1732) the learned Boerhaave (1668–1738) developed the idea that true fermentation occurred only in vegetable matter, and that although putrefaction is a true intestine motion it never produces acids or inflammable spirits and in this respect differs from fermentation. He describes a 'ferment' as that body 'which when intimately mixed with a fermenting vegetable excites, increases and promotes the fermentation'. In 1787 appeared in Florence a book on fermentation, by Fabbroni, entitled *Dell'arte di fare il vino* and this exerted a considerable influence on the ideas on fermentation. The book was translated into French by F. R. Baud (An X (1801)), and a memoir appeared on Fabbroni's work by Fourcroy in An VII (1799). There has been some mystery about the author Fabbroni in that most writers, including Poggendorff (1863), have referred to him as if he were G. V. M. Fabbroni (1752–1822), who was a well-known Florentine scientist and who, as Fourcroy

(An VII (1799)) tells us, was sent by the Italian Government to Paris to carry out some work on weights and measures. The printed Christian name on Fabbroni's work, a copy of which we consulted in the British Museum, is, however, Adamo and there is no reason for believing that the name of Adamo would have been put on a book written by Giovanni. In fact it is stated by Dryander that they were brothers. Whether this is so or not Fabbroni dealt in the first part of his book with the theory of fermentation of wine, and he believed that the process was really a decomposition of one substance by another just as a carbonate is decomposed by an acid. The substance which in fermentation brings about the decomposition of the sugar was regarded as of vegeto-animal nature and was contained in certain utricles in the grape itself. When the grape is crushed the glutinous matter, according to Fabbroni, acted on the sugar and induced fermentation. He showed that air was not necessary for fermentation to proceed as it occurred, in Fabbroni's experience, in a Toricellian vacuum. Fabbroni did not regard alcohol as a constituent of the grape nor a product of the fermentation, but is formed by the reciprocal action of the two reagents which cause fermentation.

Pasteur (1866) recognized that Fabbroni was the first who regarded the ferment as of the nature of an albumenoid substance, his 'vegeto-animal matter' being gluten or some allied substance, and Pasteur (1866) even went the length of saying that 'Fabroni peut donc être considéré à juste titre le principal promoteur des idées modernes sur la nature du ferment'. Fermentation itself continued to be almost exclusively a study for chemists, and among their notable publications must be placed in a prominent position that of Lavoisier (1789), who studied the question quantitatively in great detail. His work led him to the conclusion that the effect of the vinous fermentation upon sugar was a mere separation of the elements of the latter into two portions. One part was believed to be oxygenated at the expense of the other so as to form carbon dioxide, while the other part, being disoxygenated in favour of the former, is converted into the combustible substance alcohol. If it were possible to reunite alcohol and carbon dioxide we ought in Lavoisier's opinion to be able to reconstruct sugar. The conclusions of Lavoisier were nearly correct, but his data were not, as his analysis of the sugar and alcohol were a good deal out. Notwithstanding, Lavoisier's work on fermentation was the starting-point of many chemical investigations on fermentation. In the year VIII (1803) of the French Republic the Institut National des Sciences et Arts (Mémoires An XI) offered a prize on the question 'Quels sont les caractères qui distinguent dans les matières végétales et animales celles qui servent de ferment, de celles auxquelles elles font subir la fermentation?' i.e. What are the characters which distinguish in vegetable and animal matters those which serve as ferments from those in which they cause or excite fermentation? The prize was to be a gold medal *de la valeur d'un kilogramme* [*sic*]. The prize was again proposed in the year X of the Republic but was withdrawn in the year XII, because unexpectedly the Institut was unable to pay (Cagniard-Latour, 1838).

One of the early workers on fermentation in the nineteenth century was the French chemist Louis Jacques Thénard (1802–3). He observed that during fermentation of gooseberry juice a deposit occurs resembling brewers' yeast. Added to a fresh saccharine liquid this deposit could start fermentation. The ferment was insoluble in water. Thénard was unable to say whether the ferment was formed during fermentation or whether it was soluble at the beginning and was merely deposited from solution during the fermentation. In addition to fermentation of gooseberries he found that a similar process occurred in cherries, pears, apples, barley, and wheat. The eminent chemist L. J. Gay-Lussac (1810) was led to study fermentation on account of experiments he had made with Appert's preserves. Until recently little was known about Appert except that he was 'ancien confiseur et destillateur, élève de la bouche de maison ducale de Christian IV', as stated in his book *L'art de conserver pendant plusieurs années toutes les substances animales et végétales*, 1810. Some writers gave his Christian name as François and stated that he was brother of Benj. N. M. Appert (1797–1850), the well-known French philanthropist. It is now known that he was Nicolas Appert, who was born at Châlons-sur-Marne about 1750, and died in poverty at Massy (Seine-et-Oise) in 1841. In Appert's process the foodstuff was placed in clean bottles, well corked, and subsequently the bottles were raised to the boiling-point of water. In this way the destructible material could be kept unchanged for a long time. Taking some grape juice which had kept perfectly sweet by Appert's method for a year, Gay-Lussac found that when the bottle was opened and the contents poured into another bottle fermentation changes ensued in a few days. A control unopened bottle of Appert's grape juice remained unchanged. Pasteur (1876) (*Études sur la bière*, p. 63) found, however, that if every precaution were taken during decantation no fermentation ensued even after several months. Finding that bottles of Appert's preserves showed no oxygen Gay-Lussac concluded that this gas must play a vital part in the fermentation process. This view seemed to be supported by the results of the following experiment. Gay-Lussac introduced some small and intact grapes into a bell jar standing over mercury. He filled the jar several times with hydrogen gas to displace any oxygen and he then ruptured the grapes by means of an iron rod and watched the effect. For twenty-five days no fermentation had taken place, but it soon occurred when he admitted into the bell jar some bubbles of oxygen. The oxygen introduced was soon proved to have disappeared while carbon dioxide was evolved. From this Gay-Lussac concluded that oxygen was necessary to start fermentation but not for its continuance. Grape juice which had been preserved and poured into a fresh bottle could be re-preserved by subsequent heating. These results obtained with grape juice were also found to apply to preserved meat, fish, and mushrooms. Colin (1825) in a somewhat indefinite paper thought that too much attention had been directed to yeast as a ferment seeing that many animal substances can cause fermentation.

We now come to a period when biological research entered the field of fermentation. Three names stand prominent as the discoverers of the true nature of yeast. These are Baron Charles Cagniard-Latour (1836), Theodor Schwann (1837), and the algologist Friedrich Kützing (1837). They independently stated their opinion that yeast was a living thing. Prior to their publications yeast had been regarded merely as a chemical substance. The earliest microscopic observations on yeast date from Leeuwenhoek, who in his 32nd letter to the Royal Society, dated 14 June 1680, described what he had seen when he subjected beer yeast to his magnifier. His account is somewhat obscure and has often been misrepresented and misinterpreted. I have had the advantage of seeing the original letter and having a literal translation made from the Dutch by Mr. Clifford Dobell. These are Leeuwenhoek's exact words:

'I have made divers observations of the yeast from which beer is made and I have generally seen that it is composed of globules floating in a clear medium (which I judged to be the beer itself). Also I saw very plainly that each globule of the yeast consisted of six distinct globules of exactly the same size and shape as the corpuscles of our blood.'

It is not clear how Leeuwenhoek arrived at the idea that each globule consists of six small globules, and he made some experiments with balls of wax to try to find how the yeast globules were constituted by conglomeration of what he regarded as their constituents. He states definitely that the six globules composing a yeast globule were not sharply defined, and added that

'they looked to me as if the six globules were enclosed in a little vesicle, for when I poured the yeast into clean water (because the beer was too thick and sticky) and let the globules roll along the bottom, the six globules which together form a yeast globule did not come apart from one another. These observations I could make as plainly as if I had before my naked eye a very little and clear bladder which was filled with six other lesser bladders, all being very soft and supple, which I rolled down a sloping board'.

He also tried to make out 'the first beginnings of the air bubbles which rose in vast quantities from the bottom to the top of this beer and set themselves right on the surface, but despite all my endeavours I couldn't make out their origin'. He saw the air bubbles proceeding from particles or where the glass was a bit uneven. Although he demonstrated that yeast is composed of globules Leeuwenhoek made no suggestion that yeast was an organism or that it played any part in the fermentation process. The figures which accompany Leeuwenhoek's letter and which have been regarded by careless readers, e.g. Turpin (1840), as yeast cells are stated specifically by Leeuwenhoek to be wax balls, which he employed for trying to picture to himself the arrangement which he saw of the yeast globules. Leeuwenhoek's observations received no attention from others and were, so to speak, rediscovered after 1836, when the true nature of yeast was unfolded by Cagniard-Latour. It is true that in 1826 Desmazières, a botanist of Lille, gave an account and figured the microscopic characters of the genus *Mycoderma*, which had been created by C. J. Persoon in 1822. Desmazières (1826) differentiated five species of *Mycoderma*, viz. *M. cervisiae* [*sic*], *M. malti-cervisiae*, *M. malti-juniperini*, *M. glutini-farinulae*, and *M. vini*. Desmazières' *M. cervisiae* (loc. cit., p. 314) was found as a membrane growing on beer when it was exposed to the air in open vessels. It was in the form of a whitish pellicle which under the microscope was composed of corpuscles, hyaline gelatinous and ovoid. They were $\frac{1}{120}$ mm. in length and $\frac{1}{200}$ mm. in width. Desmazières considered that they were monads (*animalcula monadina*) which, however, were immobile for long periods, but, becoming mobile, dispose themselves in lines and branches. The figure (loc. cit., Plate VIII, Fig. 1) given of *M. cervisiae* by Desmazières renders it highly probable that he was dealing with yeast cells. He made no suggestion that they played a role in fermentation. He merely described their appearances.

The first illuminating account of the yeast cell was given in 1836 by the French physicist Baron Charles Cagniard-Latour (also written de Latour) (1777–1859). In a séance of the Société philomathique on 18 June 1836 he dealt with the nature of beer yeast and its composition of non-motile globules. He stated his conviction that the globules were organized bodies probably belonging to the vegetable kingdom ('êtres organisés lesquel sont probablement du règne végétal puisqu'on ne leur voit pas exécuter de mouvemens locomotifs'). He estimated the size of the globules as $\frac{1}{150}$ mm., and he assumed that they must have 'seminules' which must be very much smaller than the actual globules. He found similar globules in a number of wines and showed that dried yeast or that which had been cooled to $-5°$ C. in liquid CO_2 was still active. Various communications were made by Cagniard-Latour at meetings of the Société philomathique in 1836 and 1837, and they were briefly reported in L'Institut—a journal devoted to the proceedings of several learned societies of Paris at the period. A short résumé of his work was given by Cagniard-Latour and was also published in the Comptes rendus Acad. des sciences on 12 June 1837. This communication is important, for in it he stated his belief that it was by the vital activity of the yeast cells that carbonic acid and alcohol were formed from the solution of sugar. All Cagniard-Latour's work on yeast was finally summed up in his classical paper in the Annales de chimie et de physique (1838). It would appear that long before his first publication in 1836 he had occupied himself with microscopic studies of yeast, but through imperfect instruments he had been led to believe that it was composed of crystals. With a good microscope, made by Oberhauser, he however discovered the globular nature of yeast. He saw the reproduction of the yeast cell by the process of budding, as a result of which there appeared to be two attached yeast cells, one at first being smaller than the other. Finally he clearly stated that yeast is a mass of globular bodies capable of reproduction and therefore organized. It is not a simple chemical substance, as had been supposed. After 1838 Cagniard-Latour published nothing more on the subject of yeast, his energies being devoted to the study of physics and especially acoustics.

Theodor Schwann (1810–82) independently discovered the yeast cell in 1837. His paper was entitled 'A preliminary communication concerning experiments on fermentation of wine and putrefaction', after a previous communication on Schwann's behalf had been made by his teacher Johannes Müller at the Gesellschaft naturforschenden Freunde zu Berlin in February 1837. A good deal of Schwann's paper is concerned with the doctrine of spontaneous generation and has been dealt with in the chapter devoted to that subject. He examined beer yeast with the microscope and found that it consisted of granules which were often arranged in rows resembling a segmented fungus. Indeed, he regarded yeast as without doubt a plant (*ohne Zweifel eine Pflanze*). This view was also taken by Meyen the mycologist, who examined yeast at Schwann's request. Meyen was only in doubt whether the yeast plant was an alga or a fungus. Schwann failed to find yeast cells in fresh grape juice, but at a temperature of 20° R. they made their appearance in the juice within 36 hours. Schwann watched the yeast cells growing by a process of budding, after which gas development and fermentation ensued. When the process was at an end the yeast had sunk to the bottom. He pointed out differences between the yeasts of beer and wine. The connexion of the yeast plant and the process of fermentation was clearly stated by Schwann. He spoke of the yeast plant as 'Zuckerpilz' (sugar fungus), from which the term 'Saccharomyces' was derived. In addition to yeast, fermentation in a saccharine solution requires a nitrogenous substance. From the sugar solution Schwann pictured the vinous fermentation in this wise. The sugar fungus or yeast draws its food supplies from the saccharine solution and some nitrogenous substance and leaves behind the alcohol and probably other compounds. Although Schwann's paper was of the nature of a preliminary communication he had been able to make a great advance in our conception of the process of fermentation. He promised another paper on the subject but it never appeared. A note was published, however, in his famous *Mikroskopische Untersuchungen* in 1839 (p. 234), in which he supported his view that yeast granules are fungi. The grounds were that their form is that of fungi, they grow like fungi by the shooting forth of

new cells at their extremities. Like fungi they propagate
partly by the separation of distinct cells and partly by the
generation of new cells within those already present and
the bursting of the parent cells. That these fungi are the
cause of fermentation follows in the opinion of Schwann
(1) from the constancy of their occurrence during the process, (2) from the cessation of fermentation under any influence by which they are known to be destroyed, especially
boiling heat, arseniate of potash, &c., (3) because the
principle which excites the process of fermentation must
be a substance which is again generated and increased by
the process itself, a phenomenon met with only in living
organisms. Schwann did not see how any other proof could
possibly be obtained otherwise than by chemical analysis
unless it could be proved that the carbon dioxide and
alcohol are formed only at the surface of the fungi. For this
purpose he took a long test-tube filled with a weak solution
of sugar coloured blue with litmus. A very small quantity
of yeast was added so that fermentation might not begin
until several hours had elapsed, when the fungus had time
to sink to the bottom of the tube. In fact the reddening of
the litmus began at the bottom. If, at the commencement
of the experiment, a glass rod was placed in the tube so that
the fungi might settle on it, reddening began at the bottom
and on the rod—a proof at any rate that an undissolved
substance which is heavier than water gives rise to fermentation.

The third independent worker who discovered the
yeast cell was the algologist Friedrich Traugott Kützing
(1807–93), whose publication also dated from 1837. He
tells us, however, that he had begun his investigations
earlier and indeed communicated his results to Ehrenberg
and Humboldt in 1834. Delay in publication was caused
by a journey to study the marine flora of Dalmatia and
Italy in 1835, and on returning he found he had been fore-
stalled by Cagniard-Latour and Schwann. In his memoir
Kützing gives a clear account of the yeast cell and an
excellent illustration of it (loc. cit., Figs. 1–4, p. 387). He
described the nucleus of the yeast cell but does not mention

budding. He went astray in attributing to the yeast plant a relatively pleomorphic character beginning with the ordinary oval form and finishing up with a full-blown mould. Kützing (loc. cit., p. 390) also gave a description, with figures, of the living organisms in the so-called 'mother of vinegar', the scum on the surface during the formation of vinegar from alcohol. He also described and figured a large number of other microscopic living things in infusions of various plants and in solutions of organic compounds. He was clearly of opinion that yeast was not a chemical substance but a living thing, and in a discussion on the terms organic and inorganic he developed the idea that all fermentation is a vital process. Fermentation, he said, was really a conflict between organic and inorganic processes and it endured till equilibrium was restored. Kützing was perhaps the first to suggest that different fermentations were due to physiologically different organisms.

When one considers the merits of Cagniard-Latour, Schwann, and Kützing there is no doubt that they were independent workers. In point of time the French physicist was ahead of his German competitors, but a much more profound effect on the problem of the cause of alcoholic fermentation was wrought by the work of Schwann, who really established most of the fundamental points, as was admitted by Pasteur in 1878 (Fredericq (1885)). Schwann was the real founder of the germ theory of fermentation. Shortly after the appearance of the papers of Cagniard-Latour, Kützing, and Schwann came the important one of T. A. Quevenne (1838), who confirmed all the findings of these workers. He gave an accurate account of the appearances of the yeast globules and the budding formation. He showed likewise that similar globules occur during the fermentation of carrots, juniper, and gooseberries, and he showed by experiment that it was the solid parts rather than the extractives of the yeast which were the active agents in fermentation. He also studied the action of alcohol, turpentine, acids, alkalies, tannin, and various mineral salts on the activity of yeast. Fermentation was completely inhibited by acetate of copper and corrosive sublimate. According to Quevenne, yeast, during the alcoholization of sugar, undergoes a great change and loses all its nitrogen by conversion into ammonia. Although in general he supported the biological rather than the chemical view of fermentation, Quevenne considered that the former still required further corroboration. He was quite convinced however, that yeast was an organized body of new formation and undergoing vegetation in the saccharine fluid. Turpin (1838), in an elaborate investigation illustrated with excellent plates (1840), also confirmed in the main the results of Cagniard-Latour and Schwann, but was of opinion that cell germs (cytoblasts) contained in the malt or other substances being liberated from the parent structure may develop as separate individuals in a lower grade of vegetable existence. He believed, for example, that the fungus *Penicillium glaucum* may arise from milk or egg albumen—a view which Henle about this time considered to be highly improbable.

The biological theory of fermentation developed by Cagniard-Latour, Schwann, and Kützing did not, however, commend itself to the majority of scientists. The reason is not far to seek. In the first third of the nineteenth century chemistry had made enormous strides, especially under the world-wide influence of the great Swedish scientist J. J. Berzelius (1779–1848). Duclaux (1896) says that chemistry at this time 'did its best to explain everything down to the most mysterious phenomena of life by the simple play of physical and chemical forces, and behold how in a remote corner and one little known to science it sees reappear in the form of an animate cause those living forces which it had expelled little by little from the domain of physiology'. Thus it was in 1828 that F. Wöhler (1800–82), an old pupil of Berzelius, made the discovery that it was possible to prepare from inorganic materials products that it was previously believed could only be formed in the living body. Wöhler showed that urea could be produced by molecular transformation of ammonium cyanate. Justus Liebig (1803–73), another friend of Berzelius, was also imbued with the new chemical progress and throughout his life

strove to impart a purely chemical character to every so-
called vital process. It is little to be wondered then that
the chemical trio of friends, Berzelius, Wöhler, and Liebig,
should have arrayed themselves against what they regarded
as erroneous, the doctrines of Cagniard-Latour, Schwann,
and Kützing. The campaign opened with Berzelius, who, in
his famous *Jahresbericht* for 1839, treated Schwann's work
with opprobrium. He thought that yeast was no more a
living organism than a precipitate of alumina or the endless
chemical substances which are non-crystalline. He stated that
Schwann's conclusions exhibited a frivolity which had long
since been banished from the realms of science. He regarded
Schwann's experiments as worthless as proofs of his views.
Berzelius dismissed Kützing's work contemptuously, adding
that it possibly possessed some value as microscopic research
but none scientifically. Berzelius himself attributed fermentation to the 'catalytic force' doctrine which he had
suggested in 1836. Yeast, in his opinion, acted merely as a
catalyst. In 1839 a vulgar skit on the subject of the organized
nature of yeast appeared anonymously in the *Annalen der
Pharmacie* (1839, xxix. 100) under the title 'The riddle of
vinous fermentation solved'. It is now known from the
published correspondence of Wöhler to Berzelius (1901)
that the authors were none other than Liebig and Wöhler, the
editors of the *Annalen*. In the skit it was stated that the
problem of fermentation had been definitely solved by the
aid of a powerful microscope, made by Pistorius, which exhibited yeast in the form of globules and fine threads of an
albuminoid nature. When the yeast globules were brought
into a saccharine solution they swelled and gave birth to small
animals which reproduced themselves by hitherto unseen
methods and with an extraordinary rapidity. Their form
differed from any of the six hundred types which had been
figured by Ehrenberg, and could only be compared to a
Beinsdorff distilling apparatus (a most complicated piece of
mechanism figured in Liebig's *Handb. d. reinen u. angew.
Chemie*, 1837, p. 711). The animalcules were described by the
authors as possessing a suctorial snout lined by a fine brush
of bristles but devoid of teeth, or eyes. When brought into a

solution of sugar the animalcules devoured it and from the anus a stream of alcohol could be seen issuing and rising to the surface. From the enormously developed genital organs bubbles of carbon dioxide streamed forth. The presence of solanin in the fermentable fluid caused emesis among the animalcules and the passage of fusel oil. Their chemical analysis was said to have yielded remarkable results.

Whatever pleasure it had been to the authors to write their skit it was evident that the new doctrine of Cagniard-Latour and Schwann demanded a more serious answer if it was to be destroyed at its birth, and this was attempted by Liebig in a long paper which appeared in Poggendorff's *Annalen der Physik und Chemie* in the same year, viz. 1839. This is one of the principal papers in which Liebig formulated his views on the nature of fermentation. These views he maintained almost unaltered for thirty years, and at a time when he had reached his zenith. In his paper Liebig (1839) dealt with the allied processes of decomposition, eremacausis, putrefaction, and fermentation. He attributed them to the chemical instability of certain substances which were able to communicate their instability to other substances in succession. Such unstable substances he called ferments. The so-called 'ferment' was assumed by Liebig to arise as a result of change in vegetable saccharine fluids exposed to air and to be able to continue the fermentation even in the absence of air. In the ferment was to be found all the nitrogen of the nitrogenous compounds (gelatine, gluten, &c.) of the vegetable juice. Relying on the observations of Thénard and Colin he regarded 'ferment' as an exceedingly unstable substance. He did not believe that there was an actual substance 'ferment'. What had been called 'ferment' was something in a constant state of change. Neither the more solid nor the more liquid parts of 'ferment' could produce fermentation but the two together could. 'Ferment' was not a substance or material but merely a carrier of an activity which chemists call fermentation or decomposition. Liebig thought, however, that the fermentation of sugar in contact with ferment is essentially

CARL JOSEPH EBERTH
(1835–1926)

PAUL EHRLICH
(1854–1915)

ROBERT KOCH
(1843–1910)
(*the last portrait taken of him*)

different from the fermentation of plant juice or wort. In the former the ferment disappears, whereas in the latter the ferment is deposited. He did not share the opinion of Schwann and others that this deposit consisted of living cells, and although he admitted that under a good microscope the deposit consisted of globules he thought that similar globules could occur in all non-crystalline substances. Fermentation, putrefaction, and like processes were, in the opinion of Liebig, brought about by a chemical instability of a so-called 'ferment' and, once begun, this instability was transmissible to other substances. When the changes were induced, lowly organized animalcules could vegetate. These animalcules were the result, not the cause of the fermentation. Liebig pointed out analogies to fermentation in several inorganic chemical processes. These views of Liebig—which were essentially those held previously by Stahl—were reiterated in his *Organic Chemistry* (1840). In this work he says that yeast is a decomposing body, the molecules of which are in a condition of motion or in the act of losing their equilibrium. If agitated in a vessel filled with a solution of sugar the molecules of yeast communicate their condition to the particles of sugar, the result being the formation of alcohol and carbon dioxide, compounds in which the constituents are retained in combination with a greater force than in sugar. In contact with sugar, yeast disappears and none is reproduced, but when added to the gluten contained in vegetable juices new yeast is formed. Yeast, therefore, is produced from gluten. Hence it follows that a decomposing body added to a mixture, in which its component parts exist, is capable of reproducing itself in that fluid in the same manner as new yeast is formed when yeast is added to liquors containing gluten. Leibig again maintained this position in his *Chemische Briefe* from 1859 onwards. In his earlier years, at any rate, Liebig was very intolerant scientifically and had very fixed ideas. G. J. Mulder (1802–80), a Dutchman, was led to say of him that 'freedom of scientific thought has never been understood by Liebig. For years past a tribunal has been established in Giessen before which Liebig is at the same time accuser, witness, public prosecutor, advocate and judge.'

Liebig's views on fermentation were supported by one of his pupils at Giessen, Schlossberger (1844), who found yeast-like globules in several kinds of flour but never, even in yeast, had he seen any evidence of reproduction by budding. Prior to this the chemist Mitscherlich (1841) made an important contribution to the subject of fermentation. He confirmed the globular nature of yeast and showed that the globules were so large that they would not traverse fine filter-paper. Based on this observation he made an experiment (Fig. 4 A) which showed that actual contact of the yeast particles with the dissolved sugar was necessary for the occurrence of fermentation. He suspended a glass tube, the bottom of which was closed with filter-paper, in a jar containing a solution of sugar, the tube itself being also filled with saccharine solution. Yeast was now added in the tube and it was noted that for some days fermentation occurred in the tube but not in the jar. Mitscherlich never saw fermentation without yeast globules and never in any part except on the surface of

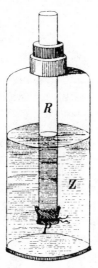

FIG. 4 A. Mitscherlich's experiment (1841) (copied from Ingenkamp, 1886).

the globules. His view was that although yeast was essential it acted not so much through its vital activity as by contact, and in this respect Mitscherlich supported the catalytic theory of Berzelius.

In the opinion of Andral and Gavarret (1843) yeast contained two elements which can be separated by dilution with water. In a few days the yeast cells were found to fall to the bottom forming a grey pulverulent deposit which is an active ferment. At the same time a scum consisting of *Penicillium glaucum*, which was inactive, grew on the surface.

Blondeau (1847), professor of physics in the College royal de Rodez (dep. Aveyron), made a study of fermentations other than the vinous. He examined particularly the lactic, butyric, acetic, urea, and what he called fatty fermentations. He concluded that they were all caused by vegetable growths, and he seems to have been the first who stated that different fermentations are due to different fungi. He studied especially four vegetable growths, viz. *Torula cerevisiae, Penicillium glaucum, P. globulosum,* and *Mycoderma vini,* but his methods did not permit of correct conclusions.

As a result of all these investigations a great deal of knowledge had been acquired on the subject of fermentation. Some idea of the extent of this knowledge may be obtained by the study of the large treatise of Balling published in 1845 in three volumes (3rd edit., 1865, 4 vols.).

These varied researches on fermentation bring us down to the middle of the nineteenth century when great advances were made through the genius and untiring industry of the French chemist Louis Pasteur (1822–95). His name has become a household word, and it is chiefly on his work that much of the fabric of the science of modern bacteriology has been raised. He was not the creator of the doctrine that fermentation is due to living organisms, but the Franklands (1898) have justly stated that he 'was an architect who built only on the solid rock of experiment, and the many mansions which he has raised on the sites where lie the shattered and almost forgotten ruins of his predecessors have invariably withstood the whips and scorns of time, becoming only more mellow with advancing age'. It may be asked why his predecessors, like Cagniard-Latour, Schwann, and Kützing, did not succeed in leaving their impress on the scientific world of their time? The reason may be sought in the remark that 'Pasteur was not only a *savant* content to seek the truth and find it, but that when he had in any matter succeeded in the difficult task of convincing himself he was impelled with almost a fanatic's zeal to force his conviction on the world, nor did he put up his sword until every redoubt of unbelief had been taken, every opponent converted or slain' (Franklands, 1898). He did a vast amount of work, he wrote a great deal, and much has been written about his work and about the man himself. Perhaps the best book about his work is that of his collaborator and successor, Emile Duclaux (1896), entitled *Pasteur: histoire d'un esprit*. Pasteur's collected works are also now in the process of appearing and will ultimately form eight large volumes. Duclaux has emphasized the point that Pasteur's

'scientific mind had an admirable unity; it was the logical and harmonious development of one and the same thought. . . . From the beginning of his studies he had before him a problem of life, and having found the road to it, from that time he always travelled in the same direction, consulting the same compass. Without doubt he has traversed many different countries leaving footprints, but he did not intend to explore them; they were merely along his pathway, and the grandeur of his discoveries makes it possible for the history of his mind, even though reduced to a report, to clothe these adventures with all the air of a romance.'

Pasteur's name will recur frequently in these pages, for there is no department of bacteriology that he did not enrich by his genius. At present, however, I confine myself to his efforts to elucidate the problems of fermentation, a subject with which his name will for ever be associated. Pasteur was a chemist and his earliest scientific researches, memorable even to-day, were entirely of a chemical character. They led him, however, directly into the question of fermentation.

It was in 1848 that Pasteur discovered the true nature of tartaric acid and showed the connexion between right- and left-handedness of crystalline form (enantiomorphism) and optical activity (Pasteur's law). It had long been known that there was a racemic or paratartaric acid which had the same composition as Scheele's tartaric acid, but while the latter rotated the plane of polarization of light, racemic acid did not. Taking up the subject here, Pasteur found that in addition to the dextro-rotatory variety there was also a laevo-rotatory form, and that a combination of the two constituted racemic acid. In the examination of the crystalline form of tartrates Pasteur noted the almost universal presence of hemihedral faces of such a character that the two hemihedral forms, which together make up the holohedral, were 'dissymmetric', i.e. could not be superposed on each other but each could be superposed on the image of the other in a mirror. For this work Pasteur was awarded the Rumford medal of the Royal Society in 1856. It was in 1854 that Pasteur, who had become Dean of the Faculty of Sciences at Lille, began to interest himself in fermentation and especially in connexion with amyl alcohol which shows rotatory powers. It was on 3 August 1857, however, that he made his first communication on lactic fermentation to the Society of Sciences, Agriculture, and Arts of Lille. This was published as a summary in the same year and fully in 1858. Duclaux tells us that Pasteur's probable reasons for working at the lactic fermentation was twofold, viz. much amylic alcohol is produced in it, and the ordinary alcoholic fermentation had already been so extensively studied by the chemists. Tracing the progress of an ordinary lactic fermentation he noticed a greyish substance form as a deposit, or lying on the surface, and that it increased in amount. Under the microscope the deposit revealed itself as composed of minute globules or very short pieces, single or in masses ('petits globules ou d'articles très courts isolés ou en amas'). Taking a trace of the grey deposit and placing it in a solution of sugar and some chalk, Pasteur saw, with great satisfaction, active lactic fermentation taking place. This process could be repeated with the same results, and each time the deposit showed the same characters. He argued that the new 'yeast', as he called it, was organized and living, and that its action on the sugar is correlated with its development and its organization.

In 1859 he showed that in a solution of pure sugar with a small quantity of ammonium phosphate and chalk, a cloudiness appears and gas is evolved. As the fermentation proceeded the ammonia disappeared, phosphates and calcium salts were dissolved, lactate of calcium was formed, and the lactic 'yeast' settled to the bottom. He showed that in all probability the lactic ferment came from the air, for when he used sterile solutions of the various ingredients and allowed only heated air to enter, neither lactic fermentation nor lactic acid nor infusoria appeared. The fluid remained barren.

The essential points of all Pasteur's work on fermentation, and indeed of bacteriology, are epitomized in his short papers on the lactic fermentation. We have here the recognition of the necessity of finding the optimum conditions for the fermentation and from this the minimal requirements on simple substances. There is the examination of the microscopic organisms which develop during the fermentation in the simplified fluids and the demonstration of their constancy of their characters. Finally, it is shown that a microscopic trace of the ferment can produce the characteristic fermentation. For an elaborate account of lactic fermentation see Hueppe (1884).

By methods such as these Pasteur worked through a large number of different fermentations during the next twenty years (1857–77) and covered an enormous field which stamped him as the greatest master in this line of inquiry. To indicate all the things which he found out has filled volumes. I can only indicate here some of his most important discoveries. His earliest papers on alcoholic fermentation date from 1857 and reach their height in the classical 'Mémoire sur la fermentation alcoolique' in 1860. In this he revealed himself as totally opposed to the doctrines of Liebig and a staunch upholder of the view that yeast is a living organism which, in the course of its life, splits sugar not only into alcohol and carbon dioxide but also into other substances. The memoir is complete and might have been written almost to-day.

In 1858 he made an important discovery in connexion with the fermentation of tartaric acid. He observed, if incompletely, the characters of the organism which causes the fermentation in ammonium tartrate, and transferring a trace of the organism to a solution of racemate of ammonia, which of course is optically inactive, he found that the fermenting solution gradually acquired a left rotatory power until finally the right-handed salt disappeared and laevo-tartaric acid was left untouched. The organism which achieved this feat was probably a *Penicillium* (1860).

In 1861 he turned his attention to the butyric fermentation and made another important discovery, viz. that this fermentation proceeds in the absence of oxygen. In the fermented material he found cylindrical rods, which he showed were the cause of the fermentation. Following the nomenclature and ideas of the time he regarded them as animal in character and named them Vibrio. The manner in which the discovery that the vibrios live without oxygen was made is simple. Examining a drop of fluid containing the butyric vibrio under a coverglass on a slide, he was astonished to see on the margin of the drop where it was in contact with air that the vibrios had ceased to move although they were actively motile in the centre. The question at once arose whether the air or oxygen was inimical to their movement and vitality. This view was tested by passing a stream of oxygen through an active butyric fermenting liquid with the result that the fermentation was inhibited. He discovered other organisms that live without air and established the idea of aerobic and anaerobic life and first used these names (1863). Aerobes were only able to live in the presence of free oxygen, whereas anaerobes or zymics, as he called them, grew in the absence of oxygen.

In 1861 he began to publish researches on the acetic fermentation and showed that it is due to organisms of the genus *Mycoderma*. These mycoderms were studied in greater detail in 1862, and his several papers on the production of vinegar appeared in 1864, and especially in 1868, under the title 'Études sur le vinaigre'.

In 1866 he published his famous *Études sur le vin*, a work of 264 pages. In it he dealt with various so-called 'diseases' of wines due to alien organisms which invade the wine and alter its chemical and physical properties. He also laid down the principles whereby these wine diseases could be minimized or abolished.

About 1865 his fermentation researches were interrupted by his studies on the formidable silk-worm disease, which was an industrial calamity in France. This work occupied him six years, and it was only in 1871 that he was able to resume his work on fermentation, this time in connexion with the subject of beer, but it was not till 1876 that he published his *Études sur la bière*, a work of nearly 400 pages. This is not, strictly speaking, a practical work on brewing, but it contains a large number of detailed researches on many points which have become fundamental in the science of bacteriology, such as the absence of germs in the normal fluids of the body, and of fruits, the question of the transformation of one bacterial and fungal species into another, the purification of commercial yeasts, and it is only at the end that there are practical methods for the manufacture of beer. The book is also remarkable for containing Pasteur's mature view on the subject of the nature of fermentative processes in general. In his opinion fermentation was essentially the result of life without oxygen. It was in 1861 that he first suggested this theory in connexion with the butyric ferment, and he recurred to it in 1863 and in 1876. His view as to the essential cause of fermentation met with a great deal of opposition and his old opponent Liebig appeared in 1869 with a slight modification of his old theory of 1839. Brefeld (1874) also challenged the doctrine of the anaerobic life of the yeast cell. There was also a growing idea that the yeast might only ferment sugar through the production of some soluble ferment or enzyme, such as were known to exist, like diastase or pepsin. Such a view of fermentation had been brought forward as early as 1858 by Moritz Traube and sustained in many subsequent papers by him.

In spite of many attempts, including those of Lüdersdorff (1846) and of Pasteur himself, no one succeeded in demonstrating any such enzyme. This would represent the state of affairs about 1880, and in this state knowledge remained till 1897 when Edouard Buchner of Tübingen, but working at the time in Munich, produced 'zymase' from yeast juice and found that it is capable of producing the alcoholic fermentation from certain sugars. A long list of workers might be mentioned who repeated Buchner's work, which was hailed as a great discovery, and as marking the end of the long-continued discussion on fermentation. The result of Buchner's work has, however, been to show that the problem is not so simple as it appeared at first, for factors other than the zymase have been found to be necessary to make the latter perform its operations. In papers from 1904 onwards Harden and Young have shown that zymase action depends on the presence of a second substance termed co-enzyme, which is coctostable, and in addition a phosphate is essential. The modern development of the problem of alcoholic fermentation cannot be followed further here but may be consulted in Harden's (1923) excellent monograph, *Alcoholic Fermentation*, which contains a very complete bibliography. See also Delbrück (1925).

IV
SPONTANEOUS GENERATION AND HETEROGENESIS

SPONTANEOUS GENERATION AND
HETEROGENESIS

ANIMALS not manifestly produced by the act of generation were considered by the ancients to come into the world spontaneously, as the result of the combined action of heat, water, air, and putrefaction. A good deal of this belief in its more concrete form was transmitted to us from Aristotle, who, in his *Historia animalium*, stated that many animals grow spontaneously and not from kindred stock. According to him some animals come from dew falling on leaves, others from decaying mud, dung, or timber. Some were supposed to develop in the fur of animals or in the excrement after it had been voided or even while still within the living body. In the opinion of Aristotle flies come from grubs that develop in the dung that farmers have gathered up in heaps. It has been pointed out by M. J. Berkeley that the true origin of larvae which appear in putrefying carcasses was known to Homer, who in the *Iliad* (xix. 23–7) says:

> Yet fear I for Menoetius' noble Son
> Lest in his spear-inflicted wounds the flies
> May gender worms and desecrate the dead
> And, life extinct, corruption reach his flesh.

Fleas, bugs, and lice were supposed to be produced from moisture and filth, and it was even believed that some fish proceed from mud, sand, or decayed matter. Eels were thought by Aristotle to come from the so-called 'Earth's guts' that grows in mud and humid ground. Varro, a writer of the time of Cicero, revived an ancient belief that bees are born in the decayed flesh of bulls, 'a sample', says Redi, 'of one of those ancient falsehoods of fabulous origin which are subsequently confirmed as truth by other writers and always with some addition'. Columella, Carystius, Ovid, and Virgil believed in the bee story from bull's flesh, as also did later writers like Aldrovandius, Cardano, Moufet, and

Thomas Bartholin. It is also related in Judges (chap. xiv) how Samson, in search of a wife, came to the vineyards of Timnah when a young lion roared against him, but he rent him as he would have rent a kid. After a while he returned and behold, there was a swarm of bees in the body of the lion—and honey.

The German Jesuit Athanasius Kircher, in the seventeenth century, considered that worms, like caterpillars, arise in the dung of oxen and putting on wings change into bees. Even the great chemist and philosopher J. B. van Helmont (1577–1644) believed that mice could be generated spontaneously. In his *Ortus medicinae* (1652) the statement is made, 'si indusium sordidum intra os vasis in quo sit triticum comprimatur: Intra paucos dies (puta 21) fermentum indusio haustum et odore granorum mutatum ipsum triticum, sua pelle incrustatum in mures transmutat', and what to van Helmont was wonderful was that such mice showed their sex and could couple with other mice born of parents' seed. It was not small sucking or abortive mice that leapt forth from the bread-corn and the shirt but mice wholly or fully formed.

Statements like these were discredited by the experiments of Francesco Redi (1626–97) of Arezzo. He was court physician to two grand dukes of Tuscany and was a member of the famous Accademia del Cimento. He was also known, and is still remembered, as a poet and lexicographer. His claim to notice here is as author of the epoch-making *Esperienze intorno alla generazione degl'insetti*, a quarto of 227 pages with 38 full-page engravings and first published in 1668.

Redi could not accept the view that 'worms' are produced in dead animals or plants but thought they were all generated by insemination. The putrefying matter merely served as a suitable nest (*un nido proporzionato*) in which animals deposit their eggs at the breeding season and in which they also find nourishment. To test this view Redi made a very large number of experiments. He placed many kinds of flesh in open boxes and left them to decay. Maggots appeared and he watched them become converted into insects. He also found ova which he considered had been dropped on the flesh by flies. He was confirmed in his belief by finding that before the flesh grew wormy, flies hovered over it of the same kind as those that later bred in it. 'Belief', says Redi, 'would be in vain without the confirmation of experiment.' He therefore put a snake, some fish, some eels, and a slice of milk-fed veal into four wide-mouthed vessels, and having well closed them with paper tied on, he prepared similar vessels in the same way except that they were left open. In the latter series the flesh rapidly teemed with maggots but he could find none in the closed series, although here and there on the paper cover he saw maggots eagerly seeking any crevice through which they could penetrate to obtain food. As the experiments had been made with closed vessels which might have hindered the circulation of air he made other experiments in which the air might freely enter but flies could not. For this purpose he placed meat and fish in large vessels closed with the finest Naples gauze (*sottilissimo velo di Napoli*) and for further protection against flies he placed the vessels in a frame covered with the same gauze. Many flies were found moving about on the frame but no maggots appeared in the meat. On occasions he saw the flies deposit their ova on the net or even in the air before they reached the net.

By such experiments Redi destroyed for ever the myth that maggots are bred spontaneously from meat. In other experiments he traced the development of the ova through maggots to fully formed flies of divers sorts. Redi described himself as the most incredulous man in the world (*incredulo uomo del mondo*) with regard to natural phenomena, and admitted that in bygone days when blinded by inexperience he believed in things 'which I am now ashamed to remember'. He studied worms in all sorts of vegetation and, strange to say, he ultimately concluded that fruits, vegetables, trees, and leaves show grubs in two ways. In the one they come from without and gnaw a path inward, whereas in the other he thought that spontaneous generation was possible and that the soul or principle which creates the flowers and fruits is the same as that which produces grubs in plants. In spite of his accurate work on the development of maggots in flesh Redi was led astray when he came to investigate the grubs in galls. Originally, he believed that oak-galls resulted from the bite of a fly which deposited its ova in the wound, but for various reasons he altered his view and maintained that grubs in trees do not originate from an ovum, and he consoled himself by saying that such a view was 'not a great sin against philosophy'.

Redi's observations on the development of flies from maggots was supported by other investigators. Leeuwenhoek, for example, could not accept the view that the edible mussel comes from sand. He showed in fact that it is derived from spawn. Swammerdam, Vallisneri (the elder), and Réaumur also published observations which cast doubt on the spontaneous generation of insects. In the seventeenth century this doctrine had been widely discussed by philosophers and divines like Sir Thomas Browne (1605–82) and Alexander Ross (1591–1654), but by degrees it was discarded so far as concerns the origin of complicated creatures.

The idea, however, persisted in regard to the vast world of microscopic creatures first revealed in water and in organic infusions by the lenses of Leeuwenhoek from 1675 onwards. Leeuwenhoek himself was an opponent of spontaneous generation and was of the opinion that the source of these microscopic creatures was the air, where, he believed, they existed in the form of seeds or germs.

Early in the field of microscopic studies was Louis Joblot (1645–1723), who in 1718 published his *Descriptions et usages de plusieurs nouveaux microscopes* . . .—a work now very rare and almost forgotten. In it, however, Joblot described certain experiments bearing on spontaneous generation, a doctrine which, he says, is inconceivable and contrary to all reason and religion. Joblot was apparently the first to carry out experiments on heated infusions to see whether they were capable of producing animalcules. His experiments date from 1711 and are described in

Chap. 15 of his book. Hay infusions made with cold water and left covered or open soon teemed with animalcules.

'On the 13th Oct. 1711', he writes, 'I boiled some similar fresh hay in ordinary water for more than one quarter of an hour. Afterwards I put equal quantities of it in two vessels of approximately the same size. One of them I closed as well as I could with parchment well soaked (du velin bien moüillé) and even before it was cooled; the other I left uncovered. In this I found animals at the end of several days but not one in the infusion which had been stoppered. I kept it thus closed for a considerable time in order that I might discover any living insects if they should have appeared in it but having found none of them I finally unclosed it and at the end of several days I then found some in it, and this shows that these animals had developed from eggs dispersed in the air since such as might have happened to be on the hay had been completely destroyed in the boiling water.'

Joblot thought that 'neither alteration, nor corruption, nor a bad odour is the cause of the generation of these animals', and he expressed the view that

'in the air near the surface of the earth there fly or float innumerable quantities of very minute animals of diverse species which settle on the plants that are agreeable to them and there come to rest, take some nourishment and bring forth their little ones while others of them lay their eggs in which new insects are enclosed. Finally, I suppose that these same animals also drop their eggs and their young in the air wherein they disport themselves.'

Despite Joblot's experiments the idea of spontaneous generation of microscopic animalcules continued to be held, and indeed received a renewed notoriety through the writings of John Turberville Needham (1713–81), and Georges Leclerc, Comte de Buffon (1707–88). About 1747 the latter was busy in preparing his monumental *Histoire naturelle*, which began to appear in 1749. He was a superior type of man, wealthy, handsome, and gifted with the power of eloquence. His only bodily defect was myopia, which in a measure unfitted him for fine and continued microscopic observations. As intendant of the Jardin du Roi he had abundant opportunities of studying natural history, and in his *Histoire* developed remarkable opinions on the nature of generation. Excellent accounts of his views have been given by John Barclay (1822) and Flourens (1844). These views had been imbibed in part from the French mathematician, P. L. Moreau de Maupertuis (1698–1759), who became Perpetual First President of the French Academy (1744), and was, at the end, the bitter enemy of Voltaire. Maupertuis published his *Vénus physique* in 1746 and among other questions he dealt with the elective attraction of molecules, a theory akin to the plastic force of the Aristotelians. Buffon developed the idea that vitality was an indestructible property of living things. He considered that all living matter was composed of organic molecules or parts; in themselves indestructible but capable of entering into the most diverse compositions which constituted vitality. He regarded spontaneous generation as a consequence of the theory of organic molecules and scoffed at the idea of pre-existing germs. 'Il n'y a point de germes préexistants,' he wrote. In his admirable and concise history of the works and ideas of Buffon, Flourens (1844) pointed out that Buffon's work on generation implied four hypotheses hanging the one upon the others. These were the idea of accumulated germs, internal moulds, organic molecules, and spontaneous generation. In Buffon's opinion the animalcules which were said to appear in organic infusions and in semen were only the molecules of animal or vegetable matter set free from their former combinations and capable of entering into new unions to constitute what appeared to be living things.

It was about this time that John Turberville Needham (1713–81) began to develop somewhat similar views. He was of Welsh origin and a Roman Catholic, and he became interested in natural science as a result of reading some of the work (1746) on infusion animalcules by the apothecary and quack who called himself 'Sir' John Hill. Between 1746 and 1749 Needham spent much of his time in London and in Paris in microscopic pursuits, the results of which he published in the *Philosophical Transactions of the Royal Society* in 1749. Some of this work was also incorporated in the early volumes of Buffon's *Histoire naturelle*. Needham supported Buffon's hypothesis of 'organical parts' which by coalition constitute the *prima stamina* of all animal and vegetable bodies as well as of every portion of aliment or nutritive juice. Needham says, 'for my own part I was then as I had been before so far of his (Buffon's) opinion as to think that there were compound bodies in Nature not rising above the condition of machines which yet might seem to be alive and spontaneous in their movements. Motion did not necessarily imply life in the common acceptance of the term.' Buffon and Needham collaborated together between 1748 and 1750, but it appears that whereas Buffon did most of the talking, Needham was the one who did the actual work. In the second volume of his *Histoire* Buffon speaks of the conjoint experiments as if they were his own, and Needham later tended to drift away from the influence of the Frenchman and published his own views, which really began where Buffon's ended.

One of Needham's fundamental experiments which he carried out 'at my own lodgings' was to take an almond germ carefully picked out between the two lobes and kernel and to place it with water in a phial closed with a cork. He saw remarkable things taking place. There was a separation or digestion of the parts, a continual flying off of the most volatile which 'offuscated' his glasses. Eight days later he began to perceive a languid motion in some of the seed particles which before seemed dead. 'It was visible', he says, 'that the motion though it had then no one characteristic of spontaneity yet sprung from an effort of something teeming as it were within the particle and not from any fermentation in the liquid or other extraneous cause.' Particles were given off and moved about 'but they did not seem to be enascent animals for the phials had been closed with corks; nay, they were the very seed or the almond germ particles themselves'. Needham tells us that these wonderful things were also seen by Buffon and that the infusions swarmed with clouds of moving atoms of extraordinary minuteness. In order to prove that these minute objects really came from the infusions, and not from without, Needham made what he considered a crucial experiment in which heat was employed.

'I took', he says, 'a quantity of mutton gravy hot from the fire and shut it up in a phial closed with a cork so well masticated that my precautions amounted to as much as if I had sealed my phial hermetically. I thus excluded the exterior air that it might not be said my moving bodies drew their origin from insects or eggs floating in the atmosphere. I neglected no precaution even so far as to heat violently in hot ashes the body of the phial that if anything existed even in that little portion of air which filled up the neck it might be destroyed and lose its productive faculty.'

In some days the phial swarmed with life. Needham examined three or four score of different infusions with like results. It mattered not. 'The phials closed or not closed, the water previously boiled or not boiled, the infusions permitted to teem and then placed on hot ashes to destroy their productions or proceeding in their vegetation without intermission, appeared to be so nearly the same that after a little time I neglected every precaution of this kind as plainly unnecessary.' Needham explained these results on the supposition that they were due to a 'vegetative force in every microscopic point of matter and every visible filament of which the whole animal or vegetable texture consists'. Studying wheat grains pounded up in a mortar and infused with water he found innumerable filaments which swelled from an internal force so active and so productive that 'even before they resolved into or shed any moving globules they were perfect zoophytes teeming with life and self moving'. In Needham's judgement these were the so-called microscopic animalcules which had been observed by naturalists.

Needham's views and experiments published in English (1749) also appeared in French with additions, in book form under the title *Nouvelles observations microscopiques . . .*, 1750, and from their nature and Needham's association with Buffon, then at the height of his fame, aroused the greatest interest among intelligent people. It was not long, however, before the theory of the two scientists was attacked, for in 1751 there appeared a work in two volumes entitled,

Lettres à un Ameriquain sur l'histoire naturelle générale et particulière de M. de Buffon, by an anonymous writer—now known to have been the French metaphysician J. A. Lelarge de Lignac (1710–62). In this work Buffon's views were very adversely criticized, 'avec une indécence et une arrogance si extraordinaire et avec si peu de connoissances' —says Needham in his notes to the Abbé Regley's translation (1769) of Spallanzani's *Saggio*—'que nous avons alors résolus ensemble de ne lui jamais répondre'. Voltaire made many most contemptuous references to Needham, e.g. in the *Histoire de Jenni* he speaks of a 'fou nommé Needham', and in *Des singularités de la nature* (1769) he refers to him as 'un jésuite irlandais nommé Needham qui voyageait dans l'Europe en habit séculier', alluding to the fact that Needham was travelling in Italy and France as tutor to the Earl of Fingal and the Hon. Charles Dillon.

The principal blow to Needham and Buffon came, however, from the great Italian naturalist, the Abbate Lazzaro Spallanzani (1729–99). His work is contained chiefly in two dissertations, the one entitled, *Saggio di osservazioni microscopiche concernenti il sistema della generazione dei Signori di Needham e Buffon*, Modena, 1765, while the other, 'Osservazioni e sperienze intorno agli animalucci delle infusioni . . .' appeared in Spallanzani's famous *Opuscoli di fisica animale e vegetabile*, Modena, 1776. The *Saggio* was translated into French by the Abbé Regley, and to his translation were added copious notes by Needham, who criticized Spallanzani's conclusions. In his *Saggio* Spallanzani dealt with the general theory of Buffon and Needham, and he described in great detail the diverse animalcules which appear in infusions made from seeds or vegetables such as pumpkins, camomile, sorrel, maize, corn, and wheat. He concluded that the 'animalcules' could not possibly be 'organical parts' as conceived by Buffon, but possessed all the attributes of animality. This was shown by the fact that they exhibited orderly motion, had definite shapes, and withstood certain degrees of heat and cold. Spallanzani could not admit that the animalcules developed spontaneously from filaments present in the infused organic material. He found animalcules in ordinary water but none in almond infusions made with well-distilled water. His numerous and well-conceived experiments indicated that in infusions the animalcules observed constant periods in which they appeared, progressed, and disappeared to be replaced by others. Needham had alleged that it was during germination of seeds that the so-called animalcules were produced, but Spallanzani found that infusions made from bruised as well as unbruised seeds gave rise to animalcules, and he also found them in twenty-five different infusions made with seeds before germination. Repeating Needham's observations on heated infusions Spallanzani showed, in hundreds of experiments, that heating prevents the appearance of animalcules in infusions although the duration of heat necessary to render an infusion barren was variable.

From various observations Spallanzani concluded that animalcules may be carried into the infusions by the air and that this was the explanation of their supposed spontaneous generation. Needham had closed his flasks with corks, and Spallanzani, in addition, used wool or paper, but was never satisfied that he had completely excluded the air. To do this effectively he resorted to hermetic sealing and found that the bigger the flask the quicker was the appearance of the animalcules. He made fresh infusions in flasks, removed the air by the pump, and then boiled the infusion. In other experiments he boiled the infusions and while they were still in ebullition he introduced them into larger flasks previously well heated. He finally sealed them hermetically. He observed that after infusions had remained barren for a long time a small crack on the neck of the flask was followed by the development of animalcules in the infusions. In an epilogue to his exhaustive research he concluded that it is not sufficient merely to seal the flasks hermetically. It is not enough merely to boil the infusions. To render an infusion permanently barren or sterile it is necessary, in addition, that the included air in the flask should contain no animalcules. If provision was made against these contingencies, and everything was done

properly, animalcules never appeared unless new air some-
how or other entered the flasks and came into contact with
the infusion.

The beautiful experimental work of Spallanzani, in
which all the pitfalls were clearly exposed, might have
ended the discussion on spontaneous generation had the
precautions which he observed been followed exactly by
others. It is necessary to state, however, that much of the
experimental work on spontaneous generation carried out
more than a century after Spallanzani's time was primitive
compared with that elaborated with so much care and
prescience by the great Italian master of experiment.

Needham's verbose notes in Regley's French translation
of Spallanzani's *Saggio* did not answer the Abbate's objections, and he was particularly ineffective in dealing with
the Italian's results with heated infusions. Needham could
only take refuge in the assertion that prolonged heating had
enfeebled or destroyed the 'vegetative force' and caused
the air to lose its elasticity owing to exhalation from the
water and from the ardour of the fire. He also advised
Spallanzani to repeat his work with this modification that
the infusions should be heated in flasks which were then
to be hermetically sealed. Afterwards the infusions should
be heated for a few minutes. If no animalcules appeared
under these circumstances Needham said, 'j'abandonne
mon système et je renonce à mes idées.'

It was to refute Needham's criticisms that Spallanzani
returned to the subject in his extensive *Opuscoli*, published
eleven years later in 1776. To settle Needham's first
objection that heat had enfeebled or destroyed the 'vegetative force' Spallanzani took infusions of haricot beans,
vetch, buckwheat, barley, indian corn, mallow, beet, and
yolk of egg, and having divided each into four portions he
heated them for $\frac{1}{2}$, 1, $1\frac{1}{2}$, and 2 hours respectively, and
found that animalcules developed in all the flasks which
were merely corked. None of the vessels hermetically
sealed and subjected to the same conditions of heat showed
any evidence of life. Extending these experiments to other
infusions, Spallanzani showed that the seeds from which the

infusions were made could be roasted even in the flame of the blowpipe without losing the power of developing animalcules. With respect to Needham's second objection that heating had diminished the elasticity of the air and rendered it incapable of vivifying the infusions, Spallanzani contracted the necks of flasks until they were of capillary dimensions and then sealed them at once. He then heated portions of such infusions for $\frac{1}{2}$, 1, 1$\frac{1}{2}$, and 2 minutes respectively and compared them with similar infusions in open flasks. While some of the sealed infusions showed no animalcules, others contained a few which were of exceedingly minute size. Spallanzani's experiments in fact revealed to him the existence of two classes of infusoria which had been indiscriminately called 'animalcules'. The one he called the 'superior order' (*ordini superiori*), which were easily destroyed by heat in half a minute at 212° F. The other class (*ultime ordine*) sometimes survived boiling for half an hour. Microscopically they were exceedingly minute compared to the superior class, and although he gave no description of their shape we may suppose that they were really bacterial in character.

In the course of his researches Spallanzani performed an enormous number of experiments on the action of heat on seeds, plants, ova, animals like silk-worms, maggots, vinegar eels, and tadpoles. He showed clearly that hot air is a much less effective means of sterilization than actual contact with hot or boiling water. Many other questions were discussed but cannot be referred to here, but it may suffice to state that he was convinced that the number of animalcules which make their appearance in organic infusions varies with the degree of communication with the air. Spallanzani's work was far ahead of its time, and his conclusions on spontaneous generation were very much the same as those expressed by Pasteur nearly a century later. Spontaneous generation cannot be disproved. All that can be given is a verdict of 'not proven'. Duclaux has put it well, that 'the debate between Needham and Spallanzani was left without a definite conclusion, each of the adversaries showing clearly that the other was wrong on some points, but not proving that he himself was right on all', and thus it was that the question of spontaneous generation continued to be the subject of discussion for many decades.

Milne-Edwards (1863) has correctly pointed out that the word 'spontaneous' as applied to generation is incongruous, and relatively early the word 'heterogenesis' was used as a substitute and to indicate the production of a living thing which was not preceded by a living thing of the same species. But the word heterogenesis, which has been used as synonymous with spontaneous generation or equivocal generation, really connotes three separate ideas, as Milne-Edwards pointed out. There is first the suggestion that a living being can be produced by the spontaneous organization of animal or dead matter without the agency of a previous living thing. This is what Milne-Edwards called 'agenetic' generation, and Charlton Bastian 'archebiosis'. Secondly, living beings may be formed as the result of dissociation of parts of previous living things now dead. This is 'necrogenic' generation, and thirdly, there is the idea that living things may be formed by the physiological action of a living organism which transmits the vital principle without its organic characters. Milne-Edwards proposed to call this 'xenogenic' generation. The word 'heterogenesis' was apparently first used by Breschet (1823) in an article entitled 'Déviation organique', and in the sense of an organic deviation with alien characters in the generation product. Burdach, in his *Physiologie* (1826), employed the words 'Homogenia' and 'Heterogenia' for two principal distinctions in the mode of origin of living things. Homogenia was the process whereby an individual results from a pre-existing living thing similar to itself in organization, whereas Heterogenia was applied to processes whereby living things arise from the matter of pre-existing organisms belonging to a totally different species. Buffon, Needham, and Pouchet were professed 'vitalists' in the sense that they held that a pre-existing 'vital force' of some kind is necessary before a new individual can make its appearance. None of these upheld what Milne-Edwards called agenetic generation. Pouchet, for example, never thought that living things can appear in solutions of mineral ingredients. They only came in organic solutions, the matter of which had been previously vivified, and which in some mysterious way still retained its vital qualities. Bastian (1872) has given the following table (l.c., p. 252) containing his ideas on how living things originate:

Origin of living things.	Archebiosis (primordial origination).		From non-living materials.
	Reproduction (from pre-existing living things).	Heterogenetic.	1. From a portion of the living matter of a pre-existing organism: a. After its death. b. Before its death.
			2. By a molecular metamorphosis of the matter of an entire organism.
			3. By the metamorphosis and fusion of many minute organisms.
		Homogenetic.	1. Indirect. Cases of 'alternate' or cyclical generation.
			2. Direct. Continuous development into the likeness of its parent.

In spite of the experiments of Spallanzani, the doctrine of spontaneous generation of animalcules in organic infusions continued to be a widely held belief and was supported by many of the leading biologists, not only at the end of the eighteenth but far into the nineteenth century. Among the distinguished men who held this belief may be mentioned O. F. Müller, Treviranus, Lamarck, Cabanis, Burdach, Kützing (1837), Dujardin (1841). A very full account of all the work on infusion animalcules will be found in the book of Ehrenberg (1838), who himself was, however, a consistent opponent of the idea of heterogenesis.

By degrees the idea of spontaneous generation of living

things became involved in the problems of fermentation and putrefaction, which were generally regarded as the result of some spontaneous chemical change in the fermentible or putrescible matter. A practical extension of Spallanzani's results was made by Appert, a retired confectioner and distiller of Paris, in the beginning of the nineteenth century. Appert's method consisted in placing the food to be preserved in bottles closed with the best corks obtainable. The bottles were placed in boiling water for considerable periods. Appert claimed that the most perishable foodstuffs could be preserved by this method and his claims were completely substantiated. Gay-Lussac (1810) believed that the keeping qualities of Appert's preserves was due to the expulsion of the air during the heating process. In this way fermentation and putrescence was avoided.

For many years the question of the influence of air on fermentation was the subject of much discussion. In 1836 Franz Schulze, working at the time in Mitscherlich's laboratory in Berlin, made an important experiment to determine whether air is necessary to vivify a heated organic infusion. In order that air might enter, Schulze took a glass flask half filled with watery vegetable infusion and closed with a good cork doubly perforated to receive two bent glass tubes fitted air-tight (Fig. 5). The flask being set upon a sand-bath was boiled, and while steam was issuing freely from the two glass tubes Schulze attached to each an absorption bulb. One bulb contained concentrated sulphuric acid and the other a solution of potassium hydrate. After cooling, the whole apparatus was placed on a window sill, the time being summer. As a control a similar but open flask was placed near by. From 28 May to the beginning of August, Schulze applied his mouth to the open end of the potash bulb several times daily and slowly aspirated air out of the flask which contained the organic infusion. To enter the flask fresh air had to traverse the bulb with sulphuric acid. Schulze watched with the microscope the edge of the fluid through the flask but nothing living could be seen nor was anything to be found when the apparatus was finally dismantled. After the infusion, hitherto barren, was exposed it rapidly teemed with confervae, infusoria, and moulds. The control infusion, which had been exposed to the air throughout, showed vibrios and monads even on the second day, and later on polygastric infusoria and rotifers made their appearance.

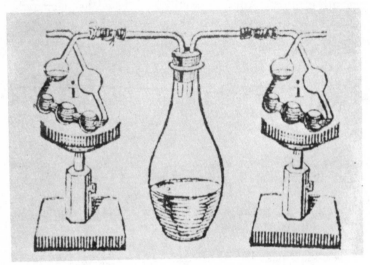

FIG. 5. Experiment of Franz Schulze (from Schulze, 1836).

Schulze's experiment is of importance in that it showed clearly that air in the flask is not the cause of the growth in the infusion, for here, fresh air which had not been heated was entering the flask daily. Further, the control infusion which early showed growths indicated that the power of the infusion to support life had not been lost by boiling.

The crucial character of Schulze's experiment has been recognized by all subsequent workers, including the heterogenists, and it has been repeated many times with equivoc results. Whether in its original form or modified technically, Pouchet (1859) always obtained growths by Schulze's method. Hughes Bennett (1868) was also unsuccessful. Tyndall (1878) pointed out that in Schulze's diagram of his experiment both the glass tubes are represented as terminating immediately below the cork, so that air entering by one tube might readily be sucked into the neighbouring tube without mixing with the general air in the flask. Tyndall also showed that even at a very slow rate of transfer germs can traverse unscathed through the sulphuric acid and potash in succession. For Schulze's experiment to succeed, the air must pass through the sulphuric acid so slowly that the floating matter up to the core of each air-bubble must come in contact with the surrounding liquid. Indeed, if the air is allowed to traverse with extreme slowness, Tyndall (1878) showed that sulphuric acid and potash may be replaced by water without detriment to the result. If the air standing over the infusion is replenished at the rate of 20 to 30 times in 24 hours the infusion will remain sterile. In Tyndall's experience, a sudden gush of air sufficed to cause the development of organisms in the infusion.

Experiments of the greatest value were carried out by Theodor Schwann (1810–82), who may be regarded as the founder of the germ theory of putrefaction and fermentation. Schwann has not always received his due reward in this connexion, but Tyndall recognized him as a precursor of Pasteur and added that Schwann was 'a man of great merit of whom the world has heard too little'. Schwann's first communication on spontaneous generation was at the Versammlung der Naturforscher und Aerzte zu Jena, on 26 September 1836. His communication was published only in the form of an abstract in *Isis*, but it contained the record of an experiment in which a small quantity of an organic infusion was placed in a large glass globe. The opening of the globe was sealed and the globe itself placed in boiling water or a Papin's digester and heated to boiling-point for a quarter of an hour. No infusoria appeared in the boiled infusion. To obviate the objection that during the boiling the organic matter of the infusion might have converted some of the oxygen in the air into carbon dioxide, Schwann modified his experiment thus (Fig. 6). The neck of the flask was bent horizontally, then into a U, at the upper end of the open limb of which was blown a small bulb. In the bend of the U was placed metallic mercury, and resting on this, and occupying the small bulb, was placed an organic infusion. The small bulb was then sealed and the whole apparatus subjected to heat, during which the infusion in the small bulb was separated from the main mass of air in the large globe. After cooling, the apparatus was inverted so as to permit the infusion to enter the flask. No infusoria developed. In the further development of his experiments Schwann made a provision for renewal of the air in the

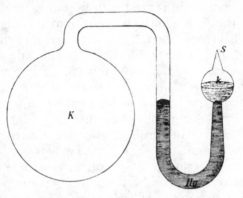

FIG. 6. Experiment of Schwann, 1836 (copied from Ingenkamp, 1886).

flask, this arrangement being the one usually associated with his name and published in 1837 in the *Ann. d. Phys. u. Chemie*. In this paper he deals with experiments in which the air was strongly heated before coming into contact with the organic infusion. In its first form Schwann's apparatus consisted of a glass flask provided with a cork through which passed two bent tubes. At a distance of about 3 inches from the flask these tubes passed into a fusible metal mixture kept at the boiling-point of mercury. As it issued from the metal bath one of the tubes was connected with a gasometer. The infusion in the flask was then well boiled so as to drive out all the air from the flask and tubes. The gasometer was then put in action and air which had been heated drawn through the apparatus. Although such heated air was allowed to traverse the apparatus during several weeks no infusoria or moulds developed in the infusion. Pieces of meat in the infusion remained unaltered. As this type of experiment required a lot of attention, Schwann simplified it in the following way. Into a 3-ounce flask (Fig. 7) about a quarter filled with meat and water, was inserted a tight-fitting cork tied in with wire. The cork was traversed by two glass tubes, one of which was bent downwards to enter a beaker containing mercury covered with

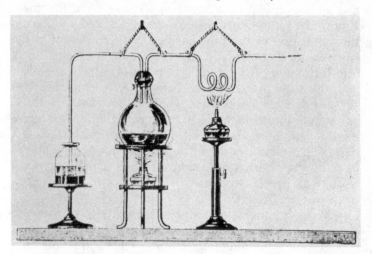

FIG. 7. Experiment of Schwann (1837). Fig. copied from Watson-Cheyne's *Antiseptic Surgery* (1882).

a layer of oil. The other tube ran at first horizontally, then downwards. It was then made to take a pair of spiral turns before it ran upwards and finally horizontally. It was drawn out to a fine opening at the end. The cork of the flask was well coated with a thick solution of rubber in linseed oil and oil of turpentine. The infusion in the flask was then boiled till the steam issuing from the tubes was so great that the mercury and oil in the beaker were unable to condense it. In order that no organisms could develop in the water between the mercury and the oil, a layer of corrosive sublimate was placed in this position. During the boiling of the infusion a spirit-lamp was placed under the spirals of the second tube and the heat was continued till the tube was on the point of softening. After the flask had been boiled for a quarter of an hour the heat was stopped, and during the cooling the air which passed through the spiral tube was still subjected to the flame of the lamp. After the flask was completely cooled the open orifice of the second tube was finally sealed in the flame. The flask thus contained an infusion which had been boiled and air which had been made to traverse a red-hot tube. From time to time this heated air was renewed, as follows: the spiral part of the tube having been well heated in the spirit-lamp, the sealed tip was broken and fresh air slowly forced through. After a time the point of entry was sealed, with the same precautions as before. At a temperature of 14°–20° R. such flasks were kept for six weeks without any sign of infusoria, moulds, or putrefaction. After opening the flask, however, the infusion became putrid in a few days. That mere heating of the air was not detrimental to its respiratory capacity was shown by Schwann by exposing frogs to air previously heated. Their growth and development was unaffected.

Schwann's crucial experiment showed that it was not air as such which brought about putrefaction in a meat extract but something in the air. That something was evidently destroyed by heat. From the standpoint of the opponents of spontaneous generation Schwann concluded that the germs or seeds of moulds and infusoria which are present in the air are destroyed when the air is heated, and his explanation of putrefaction was that these germs, on access to organic material, developed at the expense of the latter and produced the change known as putrefaction. This view was supported, in his opinion, by the fact that substances like arsenic or corrosive sublimate which are known to destroy living things prevent, for this reason, the occurrence of putrefaction. Schwann's experiments on putrefaction led him to others on the fermentation of saccharine material by yeast. Taking four flasks with sugar solution and yeast he boiled them for ten minutes, inverted them over mercury, previously heated, and passed air into them. In two instances the air was unheated, whereas in the other two the entering air had to pass a bend in the tube strongly heated. An analysis of the heated air indicated that it had suffered only a trivial diminution in the percentage of oxygen. The flasks were finally corked under mercury and removed. Those with the unheated air showed abundant fermentation in four to six weeks, whereas the two in which heated air was present remained unchanged.

Schwann did not, however, obtain constant results in this class of experiment, for occasionally none of the flasks showed fermentation or some of those with heated air did. He drew attention to the technical difficulties in the experiments, but concluded that the processes of putrefaction and fermentation were probably similar in their essence and were due to live agents which obtained their sustenance from the fermentible or putrescible materials. It was in the course of these experiments that Schwann discovered and gave an accurate account of the yeast plant and its mode of reproduction by budding. In his paper Schwann anticipated Pasteur's work when he asserted that fermentation of sugar was a chemical decomposition brought about by yeast attacking the sugar and some nitrogen containing substance necessary for its life whereby the elements not used by the yeast itself unite to form alcohol. This classical research by Schwann was described by him as 'preliminary' and at the end of it he promised to return to it. This he did, to a certain extent, in his *Mikroskopische Untersuchungen* (1839), and he added new experiments to confirm his view that alcoholic fermentation is due to the activity of the yeast plant. These experiments are referred to on p. 50.

At the British Association meeting in 1839 Ure reported that he had completely confirmed the results of Schwann, but no details of Ure's work are known except in the form of abstracts.

In 1843 Helmholtz (1821–94), afterwards the distinguished physicist, entered the discussion on heterogenesis and by experiments, similar in principle to those of Schwann, he confirmed the observations of the latter. In another type of experiment Helmholtz boiled infusions in a flask which contained a platinum spiral. An electrical current was led in and decomposed the water with liberation of oxygen. No putrefaction ensued—a result which contradicted the experience of Gay-Lussac. Nevertheless, Helmholtz considered that putrefaction might be due to exhalations from putrid substances in the air or to microbes, and this great experimenter was ultimately led to the conclusion that putrefaction can occur without the agency of living organisms. He formed this opinion, erroneous as we now know it to be, from the following experiment: taking a wide test-tube he filled it with an organic infusion and tied a piece of bladder over the end. Having sterilized the tube he inverted it in another fluid in an open vessel. Putrefaction or fermentation occurred both in the open vessel and in the closed tube, and as he could not find any infusoria or vegetable productions in the closed tube, he concluded that putrefaction is a decomposition of protein-containing substances and can be independent of living organisms, a view supported in part by the later experiments of Döpping and Struve (1847).

Further experiments of the greatest importance were carried out by H. Schröder and Th. von Dusch in 1854, and by Schröder alone in 1859 and 1861. Instead of heating the air or drawing it through sulphuric acid before permitting it to enter an infusion, they resorted to a new principle, namely, filtration of the air through cotton-wool. The use of this substance had been suggested shortly before by the chemist Loewel (1853) in connexion with the question of supersaturated solutions. In a note to one of his papers on the subject Loewel states that the air passing into supersaturated solutions could be rendered what he called 'adynamique', i.e. unable to bring about crystallization, by passing it through a long tube (40–50 cm. × 1·5–1·8 cm.) filled with cotton-wool. On the other hand supersaturated solutions such as those of sodium sulphate exposed to unfiltered air crystallized almost instantly. These experiments of Loewel suggested to Schröder and von Dusch that a freshly boiled infusion might be protected from putrefaction or fermentation if the air passing into it were filtered through cotton-wool. The arrangement of their experiment was as follows (Fig. 8): a glass flask containing the material to be tested was closed by an air-tight fitting cork dipped in hot wax previous to its insertion. The cork had two holes for the reception of two glass tubes bent at a right angle. One of these tubes was for the purpose of leading the air into the flask and the other for conducting

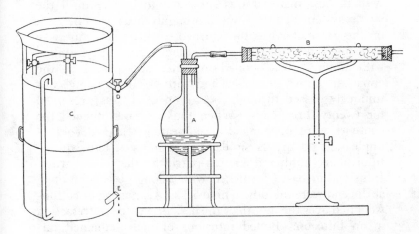

FIG. 8. Experiment of Schröder and von Dusch (1854). Fig. drawn from their description.

it away. By means of a short length of india-rubber tubing the conducting tube was connected with a glass tube, which in turn was attached to a large tube (20 in. by 1 in.) by means of a well-fitting cork, and at the other end of the large tube was another small tube open to the air. The wide tube—filtration tube—was loosely filled with cotton-wool which had been previously heated for some time in the water-bath.

The exit or suction-tube from the flask reached almost to the level of the test material in the flask, and at its other end was connected, by means of india-rubber tubing, with the upper, copper tube of an ordinary gasometer. The copper tube was provided with a stopcock. The gasometer was filled with water and by opening the oblique entering lower tube and the stopcock above the gasometer acted as an aspirator and air was drawn through the cotton-wool before it could enter the flask containing the infusion. The whole apparatus was first tested to see whether it was completely air-tight, by placing the tip of the tongue on the open end of the entry tube, when the outflow from the gasometer automatically stopped. This being demonstrated, the flask with its test material was subjected to boiling until the whole system of tubes was thoroughly heated. The cock on the gasometer was then turned so that water ran away in drops, and when empty the gasometer could be refilled with water. Meat, boiled in water and placed in the apparatus above described and with fresh air being continuously passed through the cotton-wool, was preserved for twenty-three days. When the flask was opened the contents were unchanged in appearance and devoid of unpleasant odour. A similar boiled infusion left exposed to ordinary unfiltered air was putrid by the second week. Infusions in flasks furnished with a long glass tube open at the end were mouldy in nine days. A large pad of cotton-wool tied over the mouth of the flask prevented putrefaction in an infusion. Boiled infusions of malt remained unchanged when exposed to filtered air, but when the wool was removed from the filter-tube and unfiltered air was allowed to enter the infusion, fermentation set in on the twelfth day.

Schröder and von Dusch were not successful, however, in preventing contamination of milk or of meat, without the addition of water, even when the air entering the flask had been filtered through cotton-wool. They concluded that in the case of milk and of meat without water there is a spontaneous decomposition which they attributed to the presence of oxygen in the air. In their opinion there were other substances, such as malt or meat extract (i.e. meat infused in water), which belong to a different category. In addition to air, these substances undergo putrefaction only when there is another unknown factor which is in the air, but can be removed from the air by filtration through cotton-wool.

ROBERT KOCH'S OWN LABORATORY IN INSTITUT FÜR
INFEKTIONSKRANKHEITEN, BERLIN

MAUSOLEUM OF ROBERT KOCH IN THE INSTITUT
FÜR INFEKTIONSKRANKHEITEN, BERLIN

In a paper published five years later by Schröder (1859) alone the subject of putrefaction was still further investigated. In most of his experiments Schröder relied on cotton-wool plugged directly into the neck of the flask in which the test substance was boiled. Egg albumen boiled in such a flask remained unchanged, but yolk could not be preserved, although if the yolk was first placed in a tube and heated to 160° C. in an oil-bath and then placed in a cotton-wool-plugged flask it remained sterile. Whole milk always underwent fermentative changes, whereas the casein or whey by themselves could easily be protected from infection, as also could be fibrin, fowl's blood, urine, or starch.

Schröder (1861) continued his studies on milk, yolk of egg, and meat, and now tried the effect of heat above 100° C. The substances were placed in sealed tubes and were first heated in an oil-bath at 130° C. and were then placed with water in a cotton-wool-plugged flask, which was then boiled. No fermentation or putrefaction was observed. Other refractory substances treated in the oil-bath or in a Papin's digester under 1 to 5 atmospheres could also be kept unaltered when transferred to wool-plugged flasks. Very prolonged boiling at even 100° C. also sufficed to sterilize refractory substances.

In Schröder's opinion milk, yolk, and meat contain germs which are converted into peculiar ferments. At temperatures up to 100° C. for a short time, these germs are not destroyed, but if heated for a long time at 100° C., and especially at higher temperatures, the germs are destroyed. In a series of complicated but ingenious experiments carried out under coal-gas, Schröder transferred some of the putrefactive ferment into sterile flasks containing various substances, and he found that although the putrid ferment will not grow in solutions of tartaric, or citric acids, or in sugars, dextrin or starch, it will grow when introduced into whey, casein, albumen, or egg yolk, all of which are nitrogenous in composition. Taking all the facts into consideration Schröder was of opinion that the germs that accompany putrefaction really come from the substances themselves, and are derived from the animal tissues. In spite of such a conclusion, which has turned out to be incorrect, Schröder's observations did a great deal to advance our knowledge of the principles of sterilization and the resistance of micro-organisms to heat.

Claude Bernard (1858, 1859) and Dumas (1858) again drew attention to Schwann's experiments, in a repetition of which they were successful in demonstrating that a boiled infusion supplied with calcined air remains sterile. It was at this time that F. A. Pouchet (1858) began to present to the Academy of Sciences of Paris a series of papers in which he claimed that he had proved the existence of spontaneous generation or heterogenesis. His conclusions were rejected by Milne-Edwards (1859) and by Payen, de Quatrefages, Claude Bernard, Dumas, and Lacaze Duthiers (1859). The principal experiment of Pouchet (1858) and Pouchet and Houzeau (1858) consisted in taking a large-stoppered flask filled with boiling water and inverting it over mercury. It was then filled three-quarters full with oxygen and nitrogen in the proportion to constitute an artificial air. Some hay previously heated for 20 minutes at 100° C. in the *air-bath* was introduced. The flask was then closed under mercury. At the end of some days a vegetation appeared consisting of pencillium, amoebae, monads, and vibrios. Continuing on his course Pouchet (1859) published his *Hétérogénie*, a systematic and elaborate work running to nearly 700 pages. Pouchet (1800–72) was a well-known naturalist and had previously written important historical and experimental works in various branches of biology. At the time of the publication of *Hétérogénie* he was Director of the Natural History Museum in Rouen, and an honoured member of many learned societies in France and elsewhere. Tyndall (1881) says that Pouchet was 'ardent, laborious, learned, full not only of scientific but of metaphysical fervour'. In the preface to his book Pouchet writes: 'when by meditation it was evident to me that spontaneous generation was one of the means employed by nature for the reproduction of living things I applied myself to discover the methods by which this takes place'. This was a dangerous start with a subject like spontaneous generation—so full of pitfalls. In a special chapter Pouchet attempted to destroy the doctrine of panspermism as held by Bonnet and Spallanzani. According to Pouchet organisms arise by heterogenesis. He was not of the opinion that life springs *de novo* from a fortuitous collocation of molecules. He believed in the necessary existence of a vital force coming from pre-existing living matter. Pouchet did not believe that life can be generated in solutions of mineral ingredients. It was only possible, in his opinion, in organic solutions the matter of which had formed under the influence of pre-existing life and whose properties it still retained. When the heterogenists were driven from this position by Pasteur, Charlton Bastian brought forward his theory of archebiosis, according to which a living unit may originate from a non-living fluid, i.e. life begins *de novo* owing to the occurrence of new molecular combinations. To do him justice Pouchet never went to this length. According to him the main factors which favour heterogenesis are (1) organic matter in mass or solution, (2) water, (3) access of air, and (4) suitable temperature. Variations in any one of these, even to the most trivial extent, were believed greatly to influence the kind of organisms alleged to be produced in the infusions.

To prove the occurrence of heterogenesis Pouchet brought forward experiments in which each of the supposed factors was eliminated in turn. He was satisfied that life may make its appearance even although anything previously alive had been destroyed in the putrescible material, the air, or the water. In his opinion life, therefore, must have been generated heterogenically.

Pouchet showed, as Spallanzani had shown nearly a century before, that the most intense exposure of the putrescible matter to the temperature of carbonization has no effect in preventing the appearance of infusoria when such heated substance is brought into contact with water and untreated air. He next attempted to show—although his technique was open to criticism—that the infusoria cannot come from germs in water, for when water, artificially produced by burning hydrogen in air, was boiled and subsequently added to hay which had also been heated, a development of living organisms was found to take place.

Like all the previous experimenters on the question of spontaneous generation, Pouchet admitted that the real point at issue was whether there are germs in the air or not. He repeated the experiments of Schulze (Fig. 9) and Schwann (Fig. 10) with, apparently, all the necessary

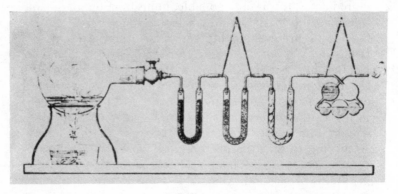

FIG. 9. Pouchet's modification of Schulze's experiment. From Pouchet (1864).

requirements, but obtained results the contrary of theirs. He devised experiments of his own which confirmed him in his belief in heterogenesis. In one experiment Pouchet drew air through a red-hot tube and then into infusions contained in Woulfe's bottles coupled up in a series of eight. In a short time the infusions teemed with life. In spite of all his precautions to destroy everything living supposed to be in the air Pouchet was constantly confronted with what he regarded as the phenomena of spontaneous generation. These phenomena, stated briefly, were that in an infusion previously clear a fermentation was soon established. An evolution of gas ensued and a whitish veil or pellicle formed on the surface and soon thickened into a membrane called by Pouchet the 'proligerous pellicle'. The microscope showed dense masses of minute organic particles named 'monads', short staff-like bodies—'bacteria'

—and large-jointed filaments called 'vibrios'. Somewhat later, animalcules of a complicated character often made their appearance and were believed by Pouchet to arise from conglomerations of monads and other particles which gradually became transformed into ova, or 'spontaneous eggs', from which the adult creatures developed by degrees into free swimming objects, the offspring of death and the embodiment of life.

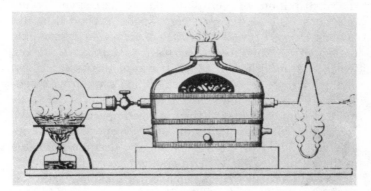

FIG. 10. Pouchet's modification of Schwann's experiment. From Pouchet (1864).

Pouchet's work on heterogenesis aroused great interest and discussion in scientific circles in France, and his conclusions were adversely criticized by men like Milne-Edwards, Claude Bernard, and Dumas. His actual experiments were criticized by Pasteur, who for several years had been making his renowned researches on fermentation. He had clearly shown that 'ferments' are really organic living beings, and that, far from fermentation being set up, as was supposed, by contact with albuminous matter, it was really an action produced by the living ferment during the course of its vital activity. The source of the living ferments had not then been worked out by Pasteur but was regarded as aerial. The study of heterogenesis, as enunciated by Pouchet, thus became imperative. The question really turned on whether the ferment germs are produced immediately from the decomposition taking place or mediately from germs or seeds for which the fermentible or putrescible substance is a suitable pabulum.

Pasteur's publications on the subject of spontaneous generation consist chiefly of communications to the Academy of Sciences of Paris and published in abstract in the *Comptes rendus de l'Acad. des Sciences* of 1860 and 1861. In these communications he dealt with the collection and demonstration of germs in the air, the origin of ferments, the distribution of germs in nature, and many other questions. These reports were finally expanded in his famous *Mémoire sur les corpuscules organisés qui existent dans l'atmosphère. Examen de la doctrine de générations spontanées*, which was published in 1861. It was of this research that Tyndall wrote: 'Clearness, strength and caution with consummate experimental skill for their minister were rarely more strikingly displayed than in this imperishable essay.' It was, indeed, the inauguration of a new epoch in bacteriology and should be studied by all who desire a schooling in the art of rigorous experimentation.

Several German writers, such as Ingenkamp, Delbrück, and Schrohe, have criticized Pasteur for his inadequate recognition and appreciation of the pioneer work of Theodor Schwann (1837), and there is something in their criticism. That Pasteur acknowledged his debt to Schwann, privately at any rate, is seen in a letter quoted by Fredericq (1885), the physiologist of Liège, in his life of Schwann. In this letter, written by Pasteur on 15 June 1878, in connexion with the fête in Liège celebrating the fortieth anniversary of Schwann's professoriate, Pasteur says: 'Depuis vingt années déjà je parcours quelques-uns des chemins que vous avez ouverts', and he signed himself, 'L'un de vos nombreux et sympathiques disciples et admirateurs L. Pasteur'.

Pasteur first dealt with an attempt to demonstrate that the air really contains germs. For this purpose he fitted up an aspirator (Fig. 11) in the form of a T-shaped tube connected above with a cistern and below with an outflow tube. The aspirator was connected with a piece of glass tubing passing through a hole in a shutter and open to the outside air. In the glass tube was inserted a plug of gun-cotton

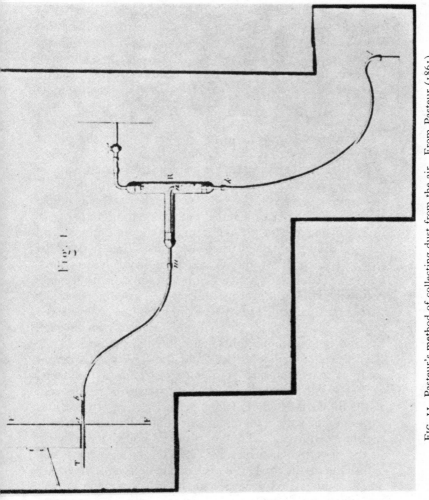

FIG. 11. Pasteur's method of collecting dust from the air.   From Pasteur (1861).

destined to act as a filter to intercept the aerial germs. At the end of the experiment the gun-cotton was removed and placed in a sedimentation tube containing a mixture of alcohol and ether to dissolve the gun-cotton. The dust being insoluble collected at the bottom of the tube and was examined under the microscope. It showed, in addition to inorganic matter, a considerable number of small, round, or oval bodies, indistinguishable from the spores of minute plants or the ova of animalcules. The number of these bodies varied according to the temperature, moisture, and movement of the air, and the distance above the soil at which the gun-cotton had been placed.

Pasteur next proceeded to repeat the experiment of Schwann (Fig. 12) in which air was passed through a red-hot tube. In a flask, the neck of which was bent almost horizontally, he placed distilled water containing 10 per cent. of sugar and 0·2 to 0·7 per cent. of the albuminoid and mineral matter obtained from beer yeast. The neck of the flask was then drawn out so that it could easily be sealed, and it was connected with a platinum tube capable of being maintained at a red heat in a furnace. The contents of the flask were then boiled for two or three minutes, and allowed to cool. Naturally, air entered but it had to pass through the red-hot platinum tube first. Finally, the neck of the flask was sealed in the flame and the flask itself was put aside at a temperature of about 30° C. for a prolonged period. With this technique no organisms ever appeared in the infusion which remained perfectly clear. In previous experiments with a different technique Pasteur's results had been equivocal. Thus in some he boiled the infusion in the flask and then merely sealed the neck in the flame. Subsequently he broke the neck of the flask under mercury and led into it oxygen gas which had been made to pass through a red-hot tube. In other experiments he introduced oxygen, prepared from the decomposition of water, or air heated or unheated. All the infusions showed growth and he was led to the conclusion that the source of the contamination was germ-laden dust from the surface of the mercury. When the mercury was discarded the infusions

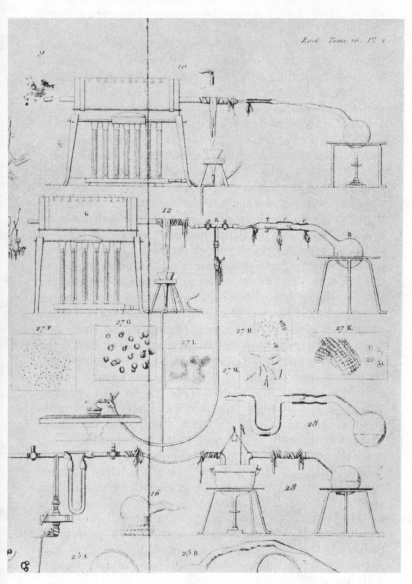

FIG. 12. Various forms of apparatus used by Pasteur in his experiments on spontaneous generation. From Pasteur (1861).

subjected to heated air by the technique described above remained sterile.

Having thus shown that germs or bodies resembling them do exist in the air, and having shown that it is possible to supply heated air to a boiled infusion without the latter being affected in any way, Pasteur then proceeded to show that a sterile infusion containing air which had been heated becomes infected if dust from the air is introduced into it. A sterile flask (Fig. 12, no. 12) with infusion was connected with a glass tube containing a smaller piece of glass tube charged with a germ-laden piece of gun-cotton through which air had been drawn. The whole was connected with a vacuum pump and a red-hot platinum tube in such a way that after the air had been pumped out of the flask the latter could be filled with air which had been heated. When this had been repeated several times the neck of the flask was broken and the gun-cotton plug was allowed to glide into the infusion. The constant result of such an experiment was that the infusion, hitherto sterile, became contaminated in from 36 to 48 hours. The same result was obtained when an asbestos plug was used instead of gun-cotton. In a modification of the experiments Pasteur also showed that if the dust-charged tube is strongly heated and then cooled it can be passed into the infusion without any change subsequently appearing in the latter. (Fig. 12, no. 28.) In addition to yeast and sugar solutions Pasteur employed urine or milk. In the case of the latter, sterilization was found to be more difficult, as had been stated by Schröder.

Pasteur also showed that infusions can be sterilized in an open flask provided that the neck of the latter is drawn out and bent down in such a way that the germs cannot ascend (Fig. 13). This type of experiment, previously used by H. Hoffmann (1860), removed all criticism on the question of air as such activating into life an organic infusion. If the bent neck of an open flask which had long remained sterile was cut off the infusion rapidly teemed with living things.

That germs are not uniformly disseminated in the air was shown by Pasteur (1860) in famous experiments.

Sterile sealed bulbs filled with infusion and opened in the cellars of the Paris Observatory, where the air was still, only rarely showed growths.

While on holiday Pasteur made 73 experiments with sterile sealed bulbs containing infusions. Twenty of these were opened on the road to Dôle, and 8 yielded growths. Proceeding to Salins, he climbed Mount Poupet (850 metres), and having opened another batch of 20 bulbs found that only 5 became infected. He then went to Chamonix (20 Septem-

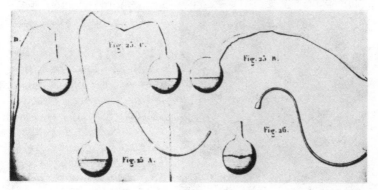

FIG. 13. Pasteur's bent-necked open flasks. From Pasteur (1861).

ber 1860) and ascended the Montanvert (6,267 feet) with a guide and with a mule carrying the remaining 33 sterile bulbs. Thirteen of these were opened on the Glacier des Bois, but difficulty was experienced in sealing them owing to an imperfection in the lamp. Sending the guide back to Chamonix to have the lamp repaired, Pasteur carried the 13 bulbs still unsealed to the little inn on the Montanvert, kept them in the room in which he slept, and sealed them the following day. Of the 13, 10 ultimately showed growth. The last batch of 20 flasks was opened on the Mer de Glace, and only 1 became contaminated. Great precautions were taken to carry out the experiments properly. Having well heated the necks of the bulbs and touched them with a file, Pasteur then elevated them high above his head in a direction opposite to the wind. The necks were then broken with long pliers and the air was allowed to enter, with the result stated. One of the original flasks exposed by Pasteur (1860) is seen in Fig. 14.

FIG. 14. One of the flasks exposed by Pasteur in 1860.

In his important *Mémoire* Pasteur also showed that growths occurred in solutions which contain no albumen at all but were composed of water (100 gm.), sugar (10 gm.), ammonium tartrate (0·2–0·5 gm.), and yeast ash (0·1 gm.).

Pouchet (1859) had also examined dust from the most diverse places, such as the Natural History Museum and the roof of the Cathedral of Rouen, the great temple of Karnak (Luxor), the banks of the Nile, the tomb of Rameses II, and the central chamber of the Great Pyramid of Gizeh. He also examined samples of air from the Mediterranean and from Greece and Constantinople. The chief organic element which he found was starch. He examined the respiratory mucous membranes of many animals and explored in particular the air cavities in the bones of birds, searching all the while for germs but finding none. 'In all our observations,' he says, 'which without exaggeration one may reckon by hundreds, we have never encountered either a single spore or a single egg of an infusorial animalcule or a single one of these in an encysted state.' In his exhaustive monograph on the microscopic examination of the air D. D. Cunningham (1873) showed this to be incorrect, as also did Miquel (1883).

Pouchet's views in general were supported by Joly and Musset at Toulouse, and by Meunier and Donné. Others took the side of Pasteur and criticized the heterogenists severely. Thus Count de Careil at a scientific congress in Paris described heterogenesis as 'une monstruosité philosophique, c'est la première abérration de la physiologie anti-chrétienne, c'est un produit de l'impiété ignorante, c'est une théorie malsaine et ceux qui la défendent sont des fils du 18ᵉ siècle égarés dans le 19ᵉ'.

Pouchet and his co-heterogenists Joly and Musset (1863) made a number of experiments which were entirely contradictory to Pasteur's teaching. They examined the air of the Pyrenees and its power to fertilize organic fluids. Starting from Luchon with a small number of sealed bulbs partly filled with boiled hay infusion, they opened 4 on the Rencluse (2,083 metres above sea-level) and 4 more on the edge of the great Maladetta glacier at a height of 3,000 metres. All Pasteur's precautions were said to have been scrupulously observed except that they cut the necks of the bulbs with a file and having shaken the contents sealed them up again. In each of the 8 bulbs exposed growths occurred. Pasteur (1863) criticized this work both with regard to the technique employed and the small number of bulbs exposed. To study the air of the Swiss Alps, and to oppose Pasteur on his own ground, so to say, Pouchet had stoppered bottles containing boiled water taken by the alpinist Kolb to the summits of Mont Blanc (15,782 feet), the Buet (10,200 feet), and Monte Rosa (15,217 feet). The bottles being unstoppered, some of the contained water was poured out and the bottles closed again. Later, in Rouen, Pouchet opened these bottles inverted in a boiling solution of clover. Animalcules appeared in a few days.

In the meantime, Pasteur had been awarded the Prix

Alhumbert for 1862 by the Academy of Sciences. The subject proposed had been 'essayer par des expériences bien faites de jeter un nouveau jour sur la question des générations dites spontanées'. In the opinion of Milne-Edwards, Flourens, Brongniart, Coste, and Claude Bernard, Pasteur's *Mémoire sur les corpuscules qui existent dans l'atmosphère* was deemed to have fulfilled all the requirements.

The renewed attack of Pouchet, Joly, and Musset on Pasteur aroused indignation in the Academy, and Flourens (1863), in reply to a further communication of Joly and Musset, expressed himself definitely and tersely on the subject.

'Les expériences de M. Pasteur sont décisives. Pour avoir des animalcules que faut-il si la génération spontanée est réelle? De l'air et des liqueurs putrescibles. Or M. Pasteur met ensemble de l'air et des liqueurs putrescibles et il ne se fait rien. La génération n'est donc pas. Ce n'est pas comprendre la question que de douter encore.'

This was, however, far from silencing the heterogenists, and in the end the Academy appointed a commission consisting of Flourens, Dumas, Brongniart, Milne-Edwards, and Balard to try to settle the question one way or another. They threw down the gauntlet with this statement: 'It is always possible in certain places to take a considerable quantity of air which has not been subjected to any physical or chemical change and yet such air is insufficient to produce any change whatsoever in the most putrescible fluid.' Pouchet and his co-workers Joly and Musset took the bait and declared that the above was erroneous. Pasteur challenged them, and they promised to supply the proof, adding, 'Si un seul de nos ballons demeure inaltéré nous avouerons loyalement notre défaite'.

The commission proposed that the respective antagonists should make similar experiments before them. To this Pouchet and co-workers demurred and insisted that they should repeat only their own experiments and not Pasteur's. They tried in fact to open up the whole question of heterogenesis *de novo*. The commission decided that each should demonstrate his results on a series of 60 infusions. At last the antagonists met for the test (22 June 1864) in Chevreul's laboratory in the Natural History Museum at the Jardin des Plantes, Paris. Pasteur first demonstrated three ballons which he had exposed on Montanvert in 1860. One was opened and, the air being analysed, was found to contain 21 per cent. of oxygen. The second was opened and left open. It was peopled with organisms within three days. The third ballon was left untouched and was subsequently exhibited at the Academy of Sciences on 20 February 1865. Pasteur then prepared his new series of 60 flasks before the commission. In each were placed 250–300 c.c. of yeast water. The neck was narrowed, the fluid was boiled for 2 minutes, and 56 out of the 60 flasks were sealed in the flame. In 4 the necks were drawn out, bent downwards, and left open. Pouchet, Joly, and Musset raised many trivial objections, refused to carry out their experiments, and finally withdrew. Pasteur completed his experiments in three sections. In the first, 19 ballons were opened in the amphitheatre of the National History Museum, with the result that 14 remained sterile and 5 became fertile. The second series of 19 ballons was opened on the highest part of the dome of the amphitheatre, when 13 remained sterile. The third set of 18 ballons was exposed in an open spot at Bellevue (near Sèvres) under some poplar trees. Two only remained sterile, and 16 developed growths. The 4 open Pasteur flasks remained sterile. As a result, the commission decided in favour of Pasteur but adjourned till the following year, when they were to submit a second report. According to Pasteur Vallery-Radot (*Œuvres de Pasteur*, Tome **2**, 647) this was never done. Pouchet in the meantime published his second book (1864) on heterogenesis, but in it he added nothing to his earlier work of 1859 and did not alter his position. He made no further communications on heterogenesis after 1864, although he lived till 1872.

In spite of the discomfiture of Pouchet, Pasteur's researches did not exterminate the Lernaean hydra of heterogenesis, for other workers appeared to continue the discussion. Among the supporters of Pouchet may be mentioned Jeffries Wyman (1862), Hughes Bennett (1868), Mantegazza (1864), Huizinga (1873, 1874), and particularly Charlton Bastian. On the side of Pasteur, Ferdinand Cohn (1877), John Tyndall (1876, 1877), William Roberts (1874), and Joseph Lister (1878) ranged themselves.

The heterogenists have always made a claim on the work of Jeffries Wyman (1862), a well-known anatomist and naturalist of Harvard, but in no part of his papers did he specifically state that he was an upholder of spontaneous generation. He merely described his experiments, which

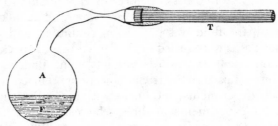

Fig. 15. Experiment of Jeffries Wyman (1862) (after Pouchet, 1864).

were as follows. The materials of the infusions were put into a flask, and a cork through which was placed a glass tube drawn out to a constriction was pushed deeply into the neck of it. The space above the cork was filled with an adhesive cement composed of resin, wax, and varnish. The glass tube was bent at right angles and inserted into an iron tube which was filled with wires closely packed. In a second series of experiments (Fig. 15) the cork of the flask was discarded and the iron tube was cemented directly into the mouth of the flask, the neck of which had been narrowed to facilitate sealing. In a third set the flask with the infusion was sealed and was afterwards submerged in boiling water.

The fluids employed by Wyman contained sugar, gelatine, and cut hay or pepper, cheese, sugar, and gelatine. After being introduced into the flask they were boiled for periods varying from 15 minutes to 2 hours, while at the same time the iron tube filled with wires was heated to redness. In 29 out of 33 experiments of this kind growths of vibrios, bacteria, spirilla, and colpoda-like bodies made their appearance. In Wyman's last series of 4 experiments ebullition took place for from 5 to 10 minutes under a pressure of 2 atmospheres in 2, and at 5 atmospheres in the remaining 2. Evidence of life consisting of monads and vibrios occurred in the first pair but not in the second.

The essentials in these experiments seem at first sight to be the same as those of Pasteur, but Watson-Cheyne and Tyndall have drawn attention to an exception which had probably an important bearing on the results. Pasteur employed flasks of a capacity of 250–300 c.c. and introduced

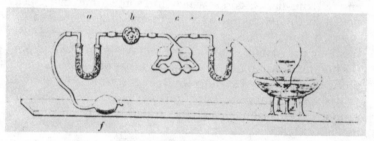

FIG. 16. Experiment of Hughes Bennett (1868).

into them 100–150 c.c. of infusion. Wyman's flasks were of 500–800 c.c. capacity and the amount of infusion was only 12–40 c.c. In some of Wyman's earlier experiments the air content in the flask was more than 30 times that of the infusion. It may now be assumed that in Wyman's experiments a large part of the flask and its aerial contents were not adequately sterilized when he imagined that they were. In a later research Wyman (1867) drew attention to the existence of algae capable of living at 208° F., and he showed that 'infusoria' might appear in hermetically sealed flasks which had been immersed in boiling water for a few hours.

Owing to the equivocal results obtained with calcined air, sulphuric acid, gun-cotton plugs, and boiling, Hughes Bennett (1868) devised experiments to try the effect of all these destructive agents at once, with the exception of the first (Fig. 16). Decoctions of liquorice-root, of tea, and of hay were maintained at the boiling-temperature in a porcelain basin over a gas flame. Flasks filled with and inverted over the boiling fluid had air pumped into them to the extent of three-fourths of their volume. Before entering the infusion the air had been made to traverse (1) a U-tube (*a*) containing liquor potassae, (2) Liebig's bulbs (*c*) with sulphuric acid, (3) a hollow glass bulb (*b*) stuffed with gun-cotton, and (4) another U-tube (*d*) filled with sulphuric acid. All the bent tubes were filled with fragments of pumice to break up the air and diminish the possibility of germs passing through in the centre of the bubbles. After the air so prepared had entered the flasks, corks which for some time had been boiled in the decoction were by means of an iron forceps inserted into the necks of the flasks and the entrance of fresh air prevented. Further, on removing the flasks from the boiling infusion the corks and the necks of the flasks were plunged into melted sealing-wax. At the same time bottles or flasks containing the same infusion but having a similar proportion of ordinary air were sealed or corked so as to be contrasted with the influence of the prepared air. All the 5 control flasks (with ordinary air) were turbid or covered with micro-organismal growths in from 6 to 12 days; 7 out of 8 infusions which had been exposed to prepared air also became turbid, but at periods varying from 4 to 9 months. The one exception was an infusion of hay. Five other series of flasks were prepared with some variations or improvements in the technique, but like results were recorded. On the ground of these experiments Hughes Bennett became a convert to the doctrine of heterogenesis. His experiments were well conceived, but according to present experience there were technical difficulties and flaws which at that time seemed incredible.

It was in 1872 that the English heterogenist Henry Charlton Bastian (1837–1915) entered the field with his large book, of more than a thousand pages, entitled *The Beginnings of Life*, and with unparalleled persistence he maintained the doctrine of heterogenesis for nearly forty years, during which he published many controversial papers and six other books on the subject. At the end he fought alone for the establishment of his cherished belief. It is not easy to account for the position he maintained for so many years unless he is regarded as a master of dialectics—a mode of reasoning which attempts to make fallacies pass for truth. The effect of Bastian's early experimental work on heterogenesis was in a way beneficial. Duclaux (1896), the greatest of Pasteur's collaborators, paid a high tribute when he said that Bastian

'has given us ideas or rather, let us say, he forced Pasteur to gain ideas the absence of which had hindered the progress of science. All our present technique has arisen from objections made by Bastian to the work of Pasteur on spontaneous generation. It was Bastian who made us see that this work which had been so vaunted abounded in false interpretations which he said invalidated its conclusions. It was Pasteur and his pupils Joubert and Chamberland who showed that even if the interpretations had sometimes been inexact the conclusions were not the less well founded.'

It was especially through Bastian's experiments with alkalized urine that it came to be known that germs may be much more thermo-resistant than had previously been supposed. The practice of heating to 115°–120° C. all liquids requiring to be sterilized dates from the repetition of Bastian's experiments by Pasteur and Chamberland. Many of Bastian's experiments were suggestive, but were often carried out with a technique now known to be defective. Bastian went much further than any of the other heterogenists and strove to establish belief in what he called archebiosis—the production of life from non-living matter. In his later years he considered that it was possible to produce highly complex living creatures from simple solutions. From the great mass which Bastian wrote it is difficult to sift out what is sound. There is no doubt that the majority who repeated his work arrived at the firm conclusion that he was deceived.

It was on 20 January 1870 that John Tyndall (1820–93), the English physicist, in a Friday evening lecture at the Royal Institution in London, first gave his views on the germs of the air, and shortly afterwards he addressed himself on the subject to the general public in *The Times* newspaper. This drew a reply from the heterogenist

Bastian, who pointed out that the matter really pertains to the biologist and physician, and he warned Tyndall that much irreparable damage might be occasioned by his (Tyndall's) methods of reasoning, which were amazing. For ten years (1859-69) before the date of his first work on germs Tyndall had been studying radiant heat and its relation to gases, and in his experiments on air he had been greatly struck with the difficulty of removing the particles that were floating in the atmosphere. It was found that although invisible to the naked eye the floating matter could be made apparent when a powerfully condensed beam of light was passed through it. Trying to remove the particles, Tyndall (1868) had recourse to the flame of a spirit-lamp, and he then found that the particles disappeared, having been burned up. The space became black, a result simply due to the absence from the track of the beam of all matter competent to scatter the light. It was what Tyndall called 'optically inactive'. Heat even far short of combustion was found to create currents passing upwards among the inert particles and dragging them after it right and left but forming between them an impassable black partition. It was thus that Tyndall began to employ a beam of light to reveal dust and not the dust to reveal light, and he began to drift into the discussion, then raging, on spontaneous generation. His researches were published in two classical memoirs in the *Philosophical Transactions* (1876, 1877). These, with other essays and lectures, were subsequently published in his illuminating book *Essays on the Floating Matter of the Air in relation to Putrefaction and Infection* (1881). Along with the researches of Pasteur the studies of Tyndall were fundamental for the progress of bacteriology. Pasteur and Tyndall, the one a chemist, the other a physicist, neither of them medical but both trained in the most exact methods of experimentation, jointly accomplished the final downfall of the doctrine of spontaneous generation. Apart from the sporadic mutterings of Bastian down to 1910, no serious objections have been brought forward to challenge the conclusions of these two masters of experiment.

Examining by chance one day an empty flask that had

been standing a long time untouched, Tyndall passed a beam of light through it and found it 'optically empty'. The motes had been deposited not only on the bottom but also on the sides, and this led him to construct a chamber in which suitable experiments could be carried out (Fig. 17).

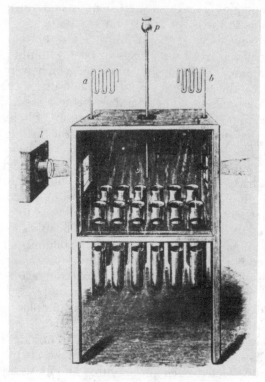

FIG. 17. Tyndall's culture chamber.

This chamber or case had a glass front, and the top, bottom, back, and sides were of wood. At the back was a small door, while two panes of glass were let in like windows in the sides. The top was perforated in the middle by a hole, 2 inches in diameter, closed air-tight by a sheet of rubber, which was pierced through the middle by a pin-hole through which was pushed a long pipette ending externally in a thistle funnel. A circular tin collar 2 inches in diameter and $1\frac{1}{2}$ inch deep surrounded the pipette, the space between being well packed with cotton-wool moistened with glycerine. Into two other smaller apertures in the top of the chamber were inserted air-tight the open ends of two narrow tubes connecting the interior of the box with the outside air. The tubes were bent up and down several times so as to intercept and retain any particles carried in the air by changes in temperature. The bottom of the chamber was pierced with holes to take a number of test tubes intended to hold the infusions which were to be exposed to the moteless air. The mode of experimentation was as follows. The chamber, being closed, was left untouched for several days and until a beam of light passed through the lateral windows showed that the floating matter of the air had been deposited on the interior surfaces, where it was retained by a layer of glycerine with which these surfaces had been previously coated. The pipette in the top of the box being manœuvred into position, the infusion was poured into the thistle funnel and allowed to enter the test-tubes until they were nearly filled. The tubes were then lowered into a bath of brine raised to the boiling-point, and allowed to boil for 5 minutes. The fluids tested were urine, infusions of mutton, beef, liver, haddock, sole, cod, turbot, herring, hare, rabbit, pheasant, grouse, and vegetable infusions of turnip or hay. Tubes filled with similar infusions but exposed to air acted as controls.

With increasing improvement in his technique Tyndall found that the 'protected' infusions in the chambers remained unaffected even for months, and it was thereby established that the power of developing bacterial life by the atmosphere and its power of scattering light go hand in hand. Tyndall also made experiments with filtered or calcined air and in hermetically sealed glass bulbs. In 130 experiments made with infusions in bulbs boiled for 3 minutes in the oil-bath, 100 remained sterile. Nearly one thousand 'exposed' tubes of infusions were examined. All showed growth. Sealed bulbs with sterile infusions were opened by Tyndall in various parts of the Royal Institution, Albemarle Street (London), by Charles Darwin in his study at Down, by Siemens at Tunbridge Wells, at Pembroke Lodge, Richmond Park, at Heathfield Park, Sussex, at Greenwich Hospital, and by J. D. Hooker at Kew Gardens. All exhibited growths. Tyndall also studied the geographical distribution of aerial germs by exposing trays containing 100 tubes filled with infusions, and he was able to show that the distribution of bacteria is not uniform.

In his second important paper Tyndall (1877) studied the question still further, especially in relation to certain remarkable results obtained by W. Roberts and F. Cohn. W. Roberts (1874) had found that when infusions of hay are carefully neutralized they can withstand three hours' boiling before they are rendered sterile. Pasteur (1861) had also shown that, whereas acid-reacting infusions had their germinal life destroyed at 100° C., a temperature over 100° C. was required to sterilize infusions that gave an alkaline reaction. Cohn confirmed this observation in part, and it was also the experience of Tyndall, although his results were at first contradictory. After a tedious inquiry involving hundreds of experiments Tyndall found that infusions made from old, dried hay were in general much more difficult to sterilize than those made with new fresh hay. The results, however, were variable. Returning to other infusions with which he had been so successful in 1876, he now found in countless experiments that he could no longer sterilize them in closed chambers by boiling for 5 minutes as he had previously been able to do. Rather than accept the suggested doctrine of spontaneous generation he laboured to make his technique more and more exact. Filtered air, calcined air, or spontaneously purified air were all tried as in his first experience, but successes now became the exception rather than the rule. Either the old substances with which he had experimented in 1876 had acquired a new generative force or some new factor enormously more heat-resistant than anything he had previously encountered was making itself felt and seemed, as he says, 'omnipresent and persistent'. Tyndall therefore prepared to leave the Royal Institution and in fact transferred all his apparatus to the Jodrell Laboratory at the

Royal Gardens at Kew. Repeating his experiments once more, he found that infusions which in the Royal Institution could not be rendered sterile after 200 minutes' boiling were completely sterilized in 5 minutes at Kew. Returning to the Royal Institution, Tyndall caused a shed to be erected on the roof and carried out further experiments there. Tinned chambers were used instead of wooden ones, but infusions of cucumber could not be sterilized after boiling for 5 minutes. All became infected. The shed was thoroughly scrubbed with carbolic acid, all the infusions were made in the shed itself, and the attendants wore overalls. At last success was achieved. Boiling for 5 minutes sufficed to sterilize the infusions. Parallel experiments made in the Laboratory of the Royal Institution and in the shed which was 24 feet away gave completely opposite results. In the laboratory, infusions withstood 5 hours' boiling and still showed life, whereas in the shed they were sterile after 5 minutes' boiling. It was impossible to believe that spontaneous generation existed in the laboratory and did not exist in the shed 8 yards away. The difference in the results was attributed, and no doubt correctly, to the fact that on the floor of the laboratory had lain bundles of old hay which when disturbed had yielded clouds of infected heat-resistant dust. Prior to the deposit of the hay in the laboratory no difficulty had been experienced in disinfecting any organic infusion by heating it to the boiling-point for 5 minutes. A simple wave of a bunch of hay in the air of the shed made it as infective as the air of the laboratory. Even the unprotected head of Tyndall's assistant when his body was carefully covered sufficed in some instances to carry the infection.

Tyndall now proceeded to test the actual limits of heat resistance of the germs which for so long had been contaminating his infusions. Most of his work was done with bulbs which were subsequently boiled and sealed. He found that the infusions could not be sterilized with certainty after even $5\frac{1}{2}$ hours' boiling. In order to prove that this resistance was not due to the germs existing in the air in the bulbs, but that the germs in the infusion itself actually survived this temperature, he devised an ingenious experiment (Fig. 18). A was the original form of the bulb. The neck was plugged with cotton-wool, *c*, and hermetically sealed at *a* (B). The lateral tube was then drawn out at *o* and *p*. The end was connected with an air-pump and

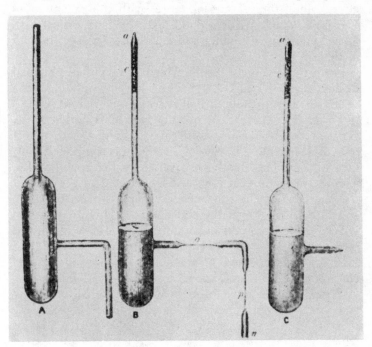

FIG. 18. Tyndall's method of determining the heat resistance of bacteria (Tyndall, 1881).

the bulb exhausted. After two or three emptyings and fillings the bulb was finally charged with one-third of an atmosphere of thoroughly filtered air. While the pump maintained this pressure within the bulb the tube was sealed at *p*. The bulb was then heated nearly to redness in a Bunsen flame, and when cooled the end *p* was introduced into the infusion, pressed against the bottom of the vessel that contained it, and was thus broken. The liquid naturally rose in the bulb, filling it to two-thirds of its capacity.

A small gas-flame was then applied at *o*, which was then sealed. The final appearance of the bulb is seen in c.

Bulbs prepared thus were exposed to boiling for periods ranging from 10 minutes to 8 hours. After 4 hours' boiling all infusions showed growth. Sterilization was attained after boiling for 5 hours or more, with the exception of one bulb which withstood the extraordinary ordeal of 8 hours' continuous boiling and was not sterile at the end of this time.

From the divergent results which Tyndall had obtained he now began to realize that bacteria have phases, one being relatively thermolabile in that it was destroyed at 100° C. in 5 minutes, whereas the other, which he regarded as the germ of the bacterium, is thermoresistant to an almost incredible extent. The idea of Tyndall was completely confirmed or independently discovered by Ferdinand Cohn, who in the same year (1877) demonstrated in the case of the hay bacillus the existence, development, germination, and heat resistance of the endospore. Tyndall and Cohn arrived at the same result by different methods of experiment, and Tyndall carried the matter further by elaborating his important method of fractional sterilization by discontinuous heating and known to-day as Tyndallization.

Observing that active bacteria are easily killed by boiling and that a certain period—the latent period—is necessary to enable the heat-resistant germ to pass into that state in which it is sensitive to heat, he devised his process of sterilization and tested it experimentally. The first account which he gave of the method was in a letter, dated 14 February 1877, to Huxley, and published in the *Phil. Trans.* (xxv. 569). He describes it thus:

'Before the latent period of any of the germs has been completed (say a few hours after the preparation of the infusion) I subject it for a brief interval to a temperature which may be under that of boiling water. Such softened and vivified germs as are on the point of passing into active life are thereby killed; others not yet softened remain intact. I repeat this process well within the interval necessary for the most advanced of those others to finish their period of latency. The number of undestroyed germs is further diminished by this second heating. After a number of repetitions which varies with the characters of the germs the infusion however obstinate is completely sterilized.'

He showed that discontinuous boiling for 1 minute on five successive occasions could render an infusion barren, whereas one single continuous boiling for 1 hour might not.

Tyndall's researches gave the final blow to the doctrine of spontaneous generation as much if not more than those of Pasteur. This was due in part to their wide diffusion by Tyndall in the form of lectures, demonstrations, books, newspapers, and magazines. Whereas the personality of Pasteur inspired something of the nature of opposition, Tyndall's magnetic personality, his exact technical methods, the logic of his interpretations, and the clarity of his literary compositions were acceptable to a large number of intelligent people. The doctrine of the germ theory of disease was then securing a foothold, and Tyndall was one of its early and one of its staunchest upholders. The medical profession owes a debt to Tyndall, and this was partly acknowledged when he was made honorary Doctor of Medicine by the University of Tübingen.

Credit must also be given to the English worker William Roberts (1874), whose observations confirmed and in some respects corrected those of Pasteur. Roberts again drew attention to the fact, known to Pasteur, that the degree of heat required to produce sterility varies greatly according to the nature of the materials tested. Roberts employed glass bulbs plugged with cotton-wool, and he found that filtered infusions of such things as beef, mutton, fish, turnip, carrot, hay, cucumber, tomato, can be readily sterilized by boiling for 5 to 10 minutes. In a second group requiring boiling for 20 to 40 minutes Roberts placed such things as chopped green vegetables, pieces of flesh meat, fish, boiled egg, milk, and turnip infusion containing cheese. In a third class by itself Roberts placed superneutralized infusions of hay, which withstand boiling in plugged bulbs for 1 to 2 hours. In skilfully devised experiments Roberts showed that this result is not due to an increase in the abiogenic aptitude of the infusions owing to the presence of the alkali, but to an increase in the vital resistance to heat of the germs in infusions. By an ingenious arrangement he showed that if a sealed and sterilized tube of potash is placed in a flask of hay infusion which is rendered sterile by heat, spontaneous generation does not occur when subsequently the alkali is allowed to mix with the infusion by breaking the glass containing the alkali without opening the flask. When, however, the sterile alkaline infusion is artificially infected by exposing it to the air it cannot be re-sterilized by merely heating it to 100° C. for 5 minutes. Roberts also made a large number of experiments to test the capacity of the juices and tissues of plants to generate bacteria without extraneous infection. These experiments are referred to later (p. 120).

In an elaborate research on the sterilization of hay infusions Ferdinand Cohn (1877) confirmed the observations of Roberts on the extreme resistance to heat of bacteria in hay, although he was not able to confirm altogether the view that there was a great difference between acid and alkaline infusions. Cohn employed ordinary test-tubes drawn out in the middle. Infusions placed in such tubes were heated in boiling water and after varying intervals were removed and then plugged with cotton-wool. All hay infusions boiled for 5 to 15 minutes showed growth with membranous formation on the surface. Sometimes the infusions became fertile even although they had been boiled for 2 hours. Cohn noticed that the flora appearing in unboiled, exposed infusions of hay was different from that which appeared in hay infusion which had been boiled. In the latter the organisms were entirely bacillary in form and of the type of *Bacillus subtilis*. He gave a very accurate description of the evolution of the hay bacillus in the boiled infusions, and showed that at a certain time the bacilli exhibit in their interiors small highly refractile bodies which he recognized as spores. He showed that when this stage is reached the vegetation of the *Bacillus subtilis* in the hay infusion ceases, although the spores are capable of germinating in a new sterile hay infusion, and he actually was fortunate enough to witness the act of germination whereby the spore gave origin to the young bacillus, which was capable of progressive vegetative growth. Cohn made an apt comparison between the evolution of *Bacillus subtilis* in a hay infusion and the course of infective disease. In each he distinguished an incubation period, a stage of progressive growth, a fastigium, and a period of remission. He showed that the difficulty of sterilizing hay infusions was closely connected with the evolution of *Bacillus subtilis*, and he was one of the first who clearly demonstrated that the spore is endowed with great heat-resisting properties. He attributed these partly to the chemical composition of the spore and partly to its physical state of dryness. Cohn also showed that the difficulty in sterilizing Bastian's turnip and cheese infusions was due to the presence of sporing bacilli in the cheese itself and had nothing to do with heterogenesis.

The result of all the work described above and carried out by so many workers sufficed to lead to the firm belief that spontaneous generation is what Pasteur called it—a chimaera.

Question of Bacteria in Fluids and Tissues of Healthy Animals and Plants

As an addendum to the experimental work on spontaneous generation as indicated above mention must be made of attempts to determine whether various normal organic fluids or substances such as blood, urine, and vegetable and animal juices and tissues contain bacteria, and if not, whether these substances can be preserved without the application of heat or chemical disinfectant. The pioneer of this kind of investigation was a Dutch professor in Utrecht, J. H. van den Broek, and he published his researches in Dutch in 1858 and in German in 1860. Along with him must be mentioned Pasteur (1863, 1876), W. Roberts (1874), Lister (1873), Watson-Cheyne (1879, 1882), and G. Meissner (see Rosenbach, 1880). All of these succeeded, often after failures, in preserving vegetable and animal tissues and juices from healthy living things and without the application of heat or chemicals. As much of our current belief rests on the work of the above-mentioned observers some account of it is necessary and will be dealt with chiefly under the different substances examined.

Grape-juice and other vegetable substances.

Van den Broek (1858, 1860) filled small beakers with mercury which had been heated almost to the boiling-point, and then placed them under the air-pump to remove all traces of air as completely as possible. When no more air could be detected he inverted the beakers over previously heated mercury and introduced into them perfectly uninjured grapes. A small slice had been cut off the skin of each grape by means of a sterile scalpel, and by pressure on the grape the contained juice was allowed to ascend into the inverted beaker. Although kept for months or years the grape-juice showed no evidence of fermentation. Van den Broek also introduced into such unaltered grape-juice, bubbles of oxygen, generated from potassium chlorate and copper oxide, or even air filtered through cotton-wool, but still the grape-juice remained unchanged. When, however, he introduced yeast, fermentation ensued early. W. Roberts (1874) had a simpler technique. He took the grape in his fingers, heated a small spot on the skin, and pushed through the heated patch a sterile cotton-wool-plugged pipette (the so-called Pasteur pipette), and by compressing the grape the juice was made to ascend into the pipette, which was then sealed in the flame. The juice of eleven grapes tested in this way remained sterile in every instance, and Roberts was equally successful in obtaining sterile juice from oranges and tomatoes. With turnip or potato tissue he had some failures, although frequently he attained success. He considered that normal tissue had no power to originate germs, and that if organisms made their appearance they were imported from without. Pasteur (1876) also obtained germ-free grape-juice by puncturing the grape through a small cauterized area in the skin.

The question was deemed to have been settled by the above experiments, but as late as 1887 Galippe stated that even after using all the necessary precautions the juices of a large number of vegetables contained bacteria. He examined, particularly, carrots, onions, parsnips, turnips, potatoes, beetroot, salsify, leeks, and Jerusalem artichokes. Bernheim (1888) also asserted that large numbers of bacteria were to be found in the interior of maize, wheat, rye, and peas, and he believed that they ascended from the soil. A re-investigation of the subject by Fernbach (1888) led to different results. In 555 cultivations made from 98 tomatoes, turnips, beets, and potatoes, the juices and tissues showed no growths except in 35 (6·3 per cent.), and Fernbach considered that these were the result of accidental contaminations.

Blood.

Theodor Schwann (1837) found that blood taken directly from the vessels remained unaltered when subjected to air which had been heated. Blood was also obtained and kept sterile by van den Broek (1860). He introduced one end of a previously heated copper tube into the carotid artery of a dog. At the other end of the copper tube was a sterilized rubber tube leading into a beaker standing over mercury. The blood remained unchanged even when oxygen or filtered air was introduced into it. A trace of putrid meat infusion, however, rapidly set up putrefaction in the blood. Pasteur (1863, 1866) also preserved unheated blood by a technique which he described in detail in 1866 and 1876. Burdon Sanderson (1871), Roberts (1874), Lister (1878), Koch (1878), and Zahn (1884) were also successful, although they also had to record occasional failures. Roberts, who attempted to take blood from the tip of the finger, failed to get it sterile 4 out of 10 times, the risk of contamination being much greater than with the technique employed by Pasteur and Lister. With very imperfect technique, Johanna Lüders (1867) and Hensen (1867) always found bacteria in fresh blood.

Urine.

Normal urine was effectively preserved from fermentation by van den Broek (1858). An animal (dog or sheep) was killed and immediately cut open. The ureters and urethra being rapidly tied, the bladder was removed and immersed under sterile mercury. An incision was made into the bladder by a sterile scalpel, and the urine ascended into a vessel standing over mercury. No ammoniacal fermentation took place. W. Roberts (1874) collected freshly passed urine in large sterile test-tubes, and he then charged 8 sterile pipettes from these samples. Seven remained sterile. Pasteur (1866, 1876), by the same method as he employed to get sterile blood, also obtained and preserved urine in a sterile state. Lister (1875) and Watson-Cheyne (1879, 1882) also obtained sterile urine. Cazeneuve and Livon (1877) showed that if the urinary bladder is cut out it can be suspended in the air for several days without the contained urine undergoing any change. If, however, the bladder is punctured, alkaline fermentation with growth of bacteria rapidly takes place. G. Meissner (Rosenbach, 1880) had no difficulty in obtaining and preserving normal urine.

Eggs.

Van den Broek (1860) introduced an egg under sterile mercury, broke the shell with a sterile scalpel, stirred up the contents with the same, and then caused them to rise in a mercury-filled vessel. The egg contents remained sterile. Gayon (1873) was partially successful in preserving egg albumen obtained directly from eggs. W. Roberts (1874) collected the contents of eggs in sterile pipettes and succeeded in keeping the contents sterile 11 times in 15 experiments.

Milk.

It has been found to be exceedingly difficult to obtain milk in a sterile state. Schröder, it will be recalled, considered that milk could not even be sterilized by heat.

W. Roberts (1874), after several fruitless attempts, succeeded with a special apparatus in collecting milk in a sterile test-tube, from which samples were aspirated into protected Pasteur pipettes. In 10 tests only 3 samples were found to be sterile. G. Meissner also had difficulty with milk, but was ultimately successful in the case of a sample from a goat. The details of interesting experiments which Meissner made with this sterile milk are given by Rosenbach (1880). Lister (1878) succeeded only twice in 24 trials to get sterile milk direct from the cow. His results, however, along with those of Meissner, Roberts, and Watson-Cheyne (1882), sufficed to show that milk is not inherently fermentable, as many had believed.

Animal tissues.

The results of preserving tissues and organs of animals just dead have been discordant. Rindfleisch (1872) was the first to attempt this, and he was successful with minute pieces of muscle from sparrows, guinea-pigs, and calves. He employed small glass capsules, and everything used in the experiments was sterilized by alcohol and burning. He showed that it was essential to sterilize the distilled water employed. Chauveau (1873) added to our knowledge of the subject by his studies on the effects of the operation of *bistournage*, which consists in the subcutaneous torsion or rupture of the spermatic cord in animals. He studied the results in rams and showed that in the natural course of events the testicle becomes separated from its blood-supply and lies loose in its tunics separated from the air. In due course it atrophies and disappears, but it does not become septic. There is thus no inherent tendency for dying or dead tissue to undergo spontaneous fermentation. If, however, an incision is made into the scrotum after *bistournage*, infection readily occurs. Chauveau found that a similar result ensues if an intravenous injection of diluted pus is made before the *bistournage*.

Paschutin (1874), Servel (1874), and Nencki and Giacosa (1879) were not successful in preserving animal tissues without contamination. Tiegel (1874) carried out experiments which have often been referred to. He removed tissues from healthy animals immediately after death and placed them in paraffin heated to 110°–150° C. It was supposed that the heated paraffin would destroy the activity of any germ dust which fell upon the piece of tissue up to the moment at which it was placed in the flask. Subsequently the tissue was placed in paraffin at 52° C., which was then allowed to set. After a number of days the enclosed tissues were examined, and many were found to contain bacteria and were putrid. This was especially the case with liver, pancreas, spleen, and to a less extent with salivary gland, testis, muscle, and blood-clot. By this method or modifications of it Koukol-Yasnopolsky (1876) and Mott and Horsley (1880–2) were unsuccessful in preserving fresh tissues. On the other hand G. Hauser (1886), following older methods, made 121 experiments on the organs of rabbits and guinea-pigs and found 99 (81·81 per cent.) sterile for various times up to 366 days.

Special reference must be made to the successful experiments of Georg Meissner (1829–1905), the physiologist of Göttingen. His work, not published by himself, was made known by Rosenbach (1880), although Meissner had shown his preparations on several occasions in the Medico-Scientific Society of Göttingen. The date of Meissner's experiments is not given by Rosenbach, but presumably it was in the seventies of last century. After a long experience and many failures Meissner was successful in preserving unchanged the organs of cats and rabbits. Whole kidneys, spleens, pancreas, and pieces of liver from these animals as well as frogs' thighs were removed immediately after death and preserved in a dust-free atmosphere for two or three years. The exact technique employed is not given, but it is known that everything had been sterilized by heat. All glassware was plugged with cotton-wool and heated to 160° C. All the water was boiled. No sponges were used, but only sterile cotton-wool swabs. The operations were carried out in a dust-proof room, and the number of assistants was reduced to a minimum. The skin of the animal was generally stripped off completely.

In other words, Meissner used high-class aseptic methods long before they were in vogue in human surgery. The general result of his experiments showed that there was no inherent tendency to putrefaction in animal tissues. Among other things, Meissner succeeded in preserving the flesh of a kestrel that had been fed for weeks on carrion.

Zweifel (1885) maintained, however, that bacteria occur in normal animal tissues, and since 1899 a good many papers have been published to show that bacteria may be found in organs taken from normal animals just killed and examined by a technique to all appearance efficient. Thus Carrière and Vanverts (1899) constantly found bacteria in the spleens of normal dogs. Ford (1901) stated that at least 70 per cent. of the internal organs of domesticated animals, such as cats, dogs, and guinea-pigs, contained bacteria provided that a sufficient time was allowed to elapse between the removal of the organs and their subsequent examination. Wrzosek (1904), Selter (1906), Conradi (1909), Amako (1910), and Bierotte and Machida (1910) have also found bacteria in normal organs, the suggestion being that they have come from the air through the lungs, or from the intestine. Morgan (1904), although he found bacterial growths in 23 out of 80 organs of rabbits and dogs, was of opinion that they were accidental contaminations. The subject would appear to require further investigation before being finally settled.

V
PUTREFACTION AND PUTRID INTOXICATION, INCLUDING PTOMAINES

V

PUTREFACTION AND PUTRID INTOXICATION, INCLUDING PTOMAINES

PUTREFACTION (putrescence, putridity) is a process which must always have been observed by mankind, and it was very early regarded as a possible cause of disease in man and animals. The exact nature of the putrid process was, however, long a matter of doubt. It was supposed that there was a putrefying agent which induced the change, and from the fact that putridity once started seemed to be progressive it was believed that putrefaction and fermentation were similar processes, and they were in fact for ages classified together. For a history of the older views on the relation of putridity to disease the reader may consult the works of C. Chisholm (1810) and C. Gussenbauer (1882), the last-named work being particularly complete. The process of putridity was intimately associated with the gradual development of the idea of sepsis as a cause of various pathological conditions such as septicaemia, pyaemia, putrid infection, and putrid intoxication. It was supposed that these conditions were due to the absorption of putrid substances into the blood-stream. This idea was the natural outcome of the observation that the injection of putrid substances such as decomposing animal and vegetable matters determined symptoms or death in animals.

The basic work on this subject was that carried out by Bernard Gaspard (1788–1871), who, beginning his observations in 1808, only published them in 1822 and 1824. He made a series of over seventy experiments on dogs, sheep, foxes, and pigs, with putrid infusions, pus, vaccine lymph, blood, bile, urine, sperma, saliva, and various relatively simple putrefactive products such as carbonic acid, hydrogen, and sulphuretted hydrogen. In some cases he administered the putrid substances first and attempted to prevent the development of symptoms by various remedies. The putrid materials were injected by intravenous, subcutaneous, intrapleural, or intraperitoneal routes, and he described with great accuracy the symptoms and lesions which ensued. In the most acute cases nervous phenomena predominated such as vomiting, excitement, convulsions, dyspnoea, and heart-failure. Diarrhoea was frequent. After death he found ecchymoses on the pleurae, pericardium, and peritoneum, which latter usually contained a serous or sanguineous exudate. The principal lesion, however, was congestion in the mucous membrane of the alimentary canal. With small doses the symptoms and signs were correspondingly less. He arrived at the conclusion that carbon dioxide, hydrogen, and sulphuretted hydrogen were inert, whereas ammonia produced symptoms resembling those following the injection of the putrid fluid itself. Large doses of pus or small doses repeatedly given also caused death, and Gaspard's opinion was that pus was a fluid already in commencing putridity. An interesting observation of Gaspard was to the effect that the blood of a dog suffering from putrid intoxication could produce similar, if slighter, symptoms when injected into another dog. The symptoms of putrid intoxication were apparently annulled, in Gaspard's experience, by subsequent treatment with decoctions of quinine or gentian, but not by ordinary water.

Gaspard's observations were confirmed in the main by the work of F. Magendie (1823), Leuret (1826), Dupuy (1826), Trousseau and Dupuy (1826), Dance (1828, 1829), and Evans (1837), and some additions were made to our knowledge. Thus Magendie observed that a dose of putrid blood, lethal when injected intravenously, was without effect *per os*. He also found that the toxicity of a putrid fluid was greatly diminished by filtration through paper. He also proved by an experiment in which pigeons, rabbits, and guinea-pigs were suspended on a grill above a putrid mass that no ill effect followed the inhalation of the putrid effluvia. From the time of these early experiments nothing was added for nearly twenty years, when Virchow (1856) reported results on dogs injected with putrid matter, and he again confirmed the main facts which had been established by Gaspard and Magendie. It was about this time

JOSEPH, LORD LISTER
(1827–1912)

ÉLIE METCHNIKOFF
(1845–1916)

that the condition now called pyaemia began to be effectively studied. This is a pathological condition associated with the occurrence of multiple abscesses and fever, in which respects it differs from the effects of ordinary putrid intoxication, or septicaemia, as it began to be called. Septicaemia is a corruption of the word *septicoémie* introduced by Piorry (1837). The nature of pyaemia and septicaemia was a great mystery, and remained so for nearly thirty years. Originally Morgagni (1761), who described multiple-abscess formation following wounds, was of opinion that the pus was carried to the viscera and deposited there. This was the current teaching till the epoch-making work of Virchow (1856) on thrombosis and embolism, which is the accepted doctrine to-day. The actual agent in the thrombo-embolic lesion which determines the occurrence of the abscesses still, however, remained in obscurity and was only settled long afterwards. The study of the effects of putrid fluids still continued. Stich (1853) made many experiments on different animals and found that there were great differences in their susceptibility to putrid poisons. The only constant effect which he noted was an inflammatory or congestive condition in the alimentary mucous membrane. Thiersch (1855–6) experimented with the putrid dejecta of cholera patients. During the progress of putridity he placed small pieces of filter-paper in the fluid and then fed them to mice. Out of 55 experiments made on 110 mice, 47 mice were made ill and 14 died, with asthenia and diarrhoea. From various theoretical considerations Thiersch concluded that putrid poisons and contagia in general were of a 'metabolic' character and analogous to ferments in that they were in a state of decomposition, this state being transferable to other substances. Of much greater significance for the doctrine of putrid intoxication was the work of P. L. Panum (1856), at that time professor in Kiel. His work, published in Danish, was extremely exact with regard to the question of dosage, and he also tried to determine the nature of the putrid or septic poison. Putrid blood, flesh, brain substance, or human faeces injected into dogs caused intoxication. In large doses this intoxication came on acutely and was associated with cyanosis and the involuntary passage of urine and faeces. After death the blood was found to be dark and incoagulable. In most of the animals there was a latent period of from 15 minutes to 2 hours, while death occurred in from 3 to 6 hours. According to Panum this was the characteristic picture of putrid intoxication properly so called. Like Gaspard, he observed a haemorrhagic condition of the alimentary mucosa, and he further noted that putrefaction set in very quickly after the death of the animal. Panum tested separately many of the known putrefactive products such as butyric acid, ammonium valerianate, ammonium carbonate and sulphide, leucin, and tyrosin. None of them produced the characteristic signs of putrid intoxication. He filtered putrid infusions clear and found that the residue amounted to only 2·962 parts per thousand, and on distilling the clear filtrate he found that the distillate was inert but that the residue was active. This residue was insoluble in absolute alcohol but soluble in water. The residue, strange to say, was still active after continuous boiling for 11 hours at 100° C., from which Panum concluded that the active agent was neither a ferment nor a living micro-organism. He was of opinion that the proteins in a putrid fluid were not in themselves active but in some way attached to themselves the active putrid principle. The action of the latter was comparable in its intensity to curare and other alkaloids, for the minimal lethal dose for a dog was approximately only 12 milligrams. O. Weber (1864, 1865) made a very careful investigation into the question of pyaemia, septicaemia, and fever, and in a long series of experiments studied the production of metastatic abscesses, the effects of putrid fluids after filtration, and the effects of many known products of putrefaction after subcutaneous, intraserous, or intravenous inoculation. His final conclusion was that putrid substances act like ferments, but whether the ferment is corporeal or humoral he was unable to say. In any case he was of opinion that bacteria played no important part in putrid intoxication and allied toxic states.

Hemmer's extensive experiments (1866) were carried out, chiefly on cats and rabbits, with small quantities of pus and putrid fluids. He regarded the putrid poison as a protein substance in a state of chemical decomposition. Like Panum, he found it insoluble in alcohol but soluble in water, and coctostable. He considered that it acted as a ferment and induced putridity of the blood.

As a result of experiments made with putrid fibrin, Schweninger (1866) was of opinion that the action of putrid infusions is not due to any single poison, but that different putrefactive products from divers fluids examined at different times produce the symptoms and lesions of putrid intoxication. The idea that putridity might be brought about by microbes had been suggested by Schwann in his classical experiments in 1837; and Pasteur in 1863 definitely stated that putrefaction was produced by organized ferments of the genus *Vibrio*, and he described the appearances in point of time of the different bacteria, aerobic and anaerobic, which bring about the putrefactive changes in organic matter. These observations of Pasteur had a considerable influence on further work on putrefaction, in that the subject was continued along one of two lines. On the one hand a number of workers pursued inquiries intended to show that the phenomena of putrid intoxication were due to definite chemical substances: on the other hand a new direction was given to inquiries by the belief that putrefaction, putrid intoxication, wound infections, pyaemia, and septicaemia are related to the development of micro-organisms in the putrid matter or wound discharges.

Of those who continued the search for chemical poisons in putrid infusions may be mentioned the surgeon Bergmann (1868) and his co-workers in Dorpat, Ravitsch (1872), Samuel (1873), Panum (1874), Kehrer (1874), Zuelzer (1878), Hiller (1879), Blumberg (1885). Bergmann's paper had a great influence on the subsequent work of others. In conjunction with the chemist Dragendorff he made experiments on putrid yeast and came to the conclusion that the symptoms produced by its injection were due to a substance which, in conjunction with Schmiedeberg,

Bergmann called sepsin. They obtained this in the form of a crystalline sulphate. The whole question was very fully dealt with in the large book published by Arnold Hiller (1879), an ardent opponent of the microbic doctrine of disease. His mature opinion, as a result of very extensive experiments, was that putrefaction was due to chemical ferments which broke up the complicated chemical compounds of organic matter into simpler substances, and it was then and then only that bacteria played any part. Hiller was in essential agreement with Panum, Hemmer, Stich, and others, that what is called putrid poison is not a single poison but a congeries of toxic bodies which vary in quantity and quality in putrefying fluids. Ravitsch also was an opponent of the germ theory of putrefaction. He thought there was no strict analogy between infectious diseases and the effects of putrid infusions. As late as 1882, Rosenberger also was a supporter of the chemical view of the action of putrid infusions. He was able, as was Panum before him, to produce putrid intoxication from the injection of putrid fluids which had been heated to 140° C. for 2 hours and were in consequence germ-free. These heated fluids apparently were able to increase their activity in the body.

In spite of all these researches, which had been going on for sixty years, the question of putrid intoxication and its cause was not definitely settled. The lethal action of putrid fluids was variable and many substances not strictly speaking in the process of putrefaction were found to be lethal when injected into the animal body. The results in general were pithily expressed by Bouley when he said 'tout ce qui pue ne tue pas, tout ce qui tue ne pue pas'. For a time the failure to find a chemical poison in putrid infusions led to a cessation of work in this direction, although a final revival was seen in the 'seventies and 'eighties in the doctrine of cadaveric alkaloids or ptomaines as a cause of disease. This doctrine was created by the Italian chemist Francesco Selmi (1817–81), whose first work was published in 1872. He demonstrated that many of the so-called ptomaines gave reactions similar to those of the vegetable alkaloids.

An early worker in this line was A. Gautier, who introduced the term leucomaine for basic substances found in the living tissues, in contradistinction to the ptomaines or basic products of putrefaction. Gautier tells us that by leucomaines he merely meant to indicate their albuminoid nature (λεύκωμα—anything whitened). The principal exact worker on ptomaines was, however, L. Brieger (1881-6), who isolated and determined the composition of a number of these bodies. The relation of ptomaines to disease was at one time believed to be an important one, but this belief has not stood the test of time. A full account of the history and details of the subject will be found in the works of Brown (1887), Vaughan and Novy (1888, 1891), and Gautier (1896). In recent years the scientific work done on ptomaines in relation to disease has been small.

VI

DISCUSSION ON PYAEMIA, SEPTICAEMIA, AND SURGICAL SEPSIS

VI

DISCUSSION ON PYAEMIA, SEPTICAEMIA, AND SURGICAL SEPSIS

THE history of the modern ideas of that which is now called sepsis, and which formerly appeared in the literature as putrid intoxication, putrid infection, septicaemia, septhaemia, sephthaemia (Virchow), haematosapie, ichorrhaemia, sapraemia, surgical fever, &c., is a complicated subject and for its proper understanding involves a brief excursion into surgery and medicine. The confusion which pervaded the whole subject down to the end of the seventies of last century came from the fact that the above terms arose from clinical and anatomical rather than from experimental and aetiological studies.

In the previous section I have indicated the ideas which were gradually formed to explain the effects following the injection of decomposing and putrid substances into the bodies of animals. For the most part the symptoms were definite, and the condition was described as 'putrid intoxication' of the blood, a term which was more concretely called *septicoémie* by P. A. Piorry before 1837. In this condition, which was regarded as a kind of putrefaction in the living blood, the lesions were noted to be chiefly of a congestive character. In man similar cases were seen but were often associated with abscess formations in the viscera, and were apparently connected with suppurative lesions in the periphery of the body. It was supposed that the latter were the cause of the former and that the essential nature of the process was an escape of pus into the blood. It was for this reason that Piorry called it *pyoémie*, which was altered to pyaemia. In the first third of last century a number of investigations were carried out on the microscopic and chemical characters of pus. The question whether or not pus globules differ from the white corpuscles of the blood was freely discussed by A. Donné (1836), Bonnet

(1837), and others. Their work has been thoroughly summarized by Bérard (1842).

In addition a great many experiments were made to test whether pus, fresh or foul, can produce pyaemia when introduced subcutaneously or into the vascular system. The works of Boyer (1834), De Castelnau and Ducrest (1846), and Sédillot (1849) may be taken as examples of this kind of investigation. In general, it was not found possible to produce the metastatic lesions of pyaemia, although some observers recorded positive results. Long before this, however, another idea had been introduced by John Hunter in 1784 in his classical paper (1793) on the inflammation of the internal coats of the veins. Hunter's doctrine of phlebitis, although at first it did not arrest attention, ultimately became a great theme in the literature, as can be seen in the works of Hodgson (1815), Velpeau (1823), Ribes (1825), Dance (1828), James Arnott (1829), and at a later period Cruveilhier (1862) and others now scarcely worthy of attention. Cruveilhier expressed the opinion that 'la phlébite domine toute la pathologie'. The question whether septicaemia and pyaemia are two diseases or one emerges for the first time in the article of Bérard (1842). He considered that the conditions were separate, and this doctrine prevailed down to the time of Birch-Hirschfeld (1873). A very full account of the views prevailing in 1849 will be found in Sédillot's treatise on pyaemia. He was convinced that the *infection purulente ou pyoémie* is always produced by the admission of the pus globules into the blood, and he supported his contention in a series of 45 experiments along with numerous clinical records. Suppuration, he said, at some site precedes the pyaemia. There is a manifest correlation between suppuration of the veins and pyaemia. Pus can be positively demonstrated in the blood, and injection of pus intravascularly in animals causes the symptoms and lesions of pyaemia.

Shortly after the date of Sédillot's work a great change of ideas began to take place through the epoch-making researches of Virchow on thrombosis and embolism. He combated the older doctrines and showed conclusively the

embolic character of certain products previously believed
to be inflammatory. All subsequent research has demonstrated the correctness of Virchow's contentions and has
constituted the basis for further studies on purulent infection
and pyaemia. The decade 1850–60 was largely occupied
in the pathological world by researches confirming Virchow's views, while clinically it still continued to be
occupied with the old sterile discussions on whether pyaemia
and septicaemia were one disease or two. Most of the
investigations on the subject appeared in French or German
and little or nothing in English. About 1860 a new line
was taken, chiefly by Billroth, on the cause of the phlogogenic and pyrogenic properties found to result from the
absorption or injection of pus. The doctrine of 'surgical
fever' dates from this period. About the same time important experimental results were reported by Coze and
Feltz of Strassburg on the occurrence of 'infusoria' in
infective diseases. Their work was published in papers in
the *Gaz. méd. de Strasb.* in 1866, 1867, and 1869, and these
with additions were collected in their important book in
1872. They made experiments on dogs and rabbits, into
which they injected various putrid substances by divers
routes.

Among other things, they found that the blood of animals
which had been injected with putrid substances also produces symptoms when injected into normal animals, and
they showed that the lethal power of the blood was increased
in successive inoculations in animals. This belief subsequently aroused great interest through the researches of
C. J. Davaine (1872) and other French observers. Coze
and Feltz found bacteria which were motile (*B. punctum*)
or in chains (*B. catenula*) in the blood both during life and
after the death of animals which had received injections of
putrid fluids. They also reported the presence of bacteria
in the blood of human beings suffering from typhoid fever,
variola, scarlet fever, measles, and puerperal fever, and they
alleged that such blood was infective for rabbits. In puerperal fever they found '*bactéries aux chaînettes*' which could
infect rabbits. They attempted to cultivate the bacteria in

sugar-water containing sodium bicarbonate and ammonium lactate, but without success. The researches of Coze and Feltz had the merit of directing attention to the bacterial or, as it was then called, the 'zymotic' nature of septic diseases, but there still remained in the minds of physicians and surgeons great uncertainty and confusion. This is clearly emphasized in the paper of W. Roser in 1867. He refers in this to five theories then prevalent on the aetiology of pyaemia. He characterized these theories as (1) the mechanical, (2) the septic, (3) the zymotic, (4) the eclectic, and (5) the sceptical. In the 'mechanical' theory pyaemia was supposed to be caused by the entry of pus cells and other inflammatory products into the blood. The 'septic' theory attributed pyaemia to decomposing pus or other putrid substances. In the 'zymotic' theory, to which Roser himself was an adherent, pyaemia was regarded as a specific partly contagious, epidemic, endemic, or sporadic infection, and included such diseases as puerperal fever and traumatic erysipelas. In the 'eclectic' theory the disease was regarded as due to various causes; whereas the sceptics maintained that there was no such thing as pyaemia at all. What had been called pyaemia was in their opinion a collection of diseases, the result of phlebitis, thrombosis, leucocytosis, or zymotic poisons. Hueter (1869), a surgical authority of this period, was also confused on the subject of pyaemia and septicaemia. The prolonged discussion on purulent infection carried on in the Academy of Medicine in Paris in 1869 will also bear testimony to the chaos of ideas prevailing in France among the surgical and medical leaders of the period.

The germ theory of disease was, however, now coming into evidence and was supported by the fundamental work of Pasteur on fermentation and that of Lister on the effect of antiseptic treatment of wounds. Semmelweis's observations on the methods of abolishing puerperal fever, although they had been made known in 1847, had been discredited and relegated to oblivion.

Fresh interest was revived on the subject of septicaemia by the experimental work of C. J. Davaine (1872), who had been long known for his discoveries in connexion with the bacteridium of anthrax and its mode of action. Taking up the work of Coze and Feltz on the effect of the injection of putrid blood, he confirmed it and made an elaborate study of the experimental disease produced, which he called septicaemia. He found that the blood of animals suffering or dead from putrid injections is capable of inducing a similar condition when introduced into a normal animal, and it can produce a lethal effect from incredibly minute doses. It appeared that in successive animals the putrid element became more and more virulent.

Starting with putrid blood, Davaine carried the injections consecutively through a series of 25 rabbits. The first rabbit died in 40 hours from the subcutaneous injection of 10 drops of putrid blood. A second rabbit was inoculated with blood from the first, and a third from the second, and so on. Already in the 5th generation the lethal dose of blood was $\frac{1}{100}$ of a drop. In the 15th generation it was $\frac{1}{40,000}$ of a drop. By the 20th generation it was $\frac{1}{100,000,000}$ of a drop. In the 25th generation the blood had become so virulent that it was lethal to rabbits after subcutaneous injection of 1 billionth (French trillionth) of a drop. Another rabbit which received a dose of 1 hundred-million-millionth of a drop survived. In Davaine's words: 'Le virus septicémique acquiert donc une plus grande activité en passant par l'économie d'un animal vivant' (l.c., p. 914). Davaine studied the rate of increase of virulence by what now came to be called 'passage' (i.e. the inoculation of animals in series), and he found that it took place much quicker than it had appeared in his earlier experiments. Thus in a series of three generations of rabbits he found that the first animal died from $\frac{1}{100}$ of a drop of putrid blood. The blood of this animal killed a second rabbit in a dose of 1 ten-millionth of a drop, and from the blood of the second rabbit a third was killed by a dose of 1 billionth (French trillionth) of a drop. It thus appeared that a maximum degree of virulence had been established after three passages.

Davaine found that the virulence of such blood disappears on keeping. He also found that the minimal lethal dose varies not with the size but with the species of the animal injected. Inside a particular species, young were more susceptible than older animals. Rabbits were found to be much more susceptible than guinea-pigs to the septicaemic blood. Rats and fowls withstood large doses of blood highly virulent for rabbits.

The virulence of putrid blood was found by Davaine to bear no relation to the degree of putridity except that as the blood progressed in putrefaction the virulent element appeared to be diminished. With regard to the nature of this virulent element Davaine held that it was identical with the 'ferment of putrefaction'. Whether Davaine meant the word 'ferment' in Pasteur's sense as a living agent or whether he used the word in its chemical sense is not apparent. The time was not yet ripe to speak freely of disease germs. In Davaine's words: 'La septicémie est donc une putréfaction, putréfaction qui s'accomplit dans le sang d'un animal vivant.' It is true, he said, that septicaemia is not accompanied by the characteristic odour of putrefaction, but this results from the fact that the odorous principles have been eliminated, whereas the ferment remains and accumulates in the animal economy, which it finally destroys. Davaine did not believe that the septicaemic virus was the only one in putrid blood, but to demonstrate others it would be necessary to have some method of isolation. 'Ce moyen ne sera peut-être pas toujours introuvable' was Davaine's surmise.

Davaine's work on septicaemia of rabbits aroused great interest in France and elsewhere, and while many supported his views others were opposed to them, at any rate in part. Among the latter may be mentioned Bouley (1872), Behier (1873), and especially Colin (1873), who found that the production of Davaine's septicaemia was not a constant result of the injection of putrid blood. Colin stated that substances like blood may be virulent before the appearance of putrefactive bacteria, and that the virulence really disappears with the onset of putrefaction. On the other hand, the essential experimental facts elicited by Davaine were confirmed by Vulpian (1872), Clementi and Thin (1873), U. Dreyer (1874), Feltz (1874), and others. Vulpian in particular expressed the view that Davaine's septicaemia was characterized by the presence of bacteria in the blood and was indeed a parasitic disease which might appropriately be called *bactériémie*. Dreyer also found bacteria in the form of small granules or, more rarely, rods in all the animals that died of septicaemia. Onimus (1873) opposed the view that bacteria play any important part in the disease on the ground that there appeared to be no relation between the virulence of the blood and the number of bacteria present. To show the harmlessness of bacteria he dialysed putrid blood through parchment. On both sides of the membrane bacteria were found in abundance, but whereas the blood itself was virulent the dialysate was not.

Laborde (1874) injected putrid matter into dogs and then coupled the femoral artery to the similar vessel in normal dogs. In the latter, symptoms regarded as septicaemic were developed, although in the blood of the donor no bacteria could be found. Notwithstanding all this experimental work, which, in the main, was a French product, the question of septic diseases continued to be discussed among clinicians, particularly on the Continent. A very full account of the whole story will be found in the masterly essay of Chauvel (1880-1). The subject was also discussed by a committee appointed by the Pathological Society of London. It was pointed out in their report that at the time (1879) septicaemia was a word without definite meaning. It appeared to include two states, the one apparently due to a chemical poison, the other a progressive infective process. Pyaemia was regarded as a similarly confused concept. The committee suggested that the various conditions should be grouped into:

(1) Septic intoxication: effect of poisoning by chemical products—non-infective.

(2) Septic infection: due to some peculiar constituent of putrid matter in the blood. By some, this peculiar constituent was regarded as a living organism; by others as a non-organized ferment.

(3) Pyaemia—a process similar to (2) but giving rise to metastases.

(4) Thrombosis—with softening of the thrombus and embolism.

The word 'sepsis' began to lose its etymological significance as being identical with putrefaction and the idea of rottenness. This change in the meaning of sepsis and septicaemia was largely the outcome of the experimental work of Davaine, who, starting with something which was, in the strict sense of the word, putrid, had produced a disease, septicaemia, which was manifestly not putrid. This altered sense of sepsis bulks largely in the writings of Lister at the end of the 'seventies and has continued to the present time. Its connexion with putridity has largely disappeared and it has become almost synonymous with infective, for the word has been used in the widest sense, notwithstanding that attempts have frequently been made to limit its meaning. The connexion of wounds with septic processes so called was an early observation, although writers like Wunderlich (1857) believed that there was such a thing as 'spontaneous' pyaemia indicating that no local lesion could be found to account for the generalized morbid process. Leube (1878) showed that a local focus could be surmised in many cases of supposed spontaneous pyaemia, and it was from him that the idea of 'cryptogenetic' septicaemia or pyaemia took its origin. When the ordinary forms of sepsis were excluded there was still a remnant of belief in putrid intoxication to which Matthews Duncan (1880) gave the name 'Sapraemia'.

Prior to this, Klebs (1871, 1872) had given a great impetus to the germ theory by his classical researches on the pathology of gunshot wounds. The material examined by him was obtained in the Carlsruhe military hospitals when, in a very short time (17 Aug.–17 Oct. 1870), he had 115 autopsies of which 73 per cent. showed the occurrence of septicaemia and pyaemia. Klebs carried out microscopic examinations in fresh and preserved specimens and found bacteria of different forms in nearly every case. Under the influence of the erroneous beliefs then prevailing, he regarded these different microbic forms as one organism to which he gave the name of *microsporon septicum*. Wrong as this turned out to be, Klebs added much to the existing knowledge of wound infections and created a basis on which stand the views now held.

These modern views really started with Robert Koch (1878) in his epoch-making work on the *Aetiology of Traumatic Infective Diseases*. This small work of eighty pages, written while Koch was still in medical practice at Wollstein and far from academic influences, was totally unlike anything that preceded it on the subject of septic diseases. His object was to determine whether the infective diseases of wounds are of parasitic origin or not. He had to confine himself exclusively to animal experiments, but he was able to show, in a manner practically conclusive, that a series of diseases, differing clinically, anatomically, and in aetiology, can be produced experimentally by the injection of putrid materials into animals. He admitted that the work of many previous investigators had rendered the parasitic nature of disease probable. Conclusive proof had not been obtained and in Koch's opinion could only be obtained by

'finding the parasitic micro-organisms in all cases of the disease in question, when we can further demonstrate their presence in such numbers and distribution that all the symptoms of the disease may thus find their explanation, and finally when we have established, for every individual traumatic infective disease, the existence of a micro-organism with well defined morphological characters'.

These were the problems that Koch set himself to solve.

In the first place he used methods which were new or at any rate had not been applied to such work before. Fluids and tissues were examined especially by the methods of staining by anilin colours (methyl violet, fuchsin, anilin brown) which had immediately before been recommended by Weigert (1878). Koch distinguished clearly the so-called 'structure picture' from the 'colour picture', and by the aid of the illuminating apparatus invented by Abbe and made by Zeiss in Jena, in association with high-class oil immersion lenses, he was able to obtain clear pictures of bacteria which were otherwise indistinct or invisible. The application of these methods to bacteriology had a great and lasting influence on all subsequent research on infective disease. Koch described six different infective diseases which he induced by the injection of putrid fluids. These diseases he called (1) septicaemia in mice, (2) progressive destruction of tissue (gangrene) in mice, (3) spreading abscesses in rabbits, (4) pyaemia in rabbits, (5) septicaemia in rabbits, and (6) erysipelas in rabbits. He compared his results on these diseases with those in anthrax. It is unnecessary to go into great detail with Koch's studies. That is the province of special bacteriology. Suffice it to say that he clearly demonstrated that none of the experimental diseases produced were identical either in symptoms, anatomical lesions, or in the morphological characters of the micro-organisms found. He showed that the distribution of the bacteria varied, and that the susceptibility of animals varies. Taking advantage of this fact he was able, in at least one instance, to separate two micro-organisms occurring together and otherwise inseparable.

In the congeries of bacteria present in putrid fluids he clearly differentiated pathogenic from non-pathogenic types. He was not fortunate enough to meet with the particular type of septicaemia described by Davaine, and he was not impressed with the theory of progressive exaltation of virus by passage. In the pyaemic disease of rabbits he could find no evidence of exaltation. He noted merely that the smaller the dose of blood injected the longer was the time which elapsed before death ensued, and he explained this on the supposition that the blood always contains a like quantity of undissolved infective particles, and that these particles must have increased to a certain number before they could produce death. In the case of mouse septicaemia, which was analogous to Davaine's septicaemia, Koch found that when he took from the second or third animal the smallest lethal quantity of blood, the maximal virulence had already been reached and remained constant. He explained this observation by the supposition that from the mass of bacteria in the original putrid fluid the mouse septicaemia bacillus is gradually emancipated or cultivated by its

escape into the blood-stream. The rapid increase of virulence in three passages had also been admitted by Davaine
(1872) and was confirmed by Gaffky (1881).

Koch's book created something of a sensation, although
there were many of the old school who could not accept his
views. As illustrating this I may say that I possess a copy
of Koch's book which was formerly in the possession of a
well-known surgical writer of the time. He had evidently
written his very free criticisms and opinions on the margins
of the various pages. These criticisms contain such remarks
as, 'all this is very strange'; 'how is this proved?'; 'all this
seems very recondite but full of fallacies'; 'the meaning
is as obscure as the grammar'; 'Koch nowhere discusses
the possibility of the diseases being the cause of what he
calls micro-organisms'; 'I see no proof that the poisonous
matter is not something chemical'; 'all sorts of contra-
dictions could be explained by hypotheses of this kind but
they leave the reader in a fog'; 'I confess I do not follow
what Koch says'. These remarks by one unable to see the
new light that had arisen will give a picture of what the
germ theory had to encounter even in 1878.

Although Koch's original intention was to determine
whether the infective diseases of wounds are of parasitic
origin or not, he was not able to get material for this
purpose, and thus there was a hiatus in his work.

This hiatus was filled up in masterly fashion by Alexander
Ogston (1880, 1881, and 1883), at that time assistant-
surgeon to the Aberdeen Royal Infirmary. Following
Koch's technical methods exactly, he examined for bacteria
100 abscesses of which some were 'cold' but many were
acute. No micro-organisms could be found in the former,
whereas in all the latter micrococci were found sometimes
in groups like the roe of fish, at other times in chains. He
showed that the difference in appearance was due to a
difference in the mode of fission. He made a very large
number of determinations by means of the haemocytometer
and found that the average number of cocci per cubic
millimetre of pus was nearly 3 million, although in individual samples it varied between a minimum of 900 and a

maximum of 45 million. In addition to cocci other micro-organisms were occasionally present. Thus he figured spirilla and fusiform bacilli in alveolar abscesses. The injection of the pus of cold abscesses, in which no micro-organisms could be found, was without pathogenic effect on mice.

A very different result occurred, however, when similar injections were made with pus containing micrococci. Ogston traced with great clearness the symptoms and lesions, and showed by the process of counting that the cocci in the experimental lesion must have increased greatly in numbers, and that the experimental inflammatory disease with abscess formation could be propagated in series. No increased virulence was noticeable on passage. Pus heated so as to kill the cocci, or treated with phenol was found to be inert. Ogston examined lesions other than abscesses. Among them were gonorrhea, soft chancre, sycosis, sputa from phthisis, discharges from wounds and ulcers. Micro-cocci were found in all. He attempted to cultivate the cocci in glass cells containing Cohn's or Pasteur's fluids, ascitic and ovarian fluids, aqueous humour or human blood. The results were variable. Aerobic and anaerobic cultures were tried and ultimately he succeeded with fresh eggs into which he introduced a small quantity of pus. After a time the cocci were found to have grown abundantly and with the cultures Ogston produced typical abscesses by inoculation of mice.

From all these observations Ogston concluded that micrococci produce inflammation and suppuration. There was microscopic evidence of their proliferation locally; the cocci invaded peripherally and might pass into the blood. In abscesses they were finally excluded by the formation of a delimiting wall of granulation tissue which arrested their invasion and led to their final extrusion among the pus corpuscles. In a subsequent paper Ogston (1882) gave a general statement of his views on 'micrococcus poisoning'. He considered that micrococci were of two kinds, the one arranged in chains (*Streptococcus*, Billroth) and the other in masses, which he named *Staphylococcus* (σταφυλή, a bunch of grapes). He connected local infections and spreading phlegmons ending in pyaemia and septicaemia as part and parcel of the same process and as being differences in degree rather than in quality. He opposed the view that pyaemia and septicaemia were blood diseases, and showed that the blood was merely the vehicle which may generalize in the body what, without its aid, would be only a local process. Ogston contended that there were no such diseases as pyaemia or septicaemia *per se*. These conditions are the sequence of local foci of micrococcus growth and are merely the expression of malign influences coming from this source. The variations in the course and termination of infective processes were attributed by Ogston to specific differences in the micro-organisms, their virulence, the organ or structure involved, and the susceptibility of the individual affected.

Ogston's publications, advanced and original as they were, soon exerted an influence on the surgical, patholo-gical, and bacteriological doctrines of suppuration and inflammation. The main criticism brought against his teaching was that micrococci are not the only organisms in the production of suppuration. By means of gelatine cultures Rosenbach (1884) and Passet (1885) split up Ogston's *Staphylococcus* into three species and added one or two organisms which have not proved to be important. At a later period the pyogenic properties of other organisms have been definitely established.

For a brief time there was an active discussion whether suppuration was necessarily the work of bacteria. A number of investigators, to name Uskoff (1881), Orthmann (1882), Councilman (1883), Grawitz and de Bary (1887), Christmas (1888), Kreibohm and Rosenbach (1888), claimed that chemical substances such as turpentine, croton-oil, or even bland substances may produce typical suppuration. These workers were opposed by Straus (1884), Scheurlen (1885), G. Klemperer (1886), who maintained the micro-organismal origin. A good account of this work will be found in Steinhaus (1889) and Janowski (1889). While it is now admitted that suppuration can be produced by various acrid chemical substances, no one doubts that these scarcely play a part in the ordinary suppurative disease processes in man, and it still remains essentially true that acute abscesses, as Ogston first showed, are mainly coccal in their aetiology. His views on the essential relation of pyaemia and septicaemia to the local phlogogenic processes have also proved to be the true teaching and have remained to a large extent unchanged. We thus come to an end of a chapter, in pathology and bacteriology, which long remained barren in its results but was ultimately clarified by the technical methods pursued with such diligence and with such mastery by Koch and his successors.

VII

SPECIFIC ELEMENT IN DISEASE

VII

SPECIFIC ELEMENT IN DISEASE

In Chap. II I traced the beginnings and development of the doctrine of animate contagion down to the end of the eighteenth century. In spite of the microscopical observations on the existence of animalcules by Leeuwenhoek and his successors the attempts to apply their results to the understanding of disease were entirely theoretical and had no immediate or even important influence on the doctrinal teaching of the day. Even towards the end of the eighteenth century the ideas on infective and contagious disease were nebulous and inexact. This was only part and parcel of the confused state of medical knowledge of the period. The idea of specificity of disease as we understand it was non-existent. It is true that Thomas Sydenham (1624–89), a century before, had clearly recognized that they were different diseases with proper and peculiar symptoms, and he was an upholder of the belief that there were species of diseases just as there were species of plants. He gave lucid accounts of the differential diagnosis of contagious and infective diseases like small-pox, dysentery, plague, and scarlet fever, and he believed that these conditions required separate cures. He admitted, however, that it was 'very hard to reduce all the species of epidemics into classes and to decipher the idiopathic character of each and to accommodate a method of healing particular to each'. It is apparent, however, from a study of Sydenham's writings that he had no clear ideas of specific causes of disease, and he had no immediate successors even to carry on what he had taught. The nosology of the eighteenth century degenerated in the hands of Boissier de Sauvages (1731, 1768), Linné (1763), and Cullen (1769) into mere catalogues of genera and species on a symptomatological basis.

A new advance, however, began to take place, especially in France, at the end of the eighteenth and the beginning of the nineteenth century, and this was possible through the important additions to knowledge from a deep study of pathological anatomy. A pioneer in this advance was Philippe Pinel (1755–1826) in his *Nosographie philosophique* (1802). His classification of inflammations (phlegmasiae) was particularly important. He recognized five orders of phlegmasiae according as they affected (1) the skin, (2) the mucous membranes, (3) the serous membranes, (4) the cellular tissue and parenchymatous organs, or (5) the muscular, fibrous, or synovial tissue. The idea of special tissues and their diseases was greatly advanced by Xavier Bichat (1771–1802), especially in his *Traité des membranes* (1800, 1802). The anatomical school attained a supreme position through the labours of G. L. Bayle (1774–1816), and particularly of Laennec (1787–1826), with their doctrines on tubercle. Diseases were no longer classified as genera and species on a symptomatic but as entities on an anatomical basis. It was the anatomical basis that Broussais (1772–1838) bent his great strength to demolish. Revolutionary in his ideas, he denounced the anatomical school and all its writings and attempted (1816) to set up in its place his doctrine of *médecine physiologique*. He believed that the differentiation of morbid forms was entirely factitious (*entités factices*). He considered that symptoms follow no rules. Tubercle was merely an inflammation and all inflammations or phlegmasiae were due to a diminished or increased degree of 'irritation'. He fanatically opposed everything that savoured of the idea of *spécificité*. By virtue of the important position he held at the Val-de-Grâce school, Broussais exerted a great influence on the medical thought of his time.

His doctrines were, however, not destined to survive, and indeed during his lifetime were completely overthrown by the highly original investigator P. F. Bretonneau (1778–1862) of Tours, who was the real founder of the doctrine of the *spécificité* of disease as we understand it to-day.

By a brilliant combination of faculties and on an extensive basis of exact clinical and pathological observations, Bretonneau asserted the doctrine of special or 'specific' diseases, but in a manner more exact than any of his predecessors. His belief, which almost amounted to proof, was that it is the *nature* of morbid causes rather than their *intensity* which explains the differences in the clinical and pathological pictures presented by diseases. Bretonneau concerned himself especially with enteric fever—*dothinentérie* as he called it—and diphtheria, and from his studies on these diseases he created the doctrine of aetiological specificity. Diphtheria and dothinentérie were specific phlegmasiae. Each was 'une affection morbide *sui generis*'. The specificity was partly anatomical but particularly aetiological. Thus, diphtheria is a lesion of structure characterized by the production of a false membrane or pellicular exudate, but as there are several pellicular exudates he claimed that there was only one true diphtheria which owed its characters to a special virus acting on the body to produce the diphtheria membrane. Bretonneau read two communications on his views to the Académie Royale de Médecine in 1821, but they remained unpublished. At the urgent request of his pupils, particularly Trousseau, he published his work in 1826. Continuing his researches, Bretonneau engaged in a profound study of what is now called enteric or typhoid fever, but which at the time lay concealed under such names as gastro-enteritis, putrid, adynamic, ataxic, or entero-mesenteric fever. Although he discussed his views freely with his pupils, and although Trousseau (1826) explained his master's views, Bretonneau himself published only one paper (1829) on the subject during his life-time. He was a habitual procrastinator and much that was known to be his soon fell into the hands of others. Indeed, Bretonneau's work on dothinentérie was only published in 1922, a century after its inception, by one of his descendants, L. Dubreuil-Chambardel. In comparison with the work of his predecessors Bretonneau's treatise on dothinentérie now appears before us as a masterpiece, and it is a great misfortune that it lay buried unknown for so long. Many of his views have appeared in works like that of P. C. A. Louis (1829), but there still remain important doctrines the significance of which Bretonneau, alone of his time, foresaw.

Discussing the question of contagion in typhoid fever, he asserted that diseases eminently contagious need not be taken as the type of all contagion. 'A disease may be transmissible without being so in the same degree or in the same manner as small-pox.' In addition to diseases essentially contagious and developing as the result of their *germe reproducteur* he recognized that other affections may up to a certain point appear to be communicable. Dothinentérie, in Bretonneau's opinion, was a case in point. He held that it was epidemic and contagious, and that it does not originate *de novo*. He traced its fluctuations and concluded that the *principe générateur* of the disease may attach itself to various bodies and preserve its properties until such time as it is enabled again to exert its deleterious effects if it is introduced into a susceptible subject.

Bretonneau's third memoir on 'specificity' is known to have been begun in 1828, but was never finished. As it appears before us now it consists of four chapters only, two of which, on diphtheria and enteric inflammations, are more or less complete. The third and fourth chapters on dysentery and malaria respectively exist only as headings without letterpress. It is clear, however, from what Bretonneau has given us in his *Traité de la spécificité* that he had grasped the whole problem with marvellous precision. All that is lacking are the specific bacteria of typhoid and diphtheria. His conception of *spécificité* was aetiological. Specific disease 'développe sous l'influence d'un principe contagieux, d'un agent reproducteur'. Again, he says,

'C'est surtout la cause de l'inflammation qui contribue à la modifier, c'est cette cause qui bien souvent lui imprime un caractère particulier durable et qui se conserve encore longtemps après que son action a cessé . . . chaque espèce de l'inflammation peut varier et varie beaucoup en son intensité en conservant les caractères qui lui sont propres.'

In Bretonneau's opinion

'une multitude d'inflammations sont déterminées par des causes matérielles extrinsèques, par les véritables êtres venus de dehors ou du moins étrangers à l'état normal de structure organique.'

It is somewhat remarkable that Bretonneau did not attempt to connect the 'reproductive agent' of infection which he so clearly foresaw with microscopic animalcules and 'infusoria', which were already well known and much discussed. Apart from theoretical fancies like those of Benjamen Marten (1720), M. A. Plenciz (1762), A. Neale (1831), H. Holland (1839), and others, few seemed to have thought of the possibility that contagia might be related to microscopic animalcules in a more exact way. Somewhat in advance of his predecessors in this direction was the Italian, Enrico Acerbi (1822), who postulated in typhus fever the existence of parasites capable of entering the body and multiplying there to produce disease (l.c., p. 362). The actual demonstration of this possibility was reserved for Agostino Bassi of Lodi in 1835. He is justly regarded as the real founder of the doctrine of pathogenic micro-organisms of vegetable origin. Bassi (1773–1856) was not a trained scientist nor even a doctor except of Laws. For a considerable period of his life he was a civil servant in Lodi, and his scientific training consisted of some courses of lectures which he attended in early life in the University of Pavia. He developed, however, a great passion for scientific pursuits and so injured his eyesight that he became half blind. Forced to give up his occupation, he betook himself to agriculture and wrote on potatoes, vines, and like subjects. Through ill health and bad luck he became greatly reduced in circumstances and at one time almost starved. It was under such conditions that he carried out his classical observations on diseases of silk-worms and especially on that disease called by the French 'muscardine' and by the Italians 'mal del segno', 'calcinaccio', 'calcino', or 'cannellino'. In Bassi's time this dreaded disease had caused extensive damage to the silk-worm industry in Lombardy. The affected worms become covered, after death, with a peculiar white efflorescence of a hard, lime-like consistence. A certain amount of work had been done on the disease which was generally looked upon as developing spontaneously as the result of unknown factors. By prolonged efforts Bassi came to the conclusion that the disease was really contagious. The characteristic efflorescence occurs only after death, but Bassi showed that long before this the worm was infective. Taking the worm in the early stage, he removed a portion of the epidermis and having passed the worm through the flame he touched the underlying tissues with a heated pin and showed that a scratch with this pin could transmit the disease to a healthy silk-worm. He worked out methods for the prevention of the disease in the silk-worm nurseries and arrived at a very correct understanding of the whole morbid process of 'calcino'.

Labouring at the subject for nearly twenty years and often suffering cruel disappointments, he ultimately was able to formulate his theory of contagion in a communication to a commission of the medical and philosophical faculties of Pavia in 1834, and in the following year he published the first and theoretical part of his great work, *Del mal del segno, calcinaccio o moscardino*, &c. In 1836 he published the practical part of his work, and in 1837 there was a second edition of the whole. By microscopic observations carried out with great difficulty on account of his defective eyesight, Bassi recognized the real agent of 'calcino' as a cryptogamic fungus of parasitic character ('una pianta del genere delle crittogame, un fungo parassito'). Further than this he was unable to go, but the cryptogam was identified by the Milanese botanist, G. Balsamo-Crivelli (1835, 1838), who found it to be *Botrytis paradoxa*, but ultimately renamed it in honour of its discoverer *B. bassiana*. Balsamo recognized the great importance of Bassi's observations and foresaw that they would lead to great developments. Bassi realized that the muscardine disease was transmitted by contact and infected food. He showed that the fungus comes to maturity only after death, and he was of opinion that dissemination occurs from 'seeds' produced by the plant. He made a great many ingenious experiments to determine the manner in which dissemination occurs and he showed that the fungus may survive as long as three years. He tried to grow the fungus outside the body but was unsuccessful. With regard to the practical problems of prevention and cure, he stated in the clearest possible manner the principles involved and in a manner formed the basis of all the doctrines on this subject held to-day. Bassi's statements were confirmed by Balsamo-Crivelli (1835), Audouin (1836), and Montagne (1836). At a later period Vittadini (1852) made important additions to knowledge and succeeded in growing *Botrytis bassiana* on such substance as honey, oil, sugar, mannitol, silk-worm blood, and isinglass. Vittadini also recognized the spore of the fungus.

During his lifetime Bassi was the recipient of many honours and distinctions. He had, however, his detractors and some attempts were made to belittle his discovery. The works of Bassi on 'calcino' are evidently exceedingly scarce and the original editions were probably very small. At any rate I have not been able to find them in any of the largest libraries. Outside Italy they are known only in a very general way. Recently (1925), however, a great service has been done by the Società medico-chirurgica di Pavia in publishing Bassi's *Opere*. It is from this large volume of 673 pages that I have derived most of my information of Bassi and his scientific works. After the publication of *Del mal del segno* in 1835, 1836, Bassi seems to have done no more microscopic work through the onset of blindness, but he continued to develop his theory of contagion from living parasites in such diseases as variola, typhus, plague, syphilis, wounds, gangrene, cholera, and pellagra. He also wrote extensively on the use of germicides such as heat, alcohol, acids, alkalies, sulphur, and chlorine. For diseases like cholera he advocated immediate isolation of the patient and disinfection of the clothes and excreta. In vaccinating children in series he strongly counselled the necessity of sterilizing the needle between each vaccination so as to prevent complications or the transference of diseases other than vaccinia. It is impossible to read the works of Bassi without realizing that he was a pioneer and that he was, as Italians like Calandruccio, Riquier, Grassi, and Monti have always maintained, the founder of the doctrine of parasitic microbes and a precursor of Schwann, Pasteur, and Koch.

Another work which exerted an influence on the doctrines of bacteriology at this period was *Recherches microscopiques sur la nature des mucus . . .*, published in 1837 by the French microscopist Alfred Donné (1801–78). In this he drew attention to certain microscopic living organisms occurring in pathological discharges, especially of the human genital organs. In cases of blenorrhoea he only found pus globules, but in chancres of the glands he found, in addition, a large number of animalcules 'ayant la forme des vibrions décrits par Müller sous le nom de Vibrio lineola'. Donné gives us a good figure of these from which it is evident that he had before him a spirochaete (probably *S. refringens*).

He saw similar 'vibrios' in a chancre of the vulva, a syphilitic foetid ulcer of the leg, and in cases of balanitis, but they were absent from pus of non-syphilitic cases and from the normal smegma. Donné went farther and endeavoured to find out by inoculation whether the vibrios bore any causal relation to syphilis. He first produced blisters artificially on the glans and on examining the exudation found it was merely purulent but devoid of vibrios. Having ascertained that vinegar exerts a deadly influence on the vibrios in pus he then made experiments on human beings, in which fresh pus containing masses of living vibrios was compared with the same pus but acidulated. The results were variable. On the right leg of a patient he introduced some of the unaltered pus, and on the left, pus treated with vinegar. Two days later a pustule appeared on the right side and contained masses of vibrios but nothing happened on the left. A repetition of this experiment gave a like result. In a third, the pus both treated and untreated produced a positive result, whereas in three other experiments neither the treated nor the untreated pus produced any result at all. He was therefore unable to reach any definite conclusions on the significance of the vibrios in syphilitic discharges.

It was in the same work that Donné, examining vaginal mucus, first described and figured the flagellate protozoon, called by him *Trico-monas vaginale*. R. Wagner (1837) examined animal secretions but could find no monads or infusoria except once in a cancer of the lip where they were numerous.

It was at this time (1836–7) that Cagniard-Latour and Schwann first showed the true nature of yeast and its relation to fermentation (see Chap. III). The works of Bassi, Donné, Cagniard-Latour, and Schwann, although not at first recognized as containing fundamental truths, ultimately exerted a profound influence on the direction of medical thought in the middle of last century. A search began for parasitic organisms in disease and this search was greatly influenced by Henle's *Pathologische Untersuchungen*, published in 1840. The interest in epidemic disease had been quickened by the advent of Asiatic cholera on European soil in 1832 during the second pandemic of that disease. Various theories were brought forward at the time to explain the cause of cholera, but it cannot be said that these theories were more lucid than others promulgated for other epidemics centuries before. Correctly to appreciate the importance of Henle's work, it will perhaps be clear if I refer to another work immediately preceding it. This was the general and special pathology, *Allgemeine und specielle Pathologie und Therapie*, published by J. L. Schönlein (1793–1864), at the time (1839) one of the leading physicians and teachers in Europe. Schönlein defines contagia as deleterious agencies (*schädliche Potenzen*) which originate in the body and excite disease, and which are communicable to another organism and excite similar disease. Contagia, in his opinion, are developed from spontaneous diseases either in single or in many individuals and they arise from acrid humours (*Schärfe; acrimonia*) which are peculiar derangements of the chemical properties of the body. Schönlein distinguished two aspects of the contagion problem, viz. (1) the basis or materies contagii, and (2) the aethereal or active principle. The nature of the latter he held to be unknown although, in his opinion, it was undoubtedly allied to free electricity (*gewiss etwas der freien Electricität verwandtes*). The *materies* of contagion occurred in the form of droplets, vapours, or gases. This may be said to have been the ordinary teaching on the subject of infection and contagion in 1839. Henle (1809–85), who was afterwards famous as an anatomist, was in 1839 assistant to Johannes Müller, the physiologist in Berlin, and the intimate friend of Schwann, who had already become famous for his work on fermentation (1837) and the cell theory (1839). It is only the first section of Henle's *Pathologische Untersuchungen* which concerns us here. It is entitled, 'Von den Miasmen und Contagien und von den miasmatisch-contagiösen Krankheiten', and occupies 82 pages. It is a theoretical not a practical work and, as Henle himself states, it contains 'few facts and many reflections'.

The main object which Henle had in view was to establish the hypothesis that 'the material of contagion is not only organic but living, endowed with individual life and standing to the diseased body in the relation of a parasitic organism'. He considered three kinds of epidemic diseases:

(1) Miasmata alone, as e.g. ague;

(2) Miasmatic-contagious diseases such as the exanthemata, cholera, plague, and influenza; and

(3) Contagious diseases such as syphilis, itch, rabies, and probably glanders and foot-and-mouth disease.

'Miasm' was, in Henle's view, a quiddity unrecognizable by the senses, but in some way related to or identical with other deleterious potencies like contagia. He regarded contagia as a kind of miasm in the second generation—a miasm which had passed through its first development in the human body. In the miasmo-contagious diseases the contagion is known to be eliminated from the body and conveyed to the healthy either by the atmosphere (volatile contagion) or by contact (fixed contagion). The discharges from the body contain the materies of infection. Variolous pus is pus plus the contagion of variola. Henle argued that contagia must be living. For example the disease does not break out at once. There is a *stadium latentis* in which the contagion is increasing. A small particle of variolous matter will ultimately render every part of the body infective. When the contagion has sustained the necessary increase in amount the effect on the body is fever and inflammation.

ALEXANDER OGSTON
(1844–1929)

LOUIS PASTEUR
(1822–1895)

CRYPT AND TOMB OF PASTEUR IN THE
INSTITUT PASTEUR, PARIS

As there are different miasmo-contagious diseases the presumption is that there are specific differences in the contagia which he concluded must be living. In support of his views Henle cited the practical observations of Bassi on muscardine, Cagniard-Latour and Schwann on yeast and fermentation. Henle clearly pointed out the difficulties of obtaining proofs that his views were correct. He foresaw that even if some agent were found in a disease it might be purely accidental and have no causal relation to the disease. To determine the latter, the agent constantly found must be isolated and tested in its isolated state to see whether it can produce the disease. This constant presence, isolation, and testing of the isolated object is the unassailable basis on which all the subsequent work on pathogenic bacteria has been built up, and Henle's statement of the proposition contains all the elements habitually referred to as the 'postulates' of Koch (1878, 1884) (see Fildes and McIntosh, 1920). It was about the middle of last century that the terms infection and contagion became interchangeable. In former times they were considered to be different although few agreed as to what constituted the differences. For the most part 'infection' was applied to communication of disease from sick to healthy by a morbid 'miasm' or exhalation diffused by the air, whereas 'contagion' was used for transmission by mediate or immediate contact. In our day the word 'infection' has largely superseded 'contagion'.

The practical work of Agostino Bassi, of Cagniard-Latour and Schwann, and the theoretical conclusions of Henle may be looked upon as the stimulus which led to the microscopic examination of a great many disease products. Within a year or two the existence of many parasitic cryptogams was recorded in diseases of the integuments and mucous membranes, both of man and animals.

Thus Schönlein (1839) discovered the fungus since known as *Achorion schönleinii* in certain 'Impetigines' or Favus, and Remak (1840) named it and showed its contagiosity. Buehlmann (1840) rediscovered the fungus *Leptothrix buccalis* previously made known by Leeuwenhoek.

Oidium albicans was independently discovered by Berg (1841) and Gruby (1842). The remarkable microscopic organism called *Sarcina ventriculi* was discovered in Edinburgh by John Goodsir (1842), who published a very full account of it with figures (Fig. 19). In the same year Müller and A. Retzius (1842) discovered a mould of the genus *Mucor* in the lungs and air cavities of an owl (*Stryx noctua*) from Lapland.

Very important practical discoveries were made by the eccentric David Gruby (1810–98)—a Hungarian long

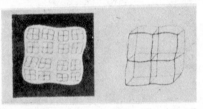

FIG. 19. *Sarcina ventriculi*. From Goodsir (1842).

resident in Paris. In 1843 he gave an accurate description of the fungus *Microsporum* (*Microsporon*) *Audouini* in Willan's *Porrigo decalvans* (a form of ringworm), and in 1844 he followed it with the discovery of a different fungus in *Teigne tondante* of Mahon (*Herpes tonsurans* of Cazenave). To this fungus Malmsten (1845) gave the name of *Trichophyton tonsurans*.

In 1844 Mayer discovered a mould (*Mucor*) in the aural passage of a child aged 8, and J. Hughes Bennett found a similar cryptogam in the sputum and lungs of a case of pneumothorax. The fungus (*Microsporon furfur*) of *Pityriasis versicolor* was discovered by Eichstedt (1846) and Sluyter (1847). A very full account of these early observations and others will be found in Robin's *Histoire naturelle des végétaux parasites*, published in 1853. The application of the fungus theory to diseases other than the above-mentioned was not so successful.

In 1846 began the third great pandemic of asiatic cholera and Europe was rapidly and extensively invaded. The time was ripe for theories and practical studies. On the theoretical side the disease was by some regarded as due to a change in the ponderable or imponderable elements of the air. Others regarded it as the result of a vegetable miasm arising from the soil. A more unlikely theory attributed cholera to volcanic or other changes in the crust of the earth. Some held it was contagious, others that it came from animalcules existent in the air. On the practical side microscopic investigations were made especially on *cholera dejecta*. In this country Brittan (1849), Swayne (1849), and W. Budd (1849), all of Bristol, recorded and figured certain 'annular bodies' which were interpreted as 'fungi'. They seemed to be peculiar to cholera and were said to be found by Budd in drinking-water from cholera districts. Baly and Gull, who were appointed by the cholera committee of the Royal College of Physicians to report on the so-called 'annular bodies', were unable to find them in air or water. They regarded them as being of divers origin and nature and that they were in part, at any rate, derived from substances taken in food or medicine. Lastly, Baly and Gull found them in conditions which had nothing to do with cholera, and they denied their fungoid nature altogether.

In France Pouchet (1849) found immense numbers of very minute 'animalcules' in the rice-water stools of cholera. He identified them as the *vibrio rugula* of Müller, but made no statement regarding their causal relation to cholera.

Although he did not deal with the ultimate cause of cholera this would seem to be the place to refer to the observations of John Snow (1813–58), the English anaesthetist who in 1849 showed, for the first time, the water-borne character of epidemic cholera. He was led to this view by the belief that cholera begins with the affection of the alimentary canal, and he considered that water might be the vehicle. He collected exact data of a large number of outbreaks and correlated them with water supplies. His theory was not received with enthusiasm, but gained ground after his exposition of the great epidemic of cholera which occurred on 31 August 1854 in Broad Street, Golden Square, London, and the adjacent streets. In a space of

250 yards from the spot where the disease began there were 500 deaths from cholera in 10 days, at the end of which time the street was deserted by the flight of the survivors. With unerring exactitude Snow traced the outbreak to contamination of the water of a particular pump in Broad Street. In this water he found evidences of contamination with organic matter. His views, as it happens, have turned out to be largely correct. Search continued to be carried out for cholera parasites but was unsuccessful, and the third pandemic of cholera subsided. The doctrine of contagion animatum had, however, received an assured place among medical ideas.

VIII
CLASSIFICATION OF BACTERIA

CLASSIFICATION OF BACTERIA

THE earlier microscopists were content to describe, often in considerable detail, the individual 'animalcules' found in organic infusions, but did not attempt any classification on the lines laid down in the first half of the eighteenth century by Linné (1707–78). The earliest attempt at arrangement in a system came from the great Danish naturalist, Otto Friderich Müller (1730–84), who in several works described, arranged, and named with exactitude a number of animalcular forms. In one of his works, entitled, *Vermium terrestrium et fluvatilium seu animalium infusoriorum . . . succinta historia*, and published in 1773 and 1774, he gave the first account (l.c., p. 25) of a micro-organism which later occupied a very prominent position in the bacteriological literature during a large part of the nineteenth century. Müller called the organism *Monas termo* (in Danish *Grendse-Monaden*), which he characterized as 'animalculum omnium quae microscopium simplex offert minimum simplicissimum punctulum gelatinosae substantiae ipsum microscopium compositum eludere videtur dum ne quidem sub hoc distinctius appareat'. Of bacteriological interest was also Müller's genus *Vibrio* including *V. lineola*, *V. bacillus*, *V. undula*, *V. proteus*. Müller's *magnum opus* was, however, his *Animalcula infusoria et marina . . .* published (1786) after his death and edited by Otto Fabricius. In this magnificent quarto of 367 pages and 50 plates Müller attempted a greatly enlarged classification and added many new species. He first divided all the animalcules into two great groups, viz. (I) those devoid of external organs, and (II) those possessing external organs. Group I is the only one of bacteriological interest and was divided into Crassiuscula and Membranaceae. Among the Crassiuscula which included five genera there were two, viz. *Monas* and *Vibrio*, which contained bacterial forms.

Müller's genus *Monas* was characterized as 'vermis

inconspicuus simplicissimus pellucidus punctiformis', and he named ten species, viz. *Monas termo, M. atomus, M. punctum, M. mica, M. lens, M. lamellula, M. ocellus, M. pulvisculus, M. tranquilla,* and *M. uva.* Müller's genus *Vibrio* was described as 'vermis inconspicuus simplicissimus teres elongatus', and he described six species, clearly of bacterial type, viz. *Vibrio linearis minutissimus, V. rugula, V. bacillus, V. undula, V. serpens,* and *V. spirillum.* Müller's nomenclature was employed by a number of microscopists early in the nineteenth century, but was extended by C. G. Ehrenberg (1795–1876) in his *Die Infusionsthierchen als vollkommene Organismen* (1838). This is a large folio of 547 pages with an atlas of 64 magnificent plates drawn by Ehrenberg himself. Like Müller and others Ehrenberg made no distinction between what would now be called Protozoa and Bacteria. He classed them all as Infusoria along with rotifers, desmids, and diatoms, and he regarded them as animals from his belief that they all possess stomachs. They constituted his 'Polygastrica' or 'Magenthiere', and he believed that they were organized in the same kind of way as larger animals. In his opinion they were perfect and complete (*vollkommene* as in the title of his ·work). He recognized 22 families of Polygastrica, and of these families the first or monadina, the second or cryptomonadina, and the fourth or Vibrionia are of special interest as they clearly comprise forms now recognized as bacterial (Fig. 20).

In his family of Monadina Ehrenberg recognized 41 species distributed in 9 genera. In the genus *Monas* there were 25 species divided according to their shape into 'Kugelmonaden' and 'Stabmonaden'. The round or Kugelmonaden were further subdivided into 9 species of 'Punktmonaden' and 8 of 'Eimonaden' according as they were mere points or oval. Ehrenberg's *Monas crepusculum* ('Dämmerungsmonade') was described as the smallest of all living things seen up to that time. *Monas termo* was called 'Schlussmonade'. Ehrenberg's fourth family, Vibrionia ('Zitterthierchen'), comprised five genera—*Bacterium, Vibrio, Spirochaeta, Spirillum, Spirodiscus.* The

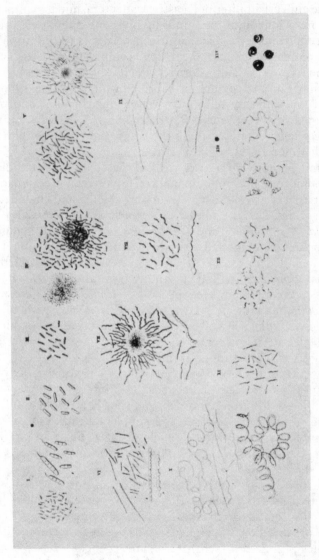

FIG. 20. Figures of micro-organisms. From Ehrenberg (1838).

general characters of the family were 'animalia filiformia distincte aut verisimiliter polygastrica, anentera, nuda, gymnica corpore Monadinorum uniformi divisione spontanea imperfecta (transversa) catenatim consociata, hinc filiformia'.

Ehrenberg's genus *Bacterium* comprised threads of filaments of rod-like forms which showed transverse division ('animal e familia Vibrioniorum divisione spontanea in catenam filiformen rigidulam abiens'), and it included three species—*B. triloculare*, *B. enchelys*, and *B. punctum*. None of these, however, can with certainty be identified to-day (Cohn, 1872).

The genus *Vibrio* was defined as 'animal e familia Vibrioniorum divisione spontanea imperfecta in catenam filiformen et anguis instar flexuosam abiens' and included five species, viz. *V. lineola*, *V. tremulans*, *V. subtilis*, *V. rugula*, *V. prolifer*, and *V. bacillus*. The third genus, *Spirochaeta*, comprised spirally wound *flexible* filaments and included only one species, *S. plicatilis*. The fourth genus, *Spirillum*, included spirally wound *inflexible* filaments and contained three species—*S. tenue*, *S. undula*, and *S. volutans*—while the fifth genus, *Spirodiscus*, with one species, *S. fulvus*, was described as compressed, inflexible, spiral filaments. Ehrenberg's principal figures of bacteria and vibrios will be found in his Plate V, figs. 1–14.

In spite of his laborious efforts at classification many of Ehrenberg's species of monads and vibrios cannot be identified with certainty to-day, and shortly after the publication of his work another and simpler classification was presented by the eminent French zoologist Félix Dujardin (1801–60) in his *Histoire naturelle des zoophytes* . . . published in 1841.

He placed (loc. cit., p. 209) the organisms that would now be called bacteria in a family 'Vibrioniens' which he characterized as 'animaux filiformes extrêmement minces sans organisation appréciable, sans organes locomoteurs visibles', and he recognized three genera—*Bacterium*, *Vibrio*, and *Spirillum*.

The genus *Bacterium* was characterized as 'corps filiforme roide devenant plus ou moins distinctement articulé par suite d'une division spontanée imparfaite; mouvement vacillante non ondulatoire' and included four species, viz. *B. termo*, *B. catenula*, *B. punctum*, and *B. triloculare*. Dujardin's genus *Vibrio* was described as 'corps filiforme plus ou moins distinctement articulé par suite d'une division spontanée imparfaite susceptible d'un mouvement ondulatoire comme un serpent'. The species of *Vibrio* were *V. lineola*, *V. rugula*, *V. serpens*, *V. bacillus*. The third genus, *Spirillum*, was defined as 'corps filiforme contourné en hélice non-extensible quoique contractile', and comprised *S. undula*, *S. volutans*, and *S. plicatile*.

Dujardin's figures are very small and the organisms which they are said to represent cannot easily be identified, although his descriptions are, in general, good.

Eleven years after the publication of Dujardin's work Maximilian Perty (1852) published a work on animalcules and suggested a new classification. He considered that some of the infusoria were not animals but what he called animal-plants or Phytozoidia, and he subdivided them into three sections, viz. (I) Filigera, (II) Sporozoidia, and (III) Lampozoidia (λάμπη, scum). The last-named contained the family Vibrionida and was further subdivided into Spirillina and Bacterina. The Spirillina, which were spirally-wound filaments, contained the genera *Spirochaetae* with one species, *S. plicatilis*, and *Spirillum* with three species, *S. volutans*, *S. undula*, *S. rufum*. The Bacterina (wavy or straight filaments) were subdivided by Perty into four genera: (1) *Vibrio* with three species, (2) *Bacterium* with one species, (3) *Metallacter* (μεταλλακτήρ, one that changes) with one species (*M. bacillus*), and (4) *Sporonema* with one species (*S. gracile*). In the last-named organism Perty recognized one or more elliptical bodies which he interpreted as spores (Fig. 21). It cannot be said that Perty's descriptions are clear and he had confused ideas with regard to the development of individual forms. Thus he regarded *Metallacter* not only as bacterial-like individuals but as elongating by progressive development to stiff or slightly flexible filaments which, under certain conditions, could lose their motility and grow like Hygrocrocis (an alga).

It was about this time that Ferdinand Cohn began to contribute his important observations, which subsequently did so much to the establishment of bacteriology as a science. Cohn (1828–98) was professor of botany in the

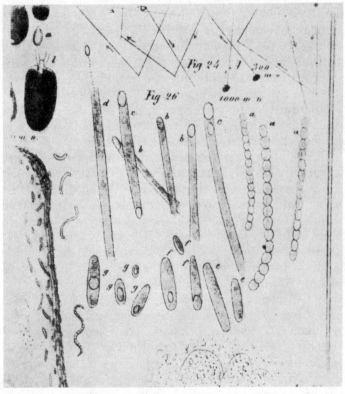

FIG. 21. Micro-organisms figured by Perty (1852).

University of Breslau and a man of great talent, learning, and industry. In 1854 he published an important memoir *On the Development of Microscopic Algae and Fungi*, and he suggested therein that Ehrenberg's family of Vibrionia should be regarded as belonging to the vegetable rather than to the animal kingdom from the fact that, apart from colour, they showed exceedingly close analogies to well-known microscopic algae. From their absence of colour, and the fact that they occur in infusions, Cohn proposed that they should be placed among the Mycophyceae or 'Wasserpilze'. In this same memoir Cohn introduced a new genus, *Zoogloea* (ζωὸν, animal, and γλοῖος, gelatin), which he defined (loc. cit., p. 123) as 'cellulae minimae bacilliformes, hyalinae, gelatina hyalina in massas mucosas globosas, uvaeformes, mox membrananaceas consociatae, dein singulae elapsae per acquam vacillantes'. As an example of this genus he cited *Zoogloea termo* (*Bact. termo*, Duj.), and he considered that it was the motile swarming form of a genus analogous to *Palmella* and *Tetraspora*. More certain in Cohn's judgement was the analogy of certain vibrios to certain of the Algae which reproduce by fission. He went so far as to state that *Spirochaeta plicatilis* is nothing more than a colourless form of *Spirulina*.

Even from this early period (1854) Cohn's views exerted a considerable influence on the opinions held on the nature of bacteria and, persisting in his studies for two more decades, he ultimately came to the front as one of the chief founders of the bacteriological doctrines held to-day. His classical *Untersuchungen über Bacterien* in 1872, 1875, and 1876 will be referred to later. Meantime, it is necessary to note that Carl Nägeli (1857), the botanist, in discussing the silk-worm disease then raging in France and Italy, described certain cells under the name of *Nosema bombycis*, and he proposed that they, along with *Umbina aceti*, *Bacterium*, *Vibrio*, *Spirillum*, *Hygrocrocis*, and *Sarcina*, should be placed in a group which he named 'Schizomycetes' (fission fungi). Although this name has persisted until to-day it is interesting to note that Nägeli was not certain whether they were plants, animals, or even elementary particles of animal or vegetable origin. He considered that the grounds for classifying them were trivial and he introduced certain errors which long persisted, for even in 1849 he had stated that microscopic fungi originate spontaneously, whereas algae originate from seeds or spores. Cohn, on the other hand, was strongly of opinion that bacteria have no genetic connexion with animals except perhaps monads. The idea that bacteria are fungi was only possible, in Cohn's opinion, on the assumption that fungi include all cellular plants devoid of cholorophyll or its equivalent, and unable to assimilate carbon dioxide. Bacteria bear no relation to typical fungi with mycelia and spore formation. On the other hand they show close analogies with the Phyco-chromaceae (Cyanophyceae) and differ only in their inability to assimilate CO_2.

The belief of Nägeli (1849) that microscopic fungi originate spontaneously had important bearings on subsequent bacteriological work and especially on systematology, for it was not supposed that, being produced spontaneously, there would be any constancy in form among the products as is known to occur in animals or plants descended from parental forms. In the middle of the nineteenth century great advances were made with regard to the life-history of certain fungi through the development of the doctrine of the 'alternation of generations'. This was mainly the outcome of the researches of the brothers Tulasne in France and the German mycologist Anton De Bary (1831-88).

The changes or alternations of form among fungi were described as di-, poly-, or pleomorphism and became an object of intensive study. The terms pleomorphism or heteromorphism were used to express the observed fact that one and the same fungus appeared in several forms, not only as regards the vegetative but also the fructifying organs, and it was soon shown that this is a widespread type of development, especially among the fungi which cause disease in cereals. Long before the time of Tulasne and De Bary, Kützing (1837) and Turpin (1838) had stated their impression that yeast cells are not constant in form but can grow into other definite types, and in the wave of excitement over the undoubted pleomorphism of certain fungi, the same doctrine came to be applied to the case of certain bacteria for a number of years.

Prominent among those who concurred in this belief may be mentioned Hallier, Lüders, Huxley, Ray Lankester, and Lister. Before dealing with their contributions, which in point of time appeared between 1865 and 1875, it is necessary to refer to the discovery of bacteria in certain diseases of man and animals. In the case of anthrax or splenic fever this discovery goes back about fifteen years. The story has often been told and in different ways. The historical facts are these. In a communication to the Société de biologie in Paris in 1850, P. F. O. Rayer (1793–1867), a well-known Paris physician and pathologist (with whom Davaine was apparently associated), reported the results obtained by inoculation of sheep with the blood of other sheep dead from anthrax, and in a microscopic examination of the blood they noted, in addition to changes in the blood corpuscles, the occurrence of 'petits corps filiformes, ayant environ le double en longueur d'un globule sanguin. Ces petits corps n'offraient point de mouvements spontanés'. Although Rayer made no other statement than that quoted it was the first dealing with the existence of such things in anthrax blood. Davaine (1875), in his historical note, tells us that the above statement was written by him and sent to Rayer for publication.

In 1855 F. A. A. Pollender (1800–79) of Wipperfürth (Westphalia) published a paper on the microscopic and microchemical examination of anthrax blood and reported that, as early as 1849, he had found certain changes in the blood of five cows dead from anthrax. He apparently discovered, quite independently, the bodies seen by Rayer (1850)—Pollender's observations being made in 1849 according to his own account—but he gave a far more exact account of them. In the spleen he found them in enormous masses as straight unbranched rods measuring $\frac{1}{400}'''-\frac{1}{200}'''$ in length and $\frac{1}{3000}'''$ in width. They were completely motionless and in their form and arrangement presented an exact analogy to *Vibrio bacillus* or *V. ambiguus*. They were not affected by acetic acid, chlorine water, alkalies, or mineral acids except nitric acid, which caused them rapidly to disappear. Iodine solutions stained them bright yellow and made them more apparent. Pollender was unable to throw any light on their origin or whether they occur during life. He could not say whether they were products of putrefaction or whether they represented the contagious element of anthrax. He was convinced they were not fibrils of the body or fibrin, but from their appearance and reactions he was inclined to regard them as of vegetable nature.

In 1857 F. A. Brauell, a veterinary professor in Dorpat, confirmed the findings of Pollender in the case of anthrax in a man who had charge of a crematorium for the destruction of anthrax carcasses. Brauell, further, carried out a number of inoculations to demonstrate the transmissibility of the disease by means of human anthrax blood to sheep in series and from equine anthrax to horses. He differed from Pollender in an important respect, in that he found that in blood kept for three days the rods showed motility, although in recent blood they were motionless. Brauell also found the rods in the blood during life in the case of a sheep, but he did not regard them as peculiar to anthrax. In a second paper Brauell (1858) gave the results of a long series of experiments on the transmission of anthrax to horses and sheep in series. Dogs and fowls were found to be resistant to the infection. He stated that anthrax blood from a horse could cause fatal anthrax, although no anthrax rods could be found in the blood.

In 1860 O. Delafond, a well-known veterinary professor at Alfort, added several new facts about the anthrax rods and attempted to cultivate them, but it was mainly through the researches of C. J. Davaine, from 1863 onwards, that the subject of anthrax and its rod-like bodies achieved wide notoriety. Reflecting on the likeness of the rods which he had seen in 1850 to the butyric ferment described by Pasteur (1861), Davaine set to work and published a long series of researches covering the pathogenesis of anthrax and on the presence of the rod-shaped bodies which he first called 'bacteria' and subsequently 'bacteridia'. He confirmed and extended the experimental work of Brauell. He discussed the relation of anthrax to the general question of putridity and stated, with considerable precision, his belief that the bacteridia must be the cause of the disease from the fact that they are constantly present in the disease, that the disease can be transmitted by inoculation, and that when the bacteridia are not present there is no anthrax.

In 1868 Davaine showed that anthrax can result from the inoculation of so small a quantity as one-millionth of a drop of anthrax blood. In his opinion the anthrax bacteridium was invariably non-motile, as indeed Rayer and Pollender had originally asserted.

Davaine's views on anthrax aroused general interest in scientific and medical circles in France. Signol (1863), Tigri (1863), Pierre (1864), and in particular Leplat and Jaillard (1864), all, however, denied the interpretation which Davaine had placed on his findings. Leplat and Jaillard (1864), two professors in the Val-de-Grâce military hospital, unable to obtain anthrax material, and being of the opinion that other bacteria, 'sans nul doute jouissent des mêmes propriétés'—so they wrote—injected, into the blood-stream of rabbits and dogs, putrid fluids teeming with bacteria. No injurious effects were seen. Pursuing their work, Leplat and Jaillard (1865[1]) then experimented with anthrax blood which was sent to them in Paris from Chartres. They inoculated a large number of rabbits, but they never found any trace of Davaine's bacteridia, not-withstanding that the fatal disease they had produced was exactly like anthrax, and was regarded as such. They considered the bacteridium of Davaine to be merely an epiphenomenon and not the cause of anthrax. In the discussion on their paper, Pasteur stated that he had had the opportunity of comparing the bacteridia in Davaine's experiments, and that they differed from the butyric ferment in that they were immobile. Davaine (1865) was also strongly of opinion that the disease induced experimentally by Leplat and Jaillard was not anthrax. It differed in being communicable to birds and the virus was conserved in spite of the onset of putrefaction. Davaine considered that the cow from which Leplat and Jaillard had obtained their original material had died, not of anthrax, but of a septic disease. To refute this Leplat and Jaillard journeyed to Chartres—the home of anthrax in France—and came into touch with the veterinary surgeon Boutet there. He subsequently sent them, through the post, a small bottle of undoubted anthrax blood. On examination it contained myriads of immobile rods and filaments, apparently the same as Davaine's bacteridia. They inoculated rabbits with the blood, and on taking samples from the femoral artery three hours after death, found no bacteridia or other alien elements. Notwithstanding, this same blood caused death in other rabbits and they were able to transmit the disease in series, although they could never observe organisms in the inoculum. Davaine, 1866 (1865), adversely criticized these results and showed that the disease produced by Leplat and Jaillard differed fundamentally from anthrax. Pasteur (1865) received some of the original blood, used by Leplat and Jaillard, but failed to find Davaine's bacteridia in it. Instead, he found putrefactive bacteria and 'butyric ferment', and he formed the opinion that Leplat and Jaillard had worked with material not of anthrax but of a disease presenting symptoms somewhat like it. I have entered into somewhat minute details of the early work on anthrax because it shows the difficulties with which the earlier workers had to contend, and because it was the precursor of the important results on which many of the bacteriological doctrines were based.

The subsequent history of the aetiology of anthrax is connected with the names of Koch and Pasteur and will be dealt with later (p. 208). The researches of Davaine on anthrax were of great importance in bringing out the doctrine which, about this time, was beginning to be called 'the germ theory of disease'. According to this, some diseases were due to the growth of 'germs' in the body, and although this seemed possible, if not probable, there were many capable persons in the scientific world who would have none of it. The theory that diseases may be due to germs was an old one and previous attempts to prove it had only been partially successful. I recall the attempts to prove that diseases were due to fungi. Between 1837 and 1850 the fungoid nature of certain diseases had been established, but for many important diseases known to be communicable such an origin could not be substantiated at the time. There were various reasons why the germ theory had a recrudescence after a period of comparative extinction. In the first place the long-continued experiments of Pasteur from 1857 onwards had shown that special fermentations have special micro-organisms as their cause. By degrees Pasteur's doctrine became firmly fixed as a belief. In the second place the old idea of a spontaneous generation of disease suffered a complete reverse with the destruction of the idea of heterogenesis of germs by Schwann, Schröder and v. Dusch and Pasteur. Thirdly, it was on the basis of germs that Lister, at that time working in Glasgow, laid the foundations of his antiseptic system about 1864. He became a convert to the germ theory as a result of the conclusive experiments of Pasteur on fermentation and spontaneous generation. His antiseptic system of the treatment of wounds had at first a chequered career, especially in England, but came back to us as a cherished belief chiefly through the experiences of German and French and Danish surgeons, among whom may be mentioned Volkmann, Nussbaum, Saxtorph, Esmarch, Hueter, and Lucas Championnière. Lister had a precursor in Jules Lemaire, who in 1860 and 1865, on the basis of the germ theory of putrefaction, made suggestions for the treatment of wounds by carbolic acid, but nothing that Lemaire did can lessen the value of Lister's own work or diminish his fame as the creator of an organized system of antiseptic treatment.

The germ theory of disease also received a great impulse owing to the researches of Pasteur (1865–70) on the silk-worm disease called 'pébrine', and culminating in his *Études sur la maladie des vers à soie* (1870). Although the disease is not due to a bacterium, and although Pasteur at first was led astray in his conclusions, he was able ultimately to show that the diseases of silk-worms are due to parasitic germs. That virus is organized and particulate was also rendered probable, if not certain, by the researches of Chauveau (1868) and Burdon Sanderson (1870) on vaccinia lymph.

Chauveau caused active vaccinia lymph to stand and allowed the leucocytes present therein to settle out. They were found not to be the carriers of the vaccinia virus.

To separate the particles from the fluids of the lymph he resorted to the process of diffusion. Active lymph was transferred with great care to small test-tubes and water was added. The whole being allowed to stand for twenty-four hours, all the soluble constituents passed into the water, while the solid particles remained unaltered. Children and heifers were vaccinated with the two portions separately. The upper aqueous layer was inactive, whereas the lower layer contained the virulent vaccine principle.

These experiments were repeated with added precautions by Sanderson (1870) and with like results, and in general he did a great deal in England to further the doctrine of the germ theory of contagion and infection. The germ theory was not, however, accepted unopposed, and apart from much writing by ordinary medical men, many persons of scientific distinction combated the theory with great ardour. The work of the chief English opponent, Henry Charlton Bastian, has already been referred to in detail, and it may be said that up to the end of his life he remained unconverted. Lionel Beale (1870, 1872), the English microscopist, was also opposed to the germ theory and published several works on the subject. His main object was to show that what was called a 'disease germ' was probably a particle of living matter derived by direct descent from the living matter of the body of man and animals. He held the view that living matter is largely composed of a hyaline structureless material which is invariably the same. This was Beale's 'germinal matter' or bioplasm, and it was believed to consist of extraordinary minute particles or bioplasts. Vegetable organisms when present in disease were regarded by Beale as merely concomitants. He believed that a poison, not of the nature of a vegetable germ, is present in the animal or solid in which the contagious properties are known to reside, and that disease germs may well be the issue of degraded but actively growing bioplasm of a diseased animal or human being. Beale was a scornful opponent of John Tyndall. Even as late as 1879 that brilliant English worker, Timothy Lewis, was not convinced that bacteria can cause disease, and in

France Ch. Robin (1879), the distinguished microscopist, was also an opponent.

This would also appear to be the proper place to introduce to the notice of the reader the now exploded doctrine of microzymas, formulated by Antoine Béchamp (1816–1908), who for many years was a thorn in the flesh of Pasteur and made it a habit to oppose everything he said, did, or wrote. For long years he was also a loquacious and trying bore in the Academy of Sciences in Paris, urging his extravagant claims on an incredulous world. The recent attempt in America (see Hume, 1923) to resuscitate Béchamp and proclaim him to have been a star of the first magnitude and the greatest biologist of all time is fantastic. Béchamp's particularly active period ranges from about 1865 to 1890 and he wrote a very large number of papers and several books, one of which (1883) alone ran to almost one thousand pages.

Béchamp was a vigorous antagonist to the germ theory in general. He proclaimed the view that life, in all its manifestations, is due to the activity of microscopic granules which he called microzyma. It is of these that protoplasm was considered to be composed. They, and not the cells, are the real agents of all the processes which appertain to life. He asserted that the microzymas produce a fluid named zymase or ferment and are surrounded by it. The microzymas nourish the body, and in a vitiated state they are the causes of all diseases. Infective diseases do not come from exogenous germs or bacteria, but are the expression of the perverted action of the microzymas. He did not believe in specific diseases or specific causes. The bacteridium of anthrax was not, in his opinion, the cause of anthrax. It merely caused a dyscrasia which led to the evolution of morbid microzymas. He claimed that he had proved the existence of these bodies in geological strata of bygone aeons. He regarded these ancient microzymas as identical with those found in the body to-day, and as the anatomical incorruptible elements of life.

Bodies die but the microzymes survive, endowed as they are with something approaching immortality. In his own words, 'Rien n'est la proie de la mort, tout est la proie de la vie', and again, 'Tout être vivant est réductible au microzyma'. There is a fatal flaw in all the superstructure which Béchamp attempted to raise. His work really rested on the comparison of objects which agreed with each other in size only—an agreement devoid of any value or interest.

In spite of the fact that the first fungoid period in the aetiology of disease had proved, for the most part, a failure, and in spite of the fact that there were many reasons for suspecting organisms other than the more complicated fungi and moulds, the latter were still believed to be potent morbid factors by several observers, and the fungus doctrine of disease was presented for a second time. This revival started with a certain J. H. Salisbury of Newark, Ohio, and he was active with his publications in the period 1862 to 1873. His earliest observations (1862) were intended to show that fungi, of the mould class, growing on damp wheat and straw, were the cause of measles. Having packed bright wheat-straw in a box and mixed it with cold well-water, he found, as must have been known to every one, that it became mouldy. He inoculated himself and his wife with some of the moulds and developed catarrhal symptoms which he considered to be akin to those of measles. During an epidemic of measles in a reformatory he inoculated twenty-seven boys with the straw fungi and believed that, by this treatment, he had prevented them from contracting the disease. Pursuing his line of thought he studied (1866) the emanations and miasms given off from a marshy place near which there had been cases of ague, and having exposed glass plates a foot above the water of stagnant pools he constantly found minute oblong cells resembling Palmellae and he interpreted them as such. To these organisms he gave the name of *Gemiasma* (earth miasm) and differentiated five species of which one, *G. rubrum*, produced intermittents 'of a congestive type'. Having filled tin boxes with the surface earth 'from a decidedly malarious drying prairie bog which was completely covered with palmellae', Salisbury afterwards exposed the boxes on the sill of a window opening into a bedroom. Two young men who slept in this room developed malaria in twelve and fourteen days respectively. In 1867 Salisbury, now professor of physiology, histology, and pathology in Cleveland, Ohio, also described two fungi (*Trichosis felinis* and *T. canis*) in skin diseases of cats and dogs. In syphilis he found (1868) minute transparent refractive algoid filaments which he named *Crypta syphilitica*. Another fungus, *Crypta gonorrhoea* (*sic*) was believed to cause gonorrhoea. In the blood of cases of erysipelas, Salisbury (1873) found another fungus, *Penicillium quadrifidum*, Salisb., and he regarded asthma and hay fever as due to an animalcule, *Asthmatos ciliaris*, Salisb., armed on one side with cilia. Horatio Wood (1868), who was opposed to the doctrine of animate contagion, criticized Salisbury's statements on malaria with disdain and himself consumed masses of palmellae intentionally but without result. Salisbury's work, inadequate as it was, soon ceased to create interest. His methods and his knowledge were manifestly insufficient for the investigation which he undertook.

If we consider the state of knowledge in regard to microorganisms and their activities about 1860–5, it is apparent that there were two views. The one, which was mainly the outcome of the labours of Pasteur on fermentation, was that fermentations owe their diversity to different germs which, however, are constant in form and are recognizable morphologically. Although, as it turned out, Pasteur was correct in his assumption, he was really unable to prove it at the time owing to the methods he employed and especially by his use of liquid media for cultivation. His very inadequate training in the microscopy of living things also led him in his nomenclature to use words in a very loose sense. Ferdinand Cohn (1872), himself a properly trained and accurate botanist, has pointed out that Pasteur used almost as synonyms 'végétaux cryptogames microscopiques', 'animalcules', 'champignons', 'infusoires', 'torulacées', 'bactéries', 'vibrioniens', 'monads', 'mucor', 'mucédinées', 'levure', &c. The word microbe now in common use was suggested by Sédillot (1878) and approved by Littré. Pasteur had no use for the established laws of botanical nomenclature, but whatever he had in his mind with regard to the above-mentioned terms, he was a firm believer in the form constancy of individual germs. Whenever it appeared that the forms altered, Pasteur inferred that he was dealing with more than one kind of 'microphyte'.

On the other hand the classical researches of Tulasne and De Bary on the polymorphism of certain fungi led to the belief that a similar state of affairs prevails among yeasts and bacteria. This view was developed in the sixties of last century in Germany, and its protagonist was Ernst Hallier (1831–1904), professor of botany in Jena. He was a most prolific writer and claimed that his discoveries had definitively settled the problem of the aetiology of many infective diseases of man, animals, and vegetables. Not content with publishing his works in established botanical or medical journals he founded a phytopathological institute in Jena and issued therefrom a special journal, *Zeitschrift für Parasitenkunde*, edited by himself and mostly containing works by himself. His active reign was brief—a matter of five years—when the edifice which he had erected with so much diligence and by such defective methods was shaken to its foundations by the drastic criticisms of Anton De Bary (1868), one of the great mycologists of his time. F. Cohn (1872[2]) also pointed out that Hallier's cultural experiments possessed not the slightest value. Any merit which Hallier had was forgotten in the general débâcle. His later years were devoted to the study of philosophy and aesthetics rather than of botany.

Hallier had, however, one merit which he shared, in an unequal way, with Pasteur. He was one of the first to direct attention to the germ theory of disease even if he was wholly incorrect in his ideas of what these germs really were. In common with others I have found the details of Hallier's numerous works very difficult to follow. His main thesis, however, was clear. It was that the microscopic forms of parasites are not in themselves genera and species, but merely stages (*Morphen*) in the development of more complicated fungi, and that the transformations are brought about through changes in medium, moisture, variations in temperature, and other less important factors. Hallier's chief aim was to study vegetable parasites under conditions altered intentionally, so that by the alterations they would be compelled to reveal their true nature and change of form. For this purpose various morbid products were inseminated on or in different media, and the results obtained were compared with similar inseminations kept under other conditions. The chief snag in Hallier's experiments was his inability to prevent contaminations with spores of common moulds (*Penicillium* and *Aspergillus*) from the air.

In his earlier work he employed test-tubes with a bent glass tube traversing a perforated cork, but, later on, he devised more complicated apparatus of two kinds, the one an *Isolir-apparat* (Fig. 22), the other a *Cultur-apparat* (Fig. 23). It must be clearly understood that the *Isolir-apparat* was not one for *isolating* but for keeping the culture *isolated*, i.e. unopened until the end of the experiment. It was intended to be a kind of control apparatus to check the results of the culture apparatus which was the one in which he followed the alleged changes. If the final result in the isolated (unopened) apparatus agreed with that in the culture apparatus, Hallier assumed that his observations were correct and his deductions justified. The *Isolir-apparat* of Hallier is seen in Fig. 22 (Hallier, 1867[2]). C is a flask with the material to be examined in a medium which has been boiled. B is a flask containing sulphuric acid. A is a tube filled with cottonwool for air filtration. D is a bell-jar connected with an air-pump by means of which the air passed through the apparatus can be renewed. The inoculum having been introduced into the sterile culture medium in C, the apparatus was sealed and not opened again. The *Cultur* apparatus is shown in Fig. 23. It consists of a bell-jar standing in a dish of water. In the centre and raised above the water-level by a small platform rests the dish in which the culture is grown. It seems needless to point out to-day that the result in such an apparatus must depend on what was present in the inoculum before it was inseminated in the culture medium, and yet it was exactly this point on which Hallier was so strangely

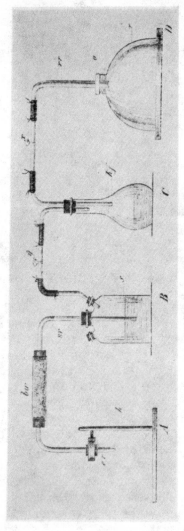

FIG. 22. Hallier's *Isolir-apparat.*

silent, and it was on this very thing which De Bary laid so much stress. He put it in simple terms when he said, 'If you wish to prove that A grows out of B it is necessary to prove that A was not present in B *before* the examination was begun', and yet it was this which Hallier never seems to have considered, and by its neglect had his whole work completely shattered, for both in his culture as well as in his isolated apparatus

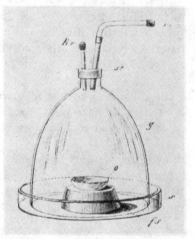

FIG. 23. Hallier's *Cultur-apparat.*

he found a succession of changes which he regarded as genetically connected. He believed that all parasitic forms belong really to only a few genera of fungi. A single instance of his type of reasoning will suffice. He alleged that if the spores of a mould are placed in a mixture of sugar and water containing an ammonium salt, the spore plasma divides into small nuclei (cocci) which multiply by division. This forms the basis of all yeast formations, and in Hallier's nomenclature is spoken of as nucleated yeast (*Kernhefe*) or micrococcus. It is the source of all the fermentations of fluids rich in nitrogen, i.e. putrid fermentations. If the nitrogen is consumed to a certain extent the nuclei become altered. They secrete a membrane which expands and lo! they are yeast cells. Reproduction is now by budding. The fermentation alters. It is now alcoholic and the fungus is in the stage of cryptococcus or sprout yeast. When the alcoholic fermentation is ended it passes into the acetic stage and the yeast is changed into arthrococcus or *Gliederhefe*. Filaments appear. It is now leptothrix or mycothrix. These micrococci, cryptococci, arthrococci, &c., are not, in Hallier's opinion, separate species, but merely forms (*morphen*) of something else which he endeavoured to discover. In disease processes of man and animals he always found micrococci, but in his apparatus he always found other things after he had inseminated his media with material containing micrococci. Thus from cholera evacuations he obtained a growth of a rice fungus. The cocci of vaccinia grew to the fungus *Eurotium herbariorum*. The micrococci of enteric fever revealed themselves as merely a stage of *Rhizopus nigricans*. Micrococci in gonorrhoea were believed to be stages of an unknown *Coniothecium*.

All this work of Hallier is now forgotten. The moral of it remains and is, perhaps, best appreciated if we remember the saying of Brefeld that if one does not work with pure cultures 'da kommt nur Unsinn und Penicillium glaucum heraus'.

From the confusion brought about by the inadequate investigations of Salisbury and Hallier, we now approach a period characterized by solid advances to knowledge. These advances we owe chiefly to Ferdinand Cohn, whose earlier work I have already referred to. He published his classical *Untersuchungen über Bacterien* in 1872, and herein the student will realize that he is on new ground which in a way was the starting-point of our modern ideas in bacteriology. His researches were the result of many years' laborious work and he was successful in disentangling almost everything that was correct and important out of a mass of confused statements on what at that time was a most difficult subject to study. His work was entirely modern in its character and expression, and its perusal makes one feel like passing from ancient history to modern times. He was clear, explicit, and fair in his judgement to other workers, and on every page it is apparent that he wrote with first-hand knowledge. In his paper of 1872 he at once raised the fundamental question whether, like other plants or animals, bacteria can be arranged in genera and species. This had been accepted as a matter of course by the earlier systematists like Müller, Ehrenberg, and Dujardin, and Cohn followed in their footsteps. He realized, however, that it was necessary to discuss the question, as even before Cohn, and after him, there were many who denied that there were bacterial genera and species. Cohn grounded his belief in the differentiation of bacteria by an accurate study of some of the larger, and especially the spiral, forms of bacteria which in his judgement and experience showed a marked constancy of form irrespective of external conditions. He clearly pointed out, however, that a purely morphological classification is insufficient, as he recognized that bacteria similar or identical in form may differ from each other in their physiological characters, and in their products.

Whether these form genera and form species are or are not natural genera and species, Cohn left for future investigators. The means at his disposal were not sufficient for the settlement of the question. Cohn has been wrongly interpreted as the founder of the doctrine that bacteria can be classified by form only. No one who has read his works can agree with such an interpretation. He realized and expressed himself clearly that we must take into account not the form alone but all the points of agreement or difference between related forms. He regarded his form genera as natural genera, but his species as largely provisional (Cohn, 1876).

Passing to the closer study of bacteria, Cohn gave an admirable account of the constitution of the bacterial cell and suggested the division of bacteria into four groups (*Tribus*), each of which contained one or more genera. Retaining, as far as possible, the older names, but with altered significance in some cases, his classification was as follows (loc. cit., p. 146):

Tribus I. Sphaerobacteria (*Kugelbacterien*).
    Genus 1. *Micrococcus* (char. emend.).
Tribus II. Microbacteria (*Stäbchenbacterien*).
    Genus 2. *Bacterium* (char. emend.).

Tribus III. Desmobacteria (*Fadenbacterien*).
    Genus 3. *Bacillus* (n.g.).
    Genus 4. *Vibrio* (char. emend.).
Tribus IV. Spirobacteria (*Schraubenbacterien*).
    Genus 5. *Spirillum* (Ehr.).
    Genus 6. *Spirochaete* (Ehrenberg.).

Among the round or Sphaerobacteria he differentiated chromogenic, zymogenic, and pathogenic species, and he gave a very full account of these species so far as they were known at the time. For the chromogenic forms he was indebted to the excellent researches of his co-worker J. Schroeter (1872), the well-known mycologist. Among the zymogenic species he placed *Micrococcus ureae*, which had become known from Pasteur's work and which Cohn confirmed. Cohn's pathogenic Sphaerobacteria included *Micrococcus vaccinae*, discovered by Cohn himself in 1872, and cocci in diphtheritis and sepsis.

Cohn's 'Microbacteria' differed from the 'Sphaerobacteria' in shape and motility and he recognized one genus—*Bacterium*. Two species, *B. termo* and *B. lineola*, were described in great detail.

The Desmobacteria or filamentous bacteria contained two genera, viz. *Bacillus* and *Vibrio*, and Cohn showed their relation to what had been previously called leptothrix forms. For the straight threads or filaments Cohn reserved the name *Bacillus*, and for the wavy forms, *Vibrio*. In the genus *Bacillus* he gave a minute description of *B. subtilis* and recognized and figured the spore which he correctly interpreted as a persisting form. Davaine's *anthrax bacteridium* was placed in the genus *Bacillus* by Cohn. Of the genus *Vibrio* he described in detail *V. rugula* and *V. serpens*.

Among the spiral bacteria (Spirobacteria) Cohn placed *Spirochaete* with flexible screw forms and *Spirillum* with inflexible screw forms. *Spirillum volutans* was described in great detail. Proceeding with his investigations, he also discussed the relation of bacteria to allied groups of plants and concluded that they constitute a pretty close group of organisms which show no relation to yeasts or moulds, but have close affinities with certain blue-green algae. He further carried out many experiments on the metabolism of bacteria and confirmed much of the work of Pasteur. In the course of this part of his investigation he devised the fluid bacterial medium since known as Cohn's fluid, which in some respects was superior to Pasteur's fluid, which contained candy sugar and yeast ash. The plate accompanying Cohn's paper is reproduced in Fig. 24. Cohn's views were not at once accepted as a correct interpretation of the observed facts and the fungoid theory of Hallier with its pleomorphism died hard, after lingering in the works of Lister (1873), Ray Lankester (1873), Huxley (1870), Klebs (1873), Warming (1875), and especially of Billroth (1874) and Zopf (1879–85). Thus Huxley regarded bacteria as the simplest change in the development of a fungus. By sowing conidia from fungi he maintained that bacteria are produced and, further, that all attempts to obtain what the Germans called 'pure cultures' must be abandoned as it was impossible to get an optically pure specimen of water. Lister, also, in 1873 (publ. 1876) described several changes which occur in the form of bacteria under varying conditions of nutrition, and he maintained that a torula arises from a filamentous fungus. He even believed that the corpuscular part of a micro-organism may differ in fermentative power from its filamentous parent. In general he regarded Cohn's classification as incorrect. It is just, however, to state that Lister's views subsequently underwent fundamental changes when the experiments were more rigorously carried out.

Ray Lankester (1873) was not a believer in the doctrine of the form constancy of bacteria and indeed was led to oppose this through his studies of *Bacterium rubescens*—a peach-coloured bacterium which he found at Oxford, in river-water that had stood for some time. He believed that the peach-coloured bacterium shows a number of phases or form species, and he grounded his belief in the genetic connexion of these phases from the fact that they all contained a definite colouring matter which he named 'bacteriopurpurin'. He described *B. rubescens* as showing sphaerous, filamentous, acicular, bacillar, serpentine, spiroid, and helicoid types, and thought they were all one and the same organism. These views he reiterated in 1876, and again with minor alterations in 1886. He did not deny that there were different species of bacteria, but that their form was not constant—a view also held later by the botanist W. Zopf (1846–1909).

Of greater importance for the subsequent development of medical bacteriology were the statements of the Viennese surgeon, Theodor Billroth (1874). In an unwieldy book he presented the result of five years' continuous labour on the bacteria of putrefaction and of infective diseases. Although the central doctrine which he submitted for future discussion has been proved to be entirely erroneous, he contributed a large number of observations, many of which were correct. His main idea was that all the round and rod-shaped bacterial forms were but stages of a plant— he regarded it as an alga—which he called *Coccobacteria septica*. He gave an elaborate description of the different bacterial elements and introduced many names, some of which have persisted till to-day. The smallest forms of *coccobacteria septica* were named coccos (κόκκος, a seed), and were distinguished either on the grounds of size or arrangement into micrococcos, diplococcos, streptococcos, gliacoccos, petalococcos (πέταλον, a plate), mesacoccos, megacoccos, ascococcos (ἀσκός, a bag). The rod-shaped forms were differentiated into gliabacteria, petalobacteria, streptobacteria, &c. In Billroth's opinion cocci changed into bacteria and were common or indeed constant in all putrefactive processes. The cocci were regarded as derived from bacterial spores and sporing bacteria were named helobacteria (ἧλος, a nail). The combination of coccos, ascococcos, and bacteria constituted the plant *coccobacteria septica*, the specific name being derived from its occurrence in putrid fluids and in septic wounds. Billroth made elaborate microscopic studies of putrid infusions of the most divers kind. In the pericardial fluids from 200 human cadavera he found *coccobacteria* in 87 instances. He also examined other fluids and the tissues of dead bodies, and extended his

Fig. 24. Cohn's first plate illustrating bacteria. From Cohn (1872).

observations to the living body in health and disease. He studied the result of inseminating *coccobacteria* in various nutrient media and made examinations of the organisms in a rich variety of surgical infections. He did not believe that inflammation and its surgical complications are due to organisms, but to a peculiar hypothetical 'phlogistic zymoid' acting as a ferment. This ferment developed autogenously by decomposition of the tissue parenchyma which was regarded as an admirable pabulum for *coccobacteria*.

The error into which, as we now know, Billroth fell was occasioned through the absence of continuous observation. His examinations occurred at intervals during which unobserved things took place, so that one form gave the impression of proceeding from another form, whereas both were present in the fluids all the time. His microscopic methods, although well carried out, were afterwards shown by methods of cultivation to be fallacious. Billroth, however, did not go nearly as far as Hallier in his views, and although supported by some observers, e.g. Klebs, his huge book now rests on library shelves unheeded except by the historian of bacteriology.

Cohn's second classical paper (1875) dealt temperately, but in detail, with Billroth's book on *coccobacteria septica*, and pointed out its defects. He also gave a description of *Ascococcus Billrothii*. Passing to the study of peach-coloured bacteria in contaminated waters, he described a number of forms which, unlike Lankester, he held to be different species rather than forms of one species, and he worked out in a very careful manner the physiology of these interesting species of 'sulphur bacteria'. In this paper Cohn also described for the first time *Cladothrix dichotoma* and *Streptothrix Foersteri*, and he also gave a minute account of the spore formation of *Bacillus subtilis*. An elaborate classification of Schizophytae was also included in Cohn's paper which, like his first paper, was illustrated with fine plates far in advance of any which had been previously published (Fig. 25).

Cohn's third research (1876) is marked IV, but this is accounted for by a paper of his co-worker, E. Eidam (1875),

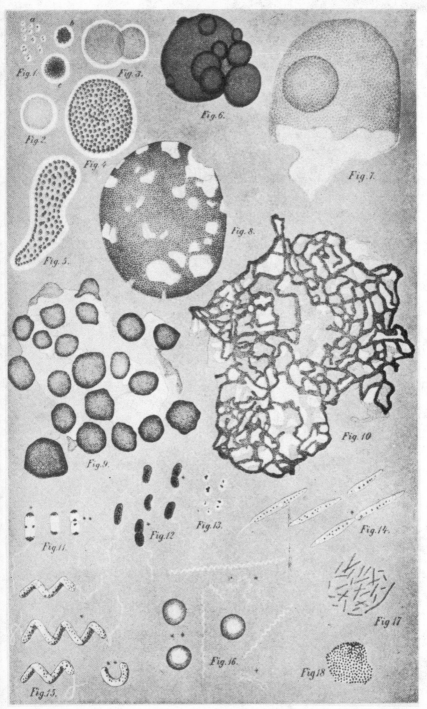

FIG. 25. Cohn's second plate illustrating bacteria. From Cohn (1875).

which is marked III and dealt with the action of heat and drying on *B. termo*. The research of Cohn (1876) is an exceedingly careful experimental inquiry into the question of the action of heat on organic liquids in relation to spontaneous generation. Next to the works of Pasteur and Tyndall, Cohn's paper was perhaps the most accurate on this subject and is referred to in detail on p. 118.

In a later work Cohn (1879) defended his ideas on the morphological unity of Schizomycetes against Nägeli's (1877) accommodation theory. A great deal has been made in bacteriological literature of the publications of the botanist, Carl von Nägeli. The special bacteriological works of this voluminous writer are two in number, and at least one of these (Nägeli, 1877) contains categoric statements rather than the presentation of observations. I have already referred to an earlier statement of his (Nägeli, 1849) that fungi develop spontaneously, and although it is probably true that he did not adhere to this belief he nevertheless stated in *Die niederen Pilze* (1877) that he could not accept the view that there were genera and species among bacteria. He said: 'For ten years I must have examined thousands of fission forms and (with the exception of Sarcina) I cannot say that I see any necessity for the differentiation of even two forms' (loc. cit., p. 20). He further stated that every 'species' of bacteria appears in several morphological and physiological forms which can be changed the one into the other by external conditions. This he called adaptation (*Anpassung* or Acclimatization). He employed this argument to write diffusely on the subject of infective diseases in general. He considered that bacteria vegetate in the soil and becoming altered they are wafted into the air, and by lighting in wounds occasion infection and contagion. He was particularly opposed to Pasteur's doctrine of specific fermentation and to Cohn's classification. Thus, for example, he states that unboiled milk becomes sour but boiled milk becomes bitter. This is not due, in his opinion, to two different agents but to the alteration of the bacteria, which cause the souring of milk, by the agency of heat. Following Nägeli closely, Hans

Buchner (1882) maintained that *Bacillus anthracis* can be produced from *Bacillus subtilis*, and, conversely, he stated that *B. subtilis* had been converted into *anthrax bacilli* by shaking at different temperatures. The change took place in the 1,500th generation in a period lasting six months. Nägeli (1882) was particularly hostile to the views of Cohn, and stated that to create, with the methods then in use, a system of fission fungi into genera and species possessed no scientific value. Time has shown that the opinions expressed by Nägeli and Hans Buchner on the subject of bacteria, at this date at any rate, were incorrect. The perusal of Nägeli's books give the impression that he was a dialectician and a verbose philosopher rather than an accurate observer in his bacteriological works, whatever he may have been in his other botanical researches.

The alleged inconstancy of bacterial forms was emphasized by Cienkowski (1877), and this view was further developed by the botanist W. Zopf in a series of works published between 1879 and 1885. He was strongly of opinion that what he called Cohn's theory possessed only historic interest and had been refuted by the work of Nägeli, Billroth, Klebs, Lankester, Warming, and others. He maintained with Nägeli that the change of one bacterial form into another is chiefly dependent on environment. In the cases which exhibit apparent constancy he considered that the developmental history was merely lacking and would in time be filled in. He considered that it was impossible to place several micro-organisms in Cohn's genera. In spite of his statements, however, Zopf classified bacteria into four groups on morphological grounds.

1. Coccaceae—known only in the globular form.
2. Bacteriaceae—consisting of coccal, rod, and thread forms but without distinction of base and apex.
3. Leptothriceae—cocci, rods, and thread forms, the latter with base and apex distinctive.
4. Cladothriceae—cocci, rods, spirals, and showing dichotomy.

Zopf's views were widely disseminated by his book *Die Spaltpilze* (1885), and raised for a considerable time active discussion on the variability of bacteria. Pleomorphism was alleged again in 1888 by Metchnikoff, but rejected by Winogradsky (1887, 1888) from his studies on sulphur bacteria. The term pleomorphism has been frequently used to denote different things. Thus it has been employed for the slight changes in bacterial form seen under different conditions of nutrition. It has also been used to denote the so-called developmental cycles of limited type which some bacteria undergo in their development. Pleomorphism in these senses is not what the name was formerly introduced to represent. The pleomorphic doctrine of Zopf and others that bacteria change from cocci to bacilli or spirals is no longer acceptable except with limitations.

While the botanists were wrangling on the subject of morphological constancy of bacteria, the medical bacteriologists were content to record the occurrence of bacteria in morbid processes. The fungus period had died down, its last representatives being Klob (1867) and Thomé (1867), who found fungi in cases of cholera, and Letzerich (1869, 1873) and Talamon (1881), who described them as causing diphtheria. From 1867 to 1877 there were many observations which showed that small bodies generally designated as 'micrococci' occurred in diphtheritis (Buhl, 1867, Tommasi and Hueter, 1868, Oertel, 1868, Nassiloff, 1870, Eberth, 1872); endocarditis (Winge, 1870, Waldeyer, 1872, Heiberg, 1872, Eberth, 1872, 1875, 1878), and in septic and putrid infections (Leyden and Jaffe, 1867, Hueter, 1868, von Recklinghausen, 1871, Cohn, 1872, Klebs, 1873, and Weigert, 1876). Of great importance for doctrine of specific pathogenic germs were the theoretical considerations of Weigert (1875). He discussed the question whether the objects called bacteria were really such or whether they were degeneration products of the tissues. He dealt with the relation of bacteria to the morbid processes or products, and whether such processes or products emanated from the bacteria or from other sources. Finally, he discussed the question whether the something that makes bacteria pathogenic is a vital bacterial product or merely something attached to the micro-organism. He endeavoured to show

RICHARD PFEIFFER
(1858–1945)

FÉLIX-ARCHIMÈDE POUCHET
(1800–1872)

that bacteria stain differently from degenerative products. In the case of small-pox, with which he was specially dealing, he showed that the observed necrotic areas in the early stages of the disease stand in close relation to the central bacterial foci. The bacteria are primary and the necrobiotic process is secondary to their action. The idea that harmless bacteria can migrate to a focus already damaged by an independent chemical poison was regarded by Weigert as improbable for the reason that foci are not found in variola before the bacteria appear. The fact that bacteria may grow without producing symptoms proves nothing, in Weigert's opinion, unless it can be shown that all bacteria are endowed with the same properties or at any rate unless in two instances, the subject of comparison, the two kinds of bacteria are the same. There is no reason to suppose that because bacteria are microscopic they are necessarily equivalent. Differences as well as agreement must be considered in estimating the question. Although Weigert's work was not experimental his theoretical conclusions carried great weight and helped on the germ theory of disease.

Towards the end of the seventies of last century it may be said that this theory had rapidly gained a firm footing in medical opinion, and bacteriology began to emerge as a definite science. People no longer spoke of 'the' bacteria, but of different bacteria. In this development both botanists and medical men played a prominent part. Among the former must be mentioned with respect the work of Cohn, whose conclusions have really remained unassailed to-day.

It is true that some of the questions which he raised are still in the melting-pot, and nowhere is this more apparent than with regard to the classification of bacteria, and the attempts of Zopf (1885), Hueppe (1886), Schroeter (1889), De Toni and Trevisan (1889), Marshall Ward (1892), Alfred Fischer (1895), Lehmann and Neumann (1896–1927), Migula (1897), down to Orla Jensen (1908, 1909), Winslow and Broadhurst (1917), and the Society of American Bacteriologists (see Bergey, 1923), cannot be said to have been entirely acceptable to all investigators.

IX
CULTIVATION OF BACTERIA

CULTIVATION OF BACTERIA

In the last chapter it was pointed out how bacteriology began to emerge as a definite branch of science. This was the outcome of the researches, particularly of Pasteur and Davaine in France, and of Ferdinand Cohn in Germany. The prolonged researches of Davaine on anthrax had rendered it highly probable that this disease was caused by the rod-shaped body which he had named bacteridium, and this view was supported by Eberth (1872), who took anthrax blood and mixed it with a large volume of water and allowed the mixture to settle. Inoculation of the supernatant liquor showed that it was inert, whereas the sediment was capable of producing anthrax. In general Davaine's views were shared by others, but there were still some lacunae which had to be filled in before the life-history of the bacteridium was completely unravelled. The infection under natural circumstances was not understood. Davaine (1870[1, 2]) showed that it may be communicated by flies, and an aerial transmission also was not denied. There still remained, however, a number of facts which the theories of the time did not explain. Why was it, for example, that the disease was endemic in certain countries or even in small localities? Why did it occur most frequently in land liable to inundation? Why was the disease more prevalent in some years than in others?

It was at this time that a new investigator appeared on the scene to throw a flood of light on the mysteries of anthrax, and to reveal the developmental cycle of Davaine's bacteridium. This investigator was Robert Koch (1843–1910), who was previously unknown in the bacteriological world, and who at the time was actually a country practitioner and district doctor at Wollstein, a small town in Posen. We are introduced to Koch by Ferdinand Cohn (1876), who tells us (loc. cit., p. 275) that it was with great pleasure that he received a letter, dated 22 April 1876, from Dr. Koch to the effect that after prolonged investigations he had discovered the complete life-history of the anthrax bacillus, and that he was prepared to come to Breslau to demonstrate his work to Cohn. The meeting took place in Cohn's institute on 30 April 1876, and lasted three days, in which time Koch completely convinced his audience of his discovery. The occasion is historic, and among those who were present were Julius Cohnheim and his assistant Weigert, L. Auerbach (of plexus fame), Moritz Traube, the chemist, Lichtheim, and Eidam, a co-worker with Cohn. What they had to see was a discovery of great moment, and one that had been overlooked by other workers. It must not be supposed that the research on anthrax was Koch's first essay in medical science. He had received a good university education at Göttingen under such men as Wöhler, the chemist, G. Meissner, and Jacob Henle, and he had studied in Berlin. He had, however, not adopted an academic career, and after his return from the Franco-Prussian war had settled in Wollstein (1872) in the capacity of a *Kreisphysikus* or district surgeon. This was, however, the beginning of a bacteriological career which stands unparalleled to-day. At the time of his anthrax work he was thirty-three years of age. In the course of his medical duties he had occasion to study anthrax in animals and he utilized his opportunities as no one had done before him. Reasoning that the facts about anthrax suggested the possibility of spore formation, he studied the disease in small animals and under primitive conditions in his home. He showed that the disease was transmissible from mouse to mouse in a series of twenty generations, and that the lesions in each member of the series were identical. He worked out the distribution of the bacilli in the bodies of different animals. Placing minute particles of fresh anthrax spleens in drops of sterile blood-serum or aqueous humour, he set himself to watch, hour after hour, what took place. His technique was simplicity itself, his apparatus was home-made. After twenty hours he saw the anthrax rods grow into long filaments especially at the edge of the cover glass, and, as he watched, he saw rounded and oval granular bodies appear in the filaments. The bodies became clearer and stood out in the filaments. He realized that they were spores which had not been seen before. In slide preparations there was every stage from Davaine's motionless bacteridia to the fully formed spore. He followed the whole process, in a primitive warm stage, at different temperatures and determined accurately the optimal thermal conditions for spore formation. By ingenious methods he disposed of the criticism that the spores belonged to some other organism which had wandered in from without, and he succeeded in showing that under suitable conditions the spores again grew into typical anthrax rods. He described and figured the act of germination of the anthrax spores, and determined that they are highly resistant to injurious influences. He grasped at once the significance of this resistance from the epidemiological standpoint. He worked at the relation of anthrax to septicaemia and showed that the two conditions differ. In putrid infusions of vitreous humour he found a bacillus almost identical in appearance as in spore production with the true anthrax bacillus, but showed that it does not produce anthrax. Tests also showed that injections of hay bacilli do not produce anthrax. From facts such as these he concluded that 'only one kind of bacillus is in position to induce the specific morbid process whereas other schizophytes are not, or, if pathogenic, they act in a manner different from anthrax'. In his own language Koch's words were: 'Es folgt heraus dass nur eine Bacillusart im Stande ist diesen specifischen Krankheitsprocess zu veranlassen, während andere Schizophyten durch Impfung gar nicht oder in anderer Weise krankheitserregend wirken' (Koch, 1876, p. 298). He thus clearly enunciated the doctrine of a specific virus.

Continuing his work, he tried to infect mice by ingestion of anthrax material but was unable to do so. He confirmed Brauell's observation that the young of pregnant animals, dead from anthrax, are non-infective, and he studied the general pathology of experimental anthrax in a complete and masterly manner. He showed that the septicaemic stage is reached only very late in anthrax of mice. Dogs, partridges, and sparrows were insusceptible to the disease. In the case of frogs he gave a very accurate description and figured the changes subsequently known as phagocytosis. Passing from the scientific to the practical aspects of anthrax, Koch showed, even at this early stage of his career, the great importance of a knowledge of prophylaxis of infective disease, and he discussed the whole question in a manner which has not been surpassed to-day. The plate accompanying Koch's (1876) classical paper on anthrax is reproduced in Fig. 26.

Koch's discovery, published (1876) under the aegis of Ferdinand Cohn, immediately became widely known, and it was at once recognized that a great investigator had arisen in the field of bacteriological research. The early hopes raised by Koch's first publication were not frustrated, for, along with Pasteur, he remains to-day the greatest exponent of bacteriological science. In connexion with his rise to fame I cannot refrain from adding a tribute to the memory of Ferdinand Cohn, who behaved towards Koch in a most generous way. Along with Cohnheim he was largely responsible for giving Koch a proper start in his scientific career, and they did everything in their power to further his worldly interests and set him free from the hum-drum of medical practice so that he could get scope for his great talents. Koch's anthrax work was practically complete, and no substantial changes have taken place in the ground which he covered. He did the work once and for all, entirely by his own efforts and by primitive means.

In France some doubt continued to be thrown on the alleged relation of Davaine's bacteridium to anthrax. Paul Bert, the physiologist, was one of the sceptics. In 1876 he stated that all living things are killed by oxygen, whereas soluble ferments or enzymes are unaffected, and he employed oxygen as a differential agent. In applying his doctrine to the special case of anthrax, he came to the conclusion that the blood contains two principles, the one—the bacteridium—which is destroyed by oxygen, and the other —a toxic substance—which remains behind unchanged.

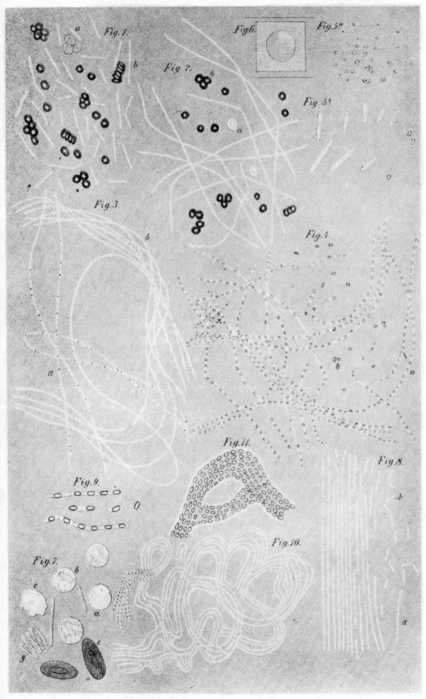

FIG. 26. Plate illustrating Koch's paper on anthrax. From Koch (1876).

This being so he could not regard the bacteridium as the cause of anthrax. Thus films of anthrax blood exposed to oxygen were found to retain their virulence although, presumably, the bacteridium had been killed. Similar results were obtained by precipitation and filtration of virulent anthrax blood. In the meantime Pasteur had begun to study anthrax, but hesitated, so he states, because he was 'étranger aux connaissances médicales et vétérinaires'. He sought for a collaborator and found one in Jules Joubert (1834–1910), a professor in the Collège Rollin in Paris. Pasteur and Joubert (1877[1]) began to investigate anthrax *de novo* and especially the question of the relation of the bacteridium of Davaine to the disease. They recognized the importance of Koch's observations and confirmed them. They also confirmed an observation already made by Tiegel (1871) that by filtering anthrax blood through plaster of Paris, the filtrate is inert whereas the filtrand is virulent. Later, Pasteur and Joubert (1877[2]), applying Bert's method, found that the anthrax spore remains unaffected by compressed oxygen. They also noted the interesting fact that the filtrate of anthrax blood not only does not produce the disease but immediately causes agglutination when added to the red blood cells even of normal animals. This agglutination of the red disks had been regarded as of diagnostic significance long before by Brauell and Pollender, but that anthrax filtrates should agglutinate normal red corpuscles was a new observation, and is probably the earliest of its kind in connexion with the subject of agglutination which became so important later on.

Pasteur and Joubert grew the anthrax bacteridia in urine rendered neutral or alkaline, and showed that such cultures reproduce the disease. They also cleared up some of the difficulties which had beset Davaine in his attempt to explain the aberrant results of Leplat and Jaillard already referred to (p. 181). Pasteur and his co-worker showed that if anthrax blood is kept, it rapidly becomes putrid and in this state may produce a mixture of anthrax and sepsis, or even of sepsis alone. In the latter case the true bacteridia were completely absent, although other micro-organisms could be found. The septic organisms were believed by Pasteur to enter the blood-stream from the alimentary canal and showed themselves as long 'vibrios' or flexuous filaments. This septic vibrio differed from the anthrax bacteridium in being anaerobic. Later, Paul Bert (1877) admitted his error and agreed with Pasteur's conclusions.

This early French excursion into the subject of anthrax really added little of importance to the clear-cut statements of Koch, but helped to fix the conviction that anthrax is a bacterial disease of special or specific character.

The study of bacteria and the diseases they may occasion then developed along two separate lines. The one direction was taken by Koch, and its aim was the thorough study and development of technical methods for the examination and cultivation of bacteria. The application of the results was intended as a basis for the rational principles of hygiene and prophylaxis. The other direction was taken by the French School led by Pasteur, who turned to the experimental analysis of how infective disease is produced in the body, and how recovery and immunity are brought about. From this work developed much that we know of preventive inoculation. The directions of the German and French schools must be considered separately and I will deal with the former first.

There was probably a reason why Koch pursued the course he did. He was working at first in a small way and animal experiments had to be economized. He therefore concentrated on technical methods, and his efforts yielded a wonderful harvest as will soon be apparent. The technical methods employed were partly the application of stains and partly attempts to obtain cultivations of bacteria in a pure state outside the body.

Staining Methods for Bacteria

The employment of stains for bacteria followed on the staining methods which had been used in ordinary histological work. This started with carmine preparations which were originally employed by Goeppert and Cohn (1849), but came into prominence through the work of Hartig (1854), and especially Gerlach (1858). The coloured extract of the logwood tree (*Haematoxylon campeachianum*) was used with indifferent success by Waldeyer (1863), but was greatly improved by Böhmer's (1865) discovery of the effects of alum added to the extract. The preparation of dyes from anilin and coal-tar took place about 1856, and substances like mauvein and fuchsin soon came into the market and were tried as tissue stains. The first who attempted to stain bacteria was Hermann Hoffmann (1819–91), professor of botany in Giessen; in 1869 he employed both carmine and fuchsin, in watery solutions. Weigert (1871) showed that carmine will colour cocci, but the staining of bacteria as an art really dates from his observations in 1875, when he showed that methyl violet can be successfully used to reveal cocci in tissues. In 1877 C. J. Salomonsen (1847–1924) successfully stained bacteria in a watery solution of fuchsin applied to a drop of fluid containing bacteria.

In his second bacteriological research, Koch (1877) greatly improved the methods of staining and laid the foundation of the technical procedures employed to-day. Realizing the importance of getting the bacteria into a non-motile state, he prepared thin films on cover glasses and dried them. To his surprise the form of the bacteria remained unchanged. He then fixed the preparations with alcohol and applied various stains, the most successful of which were methyl violet 5 B, fuchsin, and anilin brown ('new brown'). The preparations were mounted in an aqueous solution of potassium acetate or in Canada balsam. These preparations were better than any that had been seen before Koch's time, and many of them were reproduced in an excellent series of photographs taken by Koch with sunlight as an illuminant (Fig. 27). He also succeeded in staining the motile apparatus—cilia—of certain bacteria. Cilia had previously been figured by Ehrenberg (1838) in *B. triloculare*, and by Cohn (1872) in *Spirillum volutans*, but by means of logwood extract and subsequent treatment with chromic acid, Koch was the first to obtain them in the stained state. His photographs of ciliated bacteria are admirable. From now onwards staining methods were rapidly perfected. Much of this had as its basis the epoch-making work of Paul Ehrlich (1854–1915), on the staining

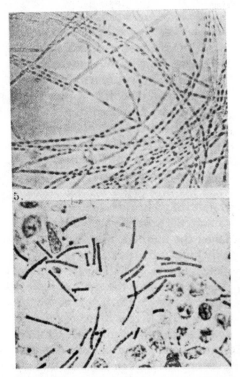

FIG. 27. Two photographs of anthrax bacilli taken by Koch. From Koch (1877).

of specific granulations in white blood corpuscles by anilin dyes. This began in 1877—at a time when he was only twenty-three years of age—and lasted till about 1881, when his blood work was practically finished. He showed that all dyes can be differentiated according as they are of basic, acid, or neutral constitution, and even at this early period he attained to astonishing perfection in stained blood-films and laid the foundations of all subsequent knowledge on the subject. His work reacted greatly on the subsequent technical methods of staining in all its branches. Bismarck brown was introduced by Weigert (1878), and methylene blue by Ehrlich (1881). The latter dye, on account of its properties, soon came to be one of the most extensively used in bacteriological work and was the dye which revealed the tubercle bacillus to Koch in 1882. He discovered that when an alkali was added to the methylene blue, the dye penetrates into the tubercle bacillus and remains there in spite of subsequent treatment with vesuvin or Bismarck brown, and it was by this process that Koch showed the tubercle bacilli in his first work on tubercle.

The alkaline addition to methylene blue was also the basis of 'Loeffler's methylene blue' (1884), which has been used ever since. It was in 1881 that heat was employed by Koch for fixing bacterial films prior to staining. On the day after Koch read his first paper on the causation of tuberculosis (24 March 1882), Ehrlich succeeded in improving the method of staining the tubercle bacillus and soon afterwards (1882) published the method, essentially the same as that now universally known by the names of Ziehl and Neelsen. Instead of potassium as the alkali he employed anilin as an addition to the dye, and the dyes he used were chiefly methyl violet or fuchsin. Ehrlich was led to the use of anilin by observing that preparations of impure violet, such as gentian violet, stain better than solutions that are pure, the enhancing element in the impure dye being probably anilin. This substance, shaken up with water and filtered, yielded the so-called anilin-water which Ehrlich then saturated with violet or fuchsin in alcoholic solution. By this means the tubercle bacillus was stained deeply and the colour was 'acid fast' in that it could not be discharged by even nitric acid in a concentration of 30 per cent. in water. As counter stains he used either vesuvin or methylene blue according as the bacillus was primarily stained with violet or fuchsin.

The use of carbolic acid instead of anilin was the contribution of F. Ziehl (1857–1926), and the use of sulphuric acid instead of nitric acid was the contribution of F. Neelsen (1854–

1894), see Johne, 1885. The whole merit of staining tubercle bacilli by the methods used up till to-day is, however, due to Ehrlich. His anilin-water solutions of methyl violet were also employed by Christian Gram (1884), a Dane working in Berlin, in the development of the universally known Gram's specific process of staining. What Gram found was that if bacteria are stained by Ehrlich's anilin-water-gentian-violet, are then treated with Lugol's solution of iodine in potassium iodide, and are finally placed in alcohol, the colour is discharged from certain bacteria but is retained in other bacteria. This Gram's method—for Gram invented no new solution—has been modified by many workers, but in its essence remains unaltered and is one of the most widely practised methods of differential staining used in bacteriology. Special methods were also discovered for staining spores, cilia, and capsules, and the number of these methods is legion and cannot be further referred to here. They are mostly trivial modifications of the original processes laid down by the early master stainers like Ehrlich, Weigert, and Koch.

Pure Cultivations of Bacteria

The conception of the existence of bacterial species, as held by Ehrenberg, Cohn, and Schroeter, early led to the idea that it might be possible to obtain what came to be called 'pure cultivations', i.e. growths of one particular kind of micro-organism unmixed with others. This was supported by the observations of Pasteur (1857) that lactic fermentation is due to a characteristic ferment, which can acidulate and ferment milk, when grown in a flask.

A pure culture in the modern sense was not at the time obtained by Pasteur, nor indeed by this rough method except perhaps in very special circumstances. The method of growing bacteria in transparent liquid media was, however, the common one employed by Pasteur, and various attempts were made to get the most suitable nutritive media. A small portion of the material to be cultivated was transferred by a sterile instrument to a sterile medium in bulk, and the latter was then said to have been inseminated or inoculated (*inoculo*, I ingraft an eye or bud from one tree into another). When growth occurred, a minute quantity was inoculated into fresh medium, and so on in series. By this means it was hoped that ultimately a 'pure' culture of one type of micro-organism might result from the original mixture. It was soon noted that media vary apparently in their nutritive characters, and efforts were made to secure media suitable to all bacteria. The early media of this type were relatively simple solutions. Thus Pasteur (1861) used a fluid consisting of water 100 parts, pure candy sugar 10 parts, ammonium tartrate 1 part, and 1 part of ash of yeast. This 'Pasteur fluid' was improved and made more suitable for many organisms by Mayer (1869), who, instead of yeast ash as such, used chemically pure solutions of the salts found in the ash of yeast, and Cohn (1872) utilized Mayer's idea to prepare a fluid medium —'Cohn's fluid'—which differed from that of Pasteur in the omission of the sugar. Cohn's fluid consisted of

Potassium phosphate	. .	0·5 gm.
Cryst. magnesium sulphate	.	0·5 ,,
Calcium phosphate	. .	0·05 ,,
Aq. destil. .	. .	100·0 ,,
Ammonium tartrate	. .	1·0 ,,

Mayer's fluid.

In the early days of bacteriology such a fluid was extensively used with variable results, as also were such fluids as Raulin's, or the various aqueous extracts and decoctions of vegetable substances (hay, turnip, carrot). Milk and urine were also employed as special media. The growths on such media were often feeble and more complicated ingredients were tried. An elaborate research on the nutritional requirements of bacteria was carried out by Nägeli (1882), who examined the effect of the presence of various carbohydrates and proteins, and found that for the assimilation of carbon and nitrogen the scale begins with sugars and peptone. Other compounds were less favourable for growth. P. Miquel (1883) suggested the use of Liebig's extract or a bouillon made from meat extract. The basis of the common broth or bouillon used everywhere to-day was worked out by F. Loeffler (1881) in connexion with his successful cultivation of the bacillus of mouse septicaemia. He grew this organism in a medium consisting of meat infusion, to which was added 1 per cent. of peptone, and 0·6 per cent. of common salt. The reaction of the final product was made very faintly alkaline with sodium phosphate.

The first attempts to obtain separate cultures of pathogenic bacteria were those of E. Klebs (1873) by what he called his 'fractional method' (*fractionirte Cultur*)—not a very happy name. The principle was to introduce by means of a capillary pipette, charged with the material to be cultivated, a minute portion into a sterile medium. After vegetation had occurred a minute fraction or quantity of the growth was placed in a second quantity of medium, and so on in series. It was hoped that any contamination would in time be eliminated, and that the organism present in the greatest quantity in the original inoculum would ultimately dominate the situation and show itself as a pure cultivation. Klebs is reticent on his results, but it is almost certain that he never obtained pure cultures by his method.

Attempts to grow mass cultures in fluid media, from mixtures of bacteria, caused the failure of Hallier, and this method was discredited. Somewhat better results were obtained by the growth on opaque solid media. This no doubt started from the observations of the peculiar blood-red stains which have at various times been observed on articles of diet sporadically or sometimes in epidemics. It was in an extensive outbreak in 1819 in Legnaro, near Padua, that Vincenzo Sette (1824) showed that what had been regarded as a portent was really something living and that it could grow on bread or polenta to which it had been transferred. He called the living agent *Zoogalactina imetrofa*, which later was named *Micrococcus prodigiosus*. Ehrenberg (1848) again saw these blood-red spots and got them to grow on various substances, and in each culture the blood-red colour was maintained.

Attempts to grow micro-organisms other than those which developed colour were made by Hermann Hoffmann (1865). He took a corked tube half filled with water, which was then boiled. A piece of bread or potato was then introduced into the tube, which was again boiled. The water being carefully poured away, a trace of yeast was applied to the surface of the medium by a long needle, and after some days growth occurred and turned out to be a mould. The apparatus of Hoffmann was described as a *Dunstrohr*, or vapour tube, and its object was to exclude aerial germs and to allow the inoculum to vegetate on a moist solid surface. Brefeld (1874) said of the method that it was on a par with a man already soaked putting on a raincoat to protect himself against the last few drops of rain.

Solid media were used with great advantage by Joseph Schroeter (1872) in his classical work on pigment bacteria. Potato, starch paste, flour paste, bread, egg albumen, and meat were all employed by him, and on them he obtained a number of bacterial growths, red, blue, green, orange, yellow, brown, or violet in colour. In the different masses or colonies the bacteria varied, but were constant in one and the same colony. No doubt Schroeter obtained pure growths.

There were limitations, however, when one had to deal with non-pigmented forms of the same colour as the medium itself. Potato, as a medium, was also employed by R. Koch (1881[1]), who, however, greatly improved the technique of preparation. This he did by soaking the raw potato in a solution of corrosive sublimate (1:1000). The potato was then sterilized in steam and finally split in two by a sterile knife. The two halves were then allowed to fall apart in a covered glass vessel which had been previously rendered sterile. The cut surface of the potato was then inoculated with the bacterial material.

Klebs (1873) attempted to get cultures by inoculating minute quantities of bacterial material in fluids which were then placed into Recklinghausen glass chambers as made by the famous Bonn glass-blower, Heinrich Geissler (1814–79). The 'cultures' were then observed under the microscope and the development watched. Klebs found difficulties with watery media and subsequently added isinglass to render the medium stiff and to keep the bacteria better in position. Notwithstanding, his method was un-satisfactory and he was led to erroneous conclusions, as has already been indicated. It may be pointed out that gelatine had also been used by Vittadini (1852) in his attempts to grow the fungus of muscardine disease.

The addition of gelatine and decoction of carragheen to liquid media was also suggested by Oscar Brefeld (1872). To this great mycologist must be given the credit of laying down with the utmost precision the principles which must be followed for obtaining pure cultures. His researches were presented in a series of eighteen volumes which appeared from 1872 onwards. These magnificent works profusely and beautifully illustrated by Brefeld himself dealt with large groups of fungi, but of especial interest bacteriologically was the series entitled *Botanische Unter-suchungen über Schimmelpilze* (1872, 1874, 1881). In addition, he wrote a number of papers, the most concise on the subject of cultivation being that published in 1875. The new principles he inculcated were (1) that the insemina-tion of the medium should be made from one single fungal spore, (2) the medium should be clear and transparent and yield optimal growth for the fungus under examination, and (3) the culture should throughout be completely pro-tected from external contamination. Brefeld worked almost entirely with the higher fungi, to which his method was peculiarly suitable. To obtain a single spore he took with a fine-pointed forceps the spore-containing organ and placed the spores in sterile water. When thorough wetting of the spore had taken place he diluted the suspension still further, and by means of a very fine needle he placed a droplet of the diluted suspension on a sterile slide. The drop being spread out he searched for a spot where the microscope showed a single spore and, this being found, the rest of the drop was removed by sterile blotting-paper. The slide with the single spore served as the lower surface of the culture, and on it he added a drop of sterile medium. He described many media and the addition of sterile gelatine for the purpose of preventing evaporation of the watery medium. He either placed the gelatinized medium

in a Recklinghausen chamber, improved to suit his particular purpose, or he suspended the drop from the under surface of a cover glass in a hollowed-out glass apparatus. This was a primitive form of the hanging drop-slide used to-day. Brefeld got excellent results with his method and was the pioneer of the single-cell method. His illustrations were so good that many of them are to be found in almost every text-book to-day.

Brefeld's technique, admirable as it was for relatively large objects like the spores of moulds, was found to be unsuitable when applied to minute objects like single bacteria. Other methods were therefore suggested and among them the principle of dilution. A fluid containing a mixture of bacteria was diluted down with sterile medium in the hope that ultimately a growth would be obtained which had taken its origin from a single microbe. The possibility of success by such a method was denied by Nägeli and Schwendener (1877), but the feat was accomplished by Joseph Lister (1878) by an ingenious method. By means of a specially constructed syringe (Fig. 28), having a graduated nut, revolving in a fine screw on the

FIG. 28. Lister's syringe. From Lister (1878).

piston rod, each degree was made to correspond to the delivery of $\frac{1}{100}$ of a minim. He found that $\frac{1}{50}$ of a minim exactly covered a cover slip of a particular size. Hence the number of bacteria under the cover glass, i.e. in $\frac{1}{50}$ minim, was equal to the number of bacteria in a microscopic field multiplied by the number of times the area of that field went into the area of the cover slip. The micrometer gave the diameter of the field in thousands of an inch across,

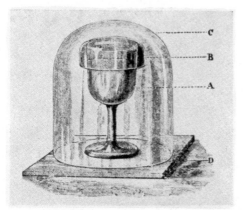

FIG. 29. Lister's culture glass. From Lister (1878).

and the areas of the circles were, of course, proportional to the squares of those diameters. All that was needful, there- fore, to calculate the number of bacteria in $\frac{1}{50}$ minim, was to form a fair estimate of the number of bacteria per field. Two different forms of bacteria were seen under the microscope.

As a result of his calculation Lister found it necessary to dilute down the bacterial fluid with one million parts of sterile water so that $\frac{1}{100}$ minim delivered should contain one single bacterium. This quantity, therefore, was delivered into each of five liqueur glasses of sterilized milk with the result that only one showed a growth, and the milk soured (Fig. 29). The growth was found to be composed of only one kind of bacterium, to which Lister gave the name of *Bacterium lactis*. He found it bred true to type and was in fact a pure culture.

Nägeli (1882) also used the dilution method for isolating bacteria from putrid urine, but he gives no exact account of how he counted the number in the original fluid.

E. C. Hansen (1842–1909), working in the Carlsberg brewery in Copenhagen, in a long series of publications

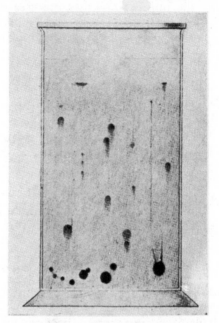

Fig. 30. Glass cylinder with putrefying blood showing dark spots of bacterial masses. From Salomonsen (1877).

dating from 1883 onwards (see also Pedersen, 1878), greatly improved the dilution technique of raising pure cultures of yeast and indeed revolutionized this aspect of the brewery industry. His method differed from the previous ones in that he partly made use of the haematimeter of G. Hayem, and partly of a special cover slip divided into squares. A drop of fluid containing yeast cells was diluted and placed on the cover glass, and the latter was placed in a Böttcher's moist chamber. The yeast cells were then counted. The diluted yeast was then introduced into sterile flasks of wort, the degree of dilution being such that only a small proportion of the flasks became infected. He was able to distinguish between the flasks which received only one cell and those that received more than one, this result being dependent on his observation that, after the yeast cells had been well distributed by agitation in the wort, each settled to the bottom and gave rise to a separate speck of yeast. Only those cultures which showed one such speck were assumed to be pure.

By means of a counting chamber and dilutions A. Fitz (1882) obtained a pure culture of *Bacillus butylicus*. He used as many as 50 flasks in his final dilutions. The mechanical separation of bacteria in putrid blood was employed by C. J. Salomonsen (1876, 1877) to obtain individual colonies. Having observed that dark spots appear in glasses containing putrid blood (Fig. 30), he found that they were really pure cultures of different kinds of bacteria, and he isolated them in a more perfect manner by aspirating the putrid blood into long capillary tubes (50–60 cm. × 0·5–1·0 mm.) which he was able to place under the microscope for continued observation (Fig. 31). Breaking the tubes where isolated spots occurred, he found that each spot contained only one sort of bacterium. In fact it was a colony of bacteria, separate spots often differing in the size and shape of the organisms. No spots or bacteria occurred in blood taken aseptically from the vessels of healthy living animals. Apart from its use in blood, Salomonsen's method could not be widely applied and was given up until modern times, when it has again been used by A. E. Wright and others.

Another early method of obtaining what were regarded as pure growths consisted of inoculating susceptible animals with some particular kind of

FIG. 31. Salomonsen's capillary culture tube showing colonies of bacteria at *b, c, d, e, f,* and *g.* From Salomonsen (1877).

infectious material. This was the method first used by L. Coze and V. Feltz (1866), and by Davaine (1872), in connexion with their work on septicaemia. Transference of the infective material from animal to animal in series was practised by these workers, and Davaine spoke of it as 'passing' through the animal body ('en passant si je puis ainsi dire dans l'économie d'un animal vivant'). The method of 'passage' in this sense was also employed by R. Koch in the case of anthrax (1876), and in traumatic infective diseases (1878). Passing anthrax material through a series of twenty mice, Koch found that the blood in the twentieth mouse showed no organism except anthrax. Again in 1878 he showed that by this method impure cultures become purified in the bodies of successive animals. It was by this means that he obtained the virus of mouse septicaemia from material which at first was grossly contaminated with other bacteria. Indeed, at this time (1878), he wrote that he regarded the successive transmission of artificial infective diseases 'as the best and surest method of pure cultivation'. He showed that the animal body can be utilized to differentiate two kinds of bacteria. From putrefying blood the only organisms which survived in the bodies of white mice were a small bacillus and a coccus in chains. They could be transferred from house mouse to house mouse. A separation, however, was possible when blood from a house mouse which contained both organisms was injected into a field mouse. The bacillus disappeared, whereas the coccus persisted and could be transferred pure in series in field mice.

The method of propagating infective diseases by inoculation has been widely practised and was one of the chief methods employed by Pasteur. It may be, and indeed usually is, the only method in cases where the particular virus cannot be cultivated *in vitro*. This is the method of propagating the unknown viruses of diseases like vaccinia, rabies, foot-and-mouth disease, fowl plague, swine fever, rinderpest, and many other diseases. Presumably, frequent transferences by 'passage' leads to purification of the virus, although this cannot be definitely proved as yet.

Apart from the method just described, the other methods of attempting pure cultivations were chiefly the result of sporadic efforts, and in all cases were laborious and very frequently ineffective. The fundamental importance of pure bacterial cultures *in vitro* was foreseen by Koch very early, and he bent all his power to attain the desired result by a simple and consistently successful method. He succeeded in grasping all the technical difficulties and necessities of the problem with wonderful clearness. He attempted to obtain a good medium which was at once *sterile, transparent,* and *solid.* After many attempts he arrived at the conclusion that it was scarcely possible to compound a kind of universal fluid which would present equal nutritive value to all bacteria. He therefore concentrated his efforts on solidifying well-tried nutritive fluid media with some clear substance, and recommended for this purpose a $2\frac{1}{2}$-5 per cent. of gelatine. The product was referred to as 'nutrient gelatine' (*Nährgelatine*). The essential ingredient in the nutritive fluid was 1 per cent. meat extract. He gave exact details how the nutrient gelatine was to be prepared. For use he poured the melted nutritive gelatine on sterile glass slides and allowed it to set under a bell jar to avoid external contaminations. The actual insemination of the gelatine was made by taking a minimal quantity of the inoculum, by means of a sterilized needle or platinum wire, and drawing it in several cross lines rapidly and lightly over the surface of the jelly. Different colonies of bacteria soon made their appearance and were then transferred to test-tubes plugged with cotton-wool and containing sterile nutrient gelatine which had been set in an upright or slanted position. In this simple way Koch had solved a problem which three years previously he had deemed to be incapable of solution. By this means he opened the door for one of the greatest advances ever made in the history of medicine. He soon gained world-wide fame by the demonstration he gave of his method and its results in the Physiological Laboratory in King's College, London, at the International Medical Congress in 1881. The audience at his demonstration comprised Lister,

Pasteur, Burdon-Sanderson, and Chauveau. Lord Lister told me that the moment Pasteur saw Koch's cultures he turned to Koch, whom he had not met previously, and said: 'C'est un grand progrès, Monsieur!' a remark which history soon verified. It was in the same International Medical Congress that the English heterogenist, Charlton Bastian, reiterated all his old views with increased ardour. In reply to him Pasteur asked whether he still believed in spontaneous generation. Receiving no negative reply to his question, Pasteur dramatically raised his hands to heaven and exclaimed: 'Mon Dieu, mon Dieu! est-ce que nous sommes encore là? Mais, mon Dieu! ce n'est pas possible!' The old doctrine was dead and received a parting kick from Pasteur; the new doctrine was born. Bastian was demolished, Koch was raised on a pedestal.

Continuing his researches with all the assiduity of his race, Koch now made many further discoveries by means of his method and strove to render it perfect. The poured plate method was published in 1883, and was recommended originally for isolating pure bacterial cultures from water. Instead of streaking bacteria on gelatine already solidified, Koch now mixed the bacterial inoculum with melted gelatine, and having well distributed the mixture by shaking he poured it, still in the melted state, on cold sterile glass plates.

The separation of adjacent colonies was very much better by this process which soon became the method of election. By this means, a large number of bacteria were obtained pure, from air, water, soil, and many other fluids and solids. Bacteriology grew at a rapid rate and students flocked to Berlin from all parts of the world to master the new technique which opened up inexhaustible fields of research. The claims which Koch put forward for his method were stated by him as follows:

'The peculiarity of my method is that it supplies a firm and, where possible, a transparent pabulum; that its composition can be varied to any extent and suited to the organism under observation; that all precautions against the possibility of after-contamination are rendered superfluous; that subsequent cultivation can be carried out by a larger number of single cultures of which of course only those cultures which remain pure are employed for further cultivation; and that, finally, a constant control over the state of the culture can be obtained by the use of the microscope.'

The subsequent history of bacteriological technique has verified to the letter the claims of Koch for his method, which in principle has remained unchanged.

The nutrient basis of the medium has been varied to an endless extent by the addition of various substances, such as sugars, alcohols, glucosides, albumen, serum, blood, and other substances to enhance the nutritive value or to reveal some particular fermentative property of the organism under examination. The titration of the reaction has been simplified and standardized by the methods of physical chemistry. The solidifying element—gelatine—in Koch's original method has suffered only one or two important substitutions. One disadvantage of its use was soon experienced, viz. that it does not remain solid at body temperature. Agar-agar soon replaced it and was introduced into bacteriological technique by Frau Hesse, the wife of Walther Hesse (1864–1911), one of Koch's early co-workers. She had obtained samples, through Dutch friends, from Batavia, where it was well known for culinary purposes and especially in the making of jam. Agar-agar is a vegetable gelatinous substance, the exact source of which is given variously by different authorities, but the principal algae from which it is derived live in the eastern Asiatic seas from Ceylon to Japan. It is said to be mainly derived from species of *Eucheuma* (*E. spinosum*) and *Gelidium* (*G. corneum*, *G. Amansii*). The peculiar virtue which has established its dominance in bacteriological culture technique is that a high temperature is required to melt it, but that once melted it can be cooled down to about 40° C. before it sets into a stiff and relatively transparent solid mass.

Blood-serum stiffened by heat also yields a solid, fairiy clear medium of great nutritive value, and was introduced by Koch (1882) for cultivating tubercle bacillus.

Koch's gelatine or agar plates were employed in their original form for some little time, but were replaced when

R. J. Petri (1887), an assistant of Koch, introduced what he called a 'slight modification' of the method. This consisted in pouring the melted medium into a covered glass dish instead of on a flat glass plate. The improvement permitted repeated examination of the plate without risk of aerial contamination at any stage of the process. Petri's dishes were destined to replace Koch's glass plates immediately and permanently, and his method has not been altered down to the present time. Taking the Koch process in principle, the results which it has given shows it to have been one by which the greatest advances ever taken in the discovery of the causation of disease have been made.

Filtration of Bacterial Liquids

The methods of bacterial filtration date from the experiments of Tiegel (1871), who filtered anthrax fluids through a porous cell of unburnt clay, by the aid of a Bunsen air pump. Eberth (1872) also used this method in connexion with experiments on diphtheria. Pasteur and Joubert (1877[1, 2]) employed plaster of Paris to separate anthrax bacilli from fluids containing them. Miquel and Benoist (1881) used a long-necked flask, the lower part of which was filled with asbestos with a layer of plaster of Paris on the top. Gautier (1884) used a somewhat similar apparatus but with a filter made of borax, silica, and red lead. In 1884 Charles Chamberland (1851–1908) gave a short description of a filtering apparatus of candle form (*bougie filtrante*), and composed of unglazed porcelain. This material apparently had come into use in Pasteur's laboratory, and the filter (Fig. 32) was improved by Chamberland, who added an end-piece of nipple shape and composed of glazed porcelain fused to the body of the *bougie*. This is the ordinary bacteriological filter so widely known as Chamberland's filter, system Pasteur. In 1891 Nordtmeyer introduced a new filtering medium made of the compressed infusorial earth known as *Kieselguhr*. Attention was directed to the filtering possibilities of this substance by the fact that the ground water in the *Kieselguhr* mine in Unterluss in the Lüneburg Heath (Hanover) was of a clear blue

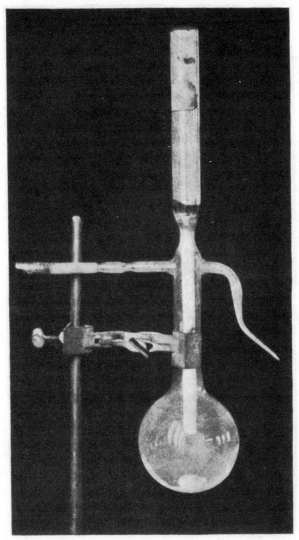

Fig. 32. One of the first bacterial filters used by Pasteur.

colour. The filter of *Kieselguhr* was called *Berkefeld* from the name of the owner of the aforesaid mine. Many other filters have been used, differing especially in size and shape and material, but the Chamberland and Berkefeld candles continue to be extensively used.

Anaerobiosis and the Technical Methods for Anaerobic Culture

Attention was drawn by Pasteur (1861) to the existence of organisms capable of living in the absence of free oxygen. His observations of this date dealt with his so-called butyric ferment. That 'animalcules' can exist without air, or at any rate in a highly rarefied atmosphere, was shown by Leeuwenhoek in 1680. In his 32nd letter (14 June 1680) to the Royal Society he described and figured an experiment in which he took a glass tube sealed at one end (Fig. 33), and filled to the level BK with powdered pepper, and up to the level CI with clear fresh rain water. By the aid of a flame the glass was then drawn out to a point which was then sealed at G. A similar glass tube was left open. In the sealed tube 'animalcules' developed although the contained air must have been in minimal quantity. Beijerinck (1913), who repeated Leeuwenhoek's experiment, considered that it was the first recorded one on anaerobic bacteria, for he found in the sealed tube *Bacterium coli*, *Azotobacter*, and *Amylobacter saccharobutyricum*, and he inferred that Leeuwenhoek's very small animalcules must have been bacterial in character. Spallanzani (1776) also saw that animalcules developed in a high vacuum and could live and move for sixteen days. These observations were, however, overlooked or forgotten and anaerobiosis was rediscovered by Pasteur (1861[1]). In a later publication (1861[2]) he showed that yeast grown in the absence of air over mercury can bring about fermentation, and still later (1863[1]) he saw typical fermentation of calcium tartrate by a micro-organism which was able to grow under a layer of oil. Pasteur (1863[2]) introduced the terms *aérobies* and *anaérobies* to indicate micro-organisms which live with or without free oxygen. An alternative word for *anaérobies* was *zymiques* (ξύμη, ferment) but it was not used subsequently. Pasteur and Joubert (1877[2]) also discovered an anaerobe which produced disease. This organism, in the form of long filaments, occurred in anthrax blood which had undergone putrefaction and was called the septic vibrio (*Vibrion septique*). Brefeld (1874) combated Pasteur's statement that yeast can ferment under anaerobic conditions, but Paschutin (1874) showed that frog's muscle undergoes putrefaction even if kept in atmospheres of hydrogen, nitrogen, carbonic oxide, or even coal gas. Hufner (1876) devised an ingenious experiment in which putrid material was brought into contact with boiled fibrin and water in a vacuum.

Pasteur (1876) also described ingenious, if complicated, methods to demonstrate fermentation in anaerobic conditions. Gunning (1877, 1878, 1879), a chemist at Amsterdam, was of opinion that strict anaerobiosis was not obtained by the methods ordinarily in use, and he clearly showed the great difficulty of freeing apparatus from small quantities of oxygen. On the other hand, M. Nencki (1879) believed that bacterial growth could occur in perfectly anaerobic conditions. With regard to yeast, Cochin (1880) found that in the continued absence of air

FIG. 33. Leeuwenhoek's tube for growing animalcules in a rarefied atmosphere. From Leeuwenhoek (1695).

the power of reproduction was lost. Hoppe Seyler (1884) considered that Pasteur's idea of life without air was highly improbable, whereas Lachowicz and Nencki (1884) saw putrefaction under conditions in which no free oxygen could be detected even by the most delicate tests. It was about this time that technical methods for cultivating anaerobes began to be used, and of these early methods may be mentioned those of Hans Buchner (1885, 1888), Hauser (1885), W. and R. Hesse (1885), Liborius (1886), Gruber (1887), Vignal (1887), Roux (1887), C. Fraenkel (1888), and Kitasato (1889). From this time down to the present day there has been one continuous stream of anaerobic apparatus of every shape and size and design, and all based on permutations and combinations of a few elementary principles. Between 1888 and 1918 no fewer than three hundred different anaerobic apparatus were described at the rate of about 10 per annum. In 1904 the new additions to the anaerobic apparatus actually numbered nineteen at least, and were probably more than this.

Sterilization in Bacteriological Technique

The necessity for the sterilization of apparatus and media was early appreciated and the methods for its production were the outcome of the work of Spallanzani, Schwann, Pasteur, Schröder, and v. Dusch, Bastian, and Tyndall. Their efforts have been already considered (Chap. IV). The modern application of the principles they inculcated developed especially out of the bacteriological work of Koch. The exact value of hot air as a sterilizating agent was determined particularly by Koch and Wolffhügel (1881). Koch, Gaffky, and Loeffler (1881) determined the limitations of steam at 100° C. The important principle of discontinuous heating was the contribution of John Tyndall and has been referred to. The extraordinary sterilizing effect of superheated steam was the outcome of Pasteur's observations of the effect of immersing hermetically sealed flasks in a bath of calcium chloride solution heated above 100° C. The disadvantages of his processes were obviated when the flasks were merely plugged with wool, and the sterilizing vessel was closed on the principle of Denys Papin's 'digester or engine for softening bones' (1681). Modern apparatus of this kind are called 'autoclaves', of which the smaller laboratory forms were made very convenient by the Parisian engineering firm of Wiesnegg under the name of Chamberland's autoclave, about 1884.

Bacteriological Technique applied to the Practice of Chemical Disinfection

The conception of chemical antiseptics or substances which prevent putrefaction is a very old one and many substances believed to possess these properties have been used in the treatment of wounds. In modern times, antiseptic treatment of wounds came into renewed prominence with Lister's antiseptic system of surgery, which was first made known publicly in 1868. An important element in this system was the use of carbolic acid in aqueous or oily solution. Lister had to estimate its virtues as an antiseptic mainly by its apparent effects in favouring the healing of wounds. Its antiseptic effect in preserving organic matter from corruption had previously been made the subject of extensive observation by J. Lemaire (1860).

Robert Koch (1881[2]) initiated a new era by comparing the action of a very large number of reputed antiseptics on certain bacteria. Pure cultures of bacteria, and amongst these especially anthrax spores, were dried on small pieces of silk threads which were then immersed in the antiseptic. After intervals of time the impregnated threads were removed from the antiseptic, were washed in sterile water or broth, and implanted in a medium to determine whether the bacteria had been killed or not. Koch was soon able to draw a distinction between the concentration of an antiseptic which prevents a micro-organism from growing and that concentration which kills the micro-organism. These limits may be, and usually are, far apart. Tested in this way, Koch found that Lister's carbolic-oil had no disinfecting power at all, and even a saturated aqueous solution of carbolic acid was much less powerful than had been supposed.

Of all the many (over 70) substances tested by Koch there was only one—perchloride of mercury—which in a high dilution and in a short space of time was able to destroy the most resistant bacterial spores.

J. Geppert (1889), in an important work, confirmed Koch's statements but showed that he had over-estimated the potency of the sublimate, for, on adding ammonium sulphide to precipitate all the mercury at the end of the experiment, he found that micro-organisms, which by Koch's technique appeared to be dead, were in fact still living but unable to develop from the traces of sublimate surrounding them. Without ammonium sulphide treatment, sublimate in a concentration of 1:1000 apparently killed all the anthrax spores in from 3 to 7 minutes, but if the mercury was precipitated as sulphide many of the spores were shown to have survived. Even after exposure to a concentration of 1:100 sublimate, anthrax spores were not destroyed with certainty in from 6 to 12 minutes. The fundamental importance of Geppert's work in determining the potency of any antiseptic was fully recognized by Krönig and Paul (1897) in their classical research, which really laid the foundations of our modern scientific knowledge on chemical disinfection. They clearly realized that disinfectants can only be accurately compared when certain conditions are fulfilled. The bacterial test object must have the same resistance and the number of bacteria must be constant. The bacteria must be brought into contact with the disinfectant without admixture of other organic matter such as nutritive medium. The temperature of the experiment must be constant. After the application of the disinfectant for a stated time its action must be promptly and completely arrested. The bacteria must then be transferred to the most favourable medium kept at optimum temperature. The result is determined by the enumeration of the living survivors in plate cultures. These general principles have been followed by subsequent workers and have yielded most of the accurate knowledge we possess to-day. The general result of an immense amount of work on the subject has been to the effect that chemical disinfection is a much more difficult thing to carry out than was formerly believed.

The simple method of pure cultivation which Koch (1881) had devised yielded, within a few years, an amazing harvest of results which fell almost exclusively to German bacteriologists. In addition to a very large number of saprophytic bacteria which were isolated from every con-

FRANCESCO REDI
(1626–1697)

ÉMILE ROUX
(1853–1933)

ceivable source such as water, air, soil, and food, the aetiological agents of a large number of human and animal diseases were revealed for the first time and an entire reconstruction of ideas on causation of disease ensued in consequence. Up till Koch's publication in 1881 it may be said that the bacterial cause of only two diseases had been made probable or certain. I refer to anthrax and to relapsing fever, in which O. Obermeier (1873) had demonstrated the constant presence of the spiral organism for long called *Spirochaete Obermeieri.* After 1881 the pure culture method disclosed the causes of many other infective diseases. It will be sufficient to name the most important only. In 1882 Robert Koch made the momentous discovery of the tubercle bacillus, and Loeffler and Schütz (*vide* Struck) discovered the cause of glanders. In 1884 Robert Koch isolated the cause of Asiatic cholera, and in the same year his assistant, F. Loeffler, isolated the diphtheria bacillus. In 1884 also G. Gaffky, another of Koch's co-workers, discovered the typhoid bacillus, and J. Rosenbach isolated pure cultures of Staphylococcus and Streptococcus. In 1885 Ernst Bumm proved that pure cultures of Neisser's gonococcus cause gonorrhoea, and Escherich isolated *Bacterium coli commune.* In 1886 Albert Fräenkel grew pneumococcus in a pure state and showed its relation to pneumonia. In 1887 David Bruce discovered the cause of Malta Fever, and in the same year A. Weichselbaum isolated the meningococcus. In 1889 Kitasato successfully cultivated the tetanus bacillus and disclosed its pathogenic properties. In 1891 Wolff and Israel succeeded in growing the fungus of actinomycosis, and in 1894 Kitasato and Yersin discovered the plague bacillus. In 1897 van Ermengem of Ghent discovered the bacillus of botulism, and in 1898 K. Shiga isolated the cause of acute dysentery. In all these instances the respective micro-organisms were constantly found, they were isolated in a pure state and the inoculation of the pure cultures determined pathological states identical or similar to those found in man or animals from which they had been isolated. Although other microbic causes were discovered in other diseases later the main aetiological period of bacteriology may be said to date from 1881 to 1900, a period of twenty years. It is remarkable that with the exception of Yersin's discovery no important pathogenic microbe was discovered by the French school, and this is all the more striking in that Pasteur had been the main pioneer in the earlier bacteriological period. French bacteriology in the hands of Pasteur and his pupils was developing along other lines and was concerned with the problems of recovery from infective disease, prophylaxis and immunity problems which early took their origin in the genius of Pasteur and later developed to a wonderful extent.

X

PASTEUR'S WORK ON ATTENUATION OF VIRUS

X

PASTEUR'S WORK ON ATTENUATION OF VIRUS

In the previous chapter attention was directed to the methods and results obtained by the pure cultivation of bacteria. It was by this means that after 1881 German bacteriologists in particular demonstrated the causes of a very large number of infective diseases. Pasteur had been the great pioneer of bacteriology in the previous two decades but his line of study was not such as to bring him in touch with problems of disease in man and animals. He now, however, began to apply himself to such questions about 1877, and his early researches on anthrax at this time have already been dealt with. He was interested not so much in the subject of technical bacteriology as in those concerning the manner in which infective disease is produced, and he suddenly struck new ground in a series of epoch-making researches dealing with the fundamental problems of prevention and recovery from infection. Although nearing his sixtieth year he entered a phase of unexampled fertility even when one remembers his previous great achievements in the domain of fermentation. In 1880 he examined cases of furunculosis, osteomyelitis, and puerperal fever, all of which he attributed to the development of micro-organisms infecting the pus of these inflammatory conditions (Pasteur, 1880[3]).

Passing from such conditions, he then turned to the study of what at that time were described in France as virus diseases. These virus diseases, which were typified by small-pox, were remarkable in that one attack left behind a lasting state of immunity. Pasteur was greatly impressed by this fact from the study of the literature of variolation and Jenner's work on vaccination. He had an instinct that a process akin to vaccination might hold for other diseases, and that such was the fact he soon disclosed in the case of fowl cholera. This disease—in France called *choléra des poules*—had been recognized as bacterial by the Italian worker Perroncito (1879), and the organism described by him had been cultivated by H. Toussaint (1879), a professor of the Lyons' Veterinary School.

Pasteur commenced his work where they left off, and in a large series of fowls showed the deadly infective character of cultivations made in chicken bouillon. He found that if grown in series in cultures the virulence for fowls was maintained, but that there were conditions in which the virulence for fowls diminished. This enfeeblement of the virus was called by Pasteur 'attenuation' (Pasteur, 1880[1]). In the same year he announced the fundamental fact that if fowls are first inoculated with the living attenuated virus of chicken cholera they withstand a subsequent inoculation which kills unprotected animals acutely. He was led to this discovery by a number of observations. He noted, for example, that the effect of chicken-cholera cultures is not always constant. The virus itself apparently was liable to fluctuate in virulence. He observed that chicken cholera is not liable to recur if the fowl has once recovered from an attack of the disease, and that where relapses do occur they are in the inverse ratio to the severity of the first attack. In some cases inoculation produces a chronic form of the disease in certain fowls, but the transference of the virus from such cases usually causes the acute form of the disease in other fowls. He noted, however, that the virulence of a culture depends on the time which elapses between successive cultures. As the period of time increases there are signs of progressive attenuation of the virus as shown by the lessened case mortality and by the delay in the development of the symptoms. If the virus is transplanted from medium to medium at intervals varying from days to a month or two no change is observed when the cultures are tested for virulence. If the interval between two successive inseminations in bouillon is carried to 3, 4, 5, or 8 months the scene changes. Instead of being active as judged by the mortality the latter becomes less and may disappear altogether (Pasteur 1880[c]).

It was apparently an inspiration which led Pasteur to apply a virulent cholera virus to fowls which had come safely through an inoculation of the attenuated virus, and this led to his establishment of the principle of prophylaxis following the inoculation of the attenuated virus—a principle which he established almost at once for anthrax, swine erysipelas, and rabies. These researches occupied him exclusively till the end of his active scientific life. The cause of the attenuation of the fowl cholera virus was believed by Pasteur to reside in the deleterious effect of air and particularly oxygen. Cultures in closed tubes of broth were found to maintain their virulence up to ten months. In open tubes plugged only with wool attenuation took place in the manner stated above. Pasteur also showed that if the fowl cholera microbe is removed from a broth culture by filtration the broth cannot be successfully re-inoculated with the microbe. Large supplies of attenuated fowl cholera virus were prepared and inoculated into fowls as a protective measure against the disease. The method as a practical proposition was not invariably a success as there was sometimes a rather high mortality from the prophylactic itself, but the soundness of the principle established by Pasteur was admitted and he attained widespread fame with his discovery. The virulence of fowl cholera bacillus led him in 1888 to recommend the use of living cultures of this microbe for the destruction of the plague of rabbits in Australia and New Zealand. A preliminary experiment to test his view was carried out on the estate of Veuve Pommery, in Rheims, and was reported as extraordinarily successful (Pasteur, 1888). Subsequent experiences by Loeffler, Danysz, and others with this method of exterminating vermin have shown that Pasteur was unduly optimistic. Turning aside temporarily from the subject of fowl cholera Pasteur returned to the study of anthrax and made a number of experiments to determine how the disease is spread under natural conditions. The extraordinary resistance of the anthrax spore had been effectively demonstrated by Koch (1876), and this was further emphasized by Pasteur (1879), who found anthrax spores in the soil above an anthrax carcass buried ten months previously. Other experiments carried out in association with his assistants, C. Chamberland and E. Roux (1880), (1881[2]), confirmed the previous ones, and they came to the conclusion that the spores in a buried anthrax carcass are gradually brought to the surface of the soil by earthworms. Earth in which there were earthworms was artificially infected with anthrax spores and when the worms were subsequently opened, anthrax spores were found in the earth-castings (Pasteur, Chamberland, and Roux, 1880).

Pasteur (1880[5]) also found anthrax in soil or in worm-casts in a field in which three cows had been buried two years previously. A paling was placed round the infected area and of four sheep made to graze in the enclosure one contracted and died of anthrax; sheep grazing in a similar but non-infective enclosure only 3 or 4 metres away remained well. In soil on a site in which anthrax cadavera had been buried twelve years previously, Pasteur, Chamberland, and Roux (1881[2]) demonstrated anthrax spores by inoculation into guinea-pigs. Pasteur's attention was then directed to the occasional recovery of domestic animals from anthrax (Pasteur and Chamberland, 1880), and he began to apply the knowledge gained. The existence of the anthrax spore in cultures rendered the problem different from that of fowl cholera. He found that anthrax will not grow above 45° C. but yielded abundant growths at 42–43° C. At the latter temperatures no spores were formed. The culture had indeed become asporogenous, and in a month had ceased to grow. When a virulent anthrax culture in broth was kept at 42–43° C. for eight days it had lost a considerable part of its virulence and was innocuous when injected into guinea-pigs, rabbits, or sheep. By longer cultivation at the above temperature it became still more attenuated. This was the method which was suggested by Pasteur, Chamberland, and Roux (1881[2, 3]) as an anthrax prophylactic or vaccine, and to demonstrate its value and emphasize its importance to farmers a large experiment was undertaken by Pasteur at the suggestion of the Agricultural Society of Melun, who placed more than sixty animals at his disposal.

In the yard of the farm Pouilly-le-Fort, a mile from

Melun, the experiment took place in the presence of a great concourse of people of all kinds, including the famous Paris correspondent of *The Times*, M. de Blowitz, who successfully mailed his copy throughout the world, and made the Pouilly-le-Fort experiment common knowledge. The experiment consisted in the inoculation, on 5 May 1881, of 24 sheep, 1 goat, and 6 cows with 5 drops of a living attenuated culture of anthrax bacillus. On 17 May 1881 all these animals were re-inoculated with a less attenuated culture of anthrax. On 31 May 1881 the crucial test was applied. All the 31 inoculated animals received a highly virulent anthrax culture which was also injected in 24 sheep, 1 goat, and 4 cows not previously inoculated, and thus serving as controls. The result was soon apparent. On 2 June 1881 a great throng had gathered, and as Pasteur arrived on the field with his assistants Chamberland, Roux, and Thuillier, he was greeted with loud acclamation. All the vaccinated sheep were well. Twenty-one control sheep and the goat were dead of anthrax, two other control sheep died in front of the spectators, and the last unprotected sheep died at the end of the day. The 6 vaccinated cows were well and showed no symptoms, whereas the 4 control cows had extensive swellings at the site of inoculation and a febrile reaction. On 3 June one of the vaccinated sheep died. It was pregnant and an autopsy suggested that it had succumbed on account of the death of the foetus. Pasteur's triumph was complete. The discovery of the anthrax vaccine constituted, he said, 'un progrès sensible sur le vaccin jennérien' (Pasteur, Chamberland, and Roux, 1881[5]).

A few weeks later Pasteur was the chief star at the International Medical Congress in London and delivered his famous address on 'Vaccination in relation to chicken cholera and splenic fever'. It was here that he explained the use of his terms vaccine and vaccination to denote the process of prophylactic inoculation in general, and as a homage 'au mérite et aux immenses services rendus par un des plus grands hommes de l'Angleterre, votre Jenner' (Pasteur, 1881).

Pasteur's anthrax vaccination, holding as it did the highest hopes for agriculture, soon became general, and already in 1882 Pasteur (1882[2]) reported the results on 85,000 animals in which it had been carried out. In 79,392 sheep the mortality from anthrax had fallen from 9·01 per cent. among uninoculated to 0·65 percent. among inoculated. The vaccine was distributed broadcast and applied in many different places, but not always so successfully as at Pouilly-le-Fort. Failures were reported in Turin (Pasteur, 1883[3]) and elsewhere, and these were seized upon by Pasteur's adversaries. In general, however, the results were favourable. A full account of these early inoculations will be found in the works of Chamberland (1883, 1894).

Among the critics of the process was R. Koch (1882), who published a reply to an address made by Pasteur (1882[3]) on the 'attenuation of virus' at the International Congress of Hygiene and Demography held at Geneva. Koch maintained that on account of the imperfections of Pasteur's method, the unfavourable results, the mortality from the inoculation itself, and the shortness of the immunity produced, anthrax vaccination was not really a practical proposition. Koch was also scathing on the purity of the vaccine and reported that he had found the French product contaminated with bacteria other than those which ought to have been met with. Pasteur (1883[1]) made a scornful reply to Koch's criticisms and proceeded quietly with his work. He seemed to have been especially hurt by Koch's suggestion that his anthrax cultures were not pure. By 1894 Chamberland (1894) reported that over 3½ million sheep had been inoculated.

In association with L. Thuillier, Pasteur now proceeded to study the disease known in France as 'rouget des porcs' (Eng. Swine erysipelas). Thuillier was sent to Peux in the department of Vienne, and there he found a well-defined microbe in the blood. Pasteur himself joined Thuillier at Bollène (dep. Vaucluse), and succeeded in obtaining cultures of the microbe of *rouget* (Pasteur and Thuillier, 1882[1, 2]). Attempts were made by Pasteur and Thuillier (1883) to attenuate the *rouget* virus. Difficulties were experienced owing especially to the varying susceptibilities of the races of pigs, but the study proceeded, and the remarkable fact was elicited that the virus of *rouget* becomes attenuated for the pig by passage through the bodies of rabbits, and this was the source of the vaccine they now recommended for inoculation of the pig, and the correctness of their conclusions has not been a serious matter for dispute. Between 1886 and 1892 over 100,000 pigs had been inoculated in France, while in Hungary from 1889–94 the number exceeded 1 million. Shortly after the completion of his work on *rouget* Thuillier went to Egypt with the French Cholera Commission and died in Alexandria (1883) of the most acute form of cholera, to Pasteur's great sorrow. He was only 27 years old.

The work of Pasteur on vaccination against fowl cholera, swine erysipelas, and anthrax had aroused world-wide interest and his fame was higher than ever before. He was now about to enter the last phase of his great scientific career, and for it he chose the mysterious and inscrutable disease rabies, which had long had a firm hold on the public imagination. Possibly it was this very fact that impelled Pasteur to its study, for it cannot be said that the disease was of any great importance in the general mortality of man or animals. Pasteur's first introduction to rabies was the case of a child aged 5 in the Sainte Eugénie hospital in Paris. Four hours after its death saliva was collected and sent to Pasteur, who injected it into two rabbits. They died in 36 hours of a disease which could be transmitted in rabbits in series. In the blood he found a capsulated minute organism in the form of a figure of 8 and surrounded by a kind of capsule (Pasteur, Chamberland, and Roux, 1881[1]), but the fact that it was later found in normal saliva rendered the view that it had a causal connexion with rabies untenable.

It is indeed now well known that the microbe of saliva was merely the pneumococcus. The transmission of rabies to rabbits was shown about this time by V. Galtier (1881), of Lyons, and this no doubt arrested Pasteur's attention. From now onward and in association with Chamberland

and Roux, Pasteur began a series of epoch-making experiments in which the effects of rabies on animals were analysed in a masterly way and on a very large mass of material. It was shown that the infective matter of rabies occurs not only in the saliva of rabid dogs, as had been pointed out by Magendie (1821) and Galtier (1881), but also in the central nervous system. By inoculation directly under the *dura mater* of dogs it was shown that the lengthy incubation period following subcutaneous inoculation of ordinary rabies virus can be shortened so that it lasts only one or two weeks (Pasteur, Chamberland, Roux, and Thuillier, 1881, 1882).

It was also shown that the virus passed through the nervous system of rabbits becomes greatly exalted. The incubation period of the inoculated disease becomes shorter but a limit is reached beyond which it is impossible to go. The virus has become 'fixed', as Pasteur expressed it. This *virus fixe* acts with extraordinary regularity. After intracerebral inoculation in rabbits there is an incubation period of seven days, at the end of which the virus begins to show its lethal effect (Pasteur, Chamberland, and Roux, 1884[2]). As early as 1882 Pasteur and his collaborators had noted spontaneous cure of canine rabies in certain rare instances after the first symptoms had shown themselves. In one dog which exhibited this phenomenon a subsequent intracerebral inoculation made on two occasions remained without effect. Four other dogs behaved in a similar fashion, and it was from this point that Pasteur now exerted all his powers to produce an effective prophylactic or vaccine against rabies. Large kennels were constructed at Villeneuve l'Etang, near St. Cloud, and experiments proceeded, and it was shown that the rabies virus is attenuated for the dog, rabbit, and guinea-pig, by passage through a series of monkeys. At the International Congress of Medicine in Copenhagen, Pasteur (1884) gave an extensive account of his researches on rabies and announced the results of protective inoculation in dogs. These results were confirmed by a commission which included Paul Bert, Villemin, and Vulpian.

Having proved that animals can be protected *before* the application of virulent rabies virus, Pasteur (1885), in a sixth communication, described a method for protecting dogs *after* they had received the virus from a bite of a rabid dog. The method consisted in suspending in dry, sterile and still air the spinal cords of rabbits which had died from the intradural injection of *virus fixe*. In the course of about two weeks the cord, at first highly virulent, becomes almost non-virulent. By inoculating emulsions of the attenuated cord it was possible to proceed to emulsions of less attenuated virus, and finally *virus fixe* could be applied with impunity to the brains of dogs thus protected. These remarkable results on dogs suggested that the method might be applicable to human beings bitten by rabid animals and still in the incubation period of the disease. Pasteur had not long to wait for an opportunity of testing this idea, for a boy, Joseph Meister, aged 9, was brought to him on the sixth of July 1885 from Alsace suffering from bites on the hand, legs, and thighs from a dog certainly rabid. On 7 July, sixty hours after the bites had been inflicted, Meister, the first human being treated by Pasteur's method, was injected with attenuated rabbit marrow, fourteen days old. In a further twelve inoculations he received virus stronger and stronger until on 16 July he received an inoculation of virulent marrow only one day after it had left the body of a rabbit dead from *virus fixe*.

The virulence of the material used in the whole of Meister's thirteen inoculations was controlled on rabbits, and it was shown that the boy in his last two inoculations had withstood a living virus which in rabbits was shown to be of maximal virulence with a seven-day incubation period. The boy remained well, became Concierge de l'Institut Pasteur, and was alive in 1935 (*Ann. Pasteur*, Suppl. no., 1935, p. 4).

Pasteur (1885) also gave an account of the second case treated by him. He was Jean Baptiste Jupille, aged 15, a shepherd boy of Villers-Farlay in the Jura. Seeing a suspicious-looking dog about to attack some children Jupille seized his whip and attempted to drive it off. He was severely bitten, but having closed with the brute he wound the thong of his whip round its muzzle and beat in its skull with his sabot. The dog was subsequently declared to be rabid, and six days after being bitten Jupille was brought

FIG. 34. Statue of Jean Baptiste Jupille in the Pasteur Institute, Paris.

to Paris and treated by Pasteur (1886). The result was successful and Jupille's deed was commemorated in a bronze statue which to-day stands in front of the Institut Pasteur in Paris (Fig. 34). After these early and dramatic successes numerous patients bitten by animals rabid or presumed to be so, arrived in Paris to receive Pasteur's treatment. By October 1886 no fewer than 2,490 persons had been inoculated. Analysing the results, Pasteur concluded that of 1,726 bitten persons of French nationality and subjected to treatment by inoculation, there were only 10 failures (1 in 170). Up to 1905 Bernstein (1905) estimated that all over the world at least 104,347 persons had received the Pasteur treatment for rabies. Up to 1935 51,057 persons had been inoculated in the Institut Pasteur, Paris, with 151 deaths—a mortality of 0·29 per cent. (*Ann. de l'Inst. Pasteur*, Suppl. no., 1935, p. 30).

By his researches on rabies Pasteur had now attained world fame. The number of persons who came for treatment was much larger than could be dealt with, and it was suggested by a commission of the Academy of Sciences in Paris that a scientific institute should be erected where his protective inoculations could be fitly given, and where scientific researches on the subjects in which he had long been interested could be further developed. The suggestion roused the greatest enthusiasm and generosity not only in France but all over the world. Money poured in to support the scheme, and the total amounted to 2,586,680 francs. A plot of land was purchased in the rue Dutot and on it was erected in 1888 the Institut Pasteur, which was inaugurated on 14 November 1888 before an audience including the most distinguished persons in France. Pasteur was there but his own life-work was at an end. He was feeble and ill, but lingered on for another seven years when he died on 28 September 1895. His last scientific communication was in 1889. He had then been before the scientific world for forty-five years and had created for himself a name which will resound for ages. Few men during their lifetime have been accorded greater acclamation, and the cause of this must be sought not only in his genius but also in the practical outcome of his work. Throughout his life he was impelled by a continuous desire to work, and he had an unexampled flair for keeping on the right path, never diverted from his aims. His scientific work is a wonderful illustration of direct continuity, one piece fitting into another almost as if guided by an inspiration. Not only were Pasteur's own contributions to science of the highest class but he had the ability to single out collaborators of rare merit, and of these particular mention must be made of Duclaux, Chamberland, Roux, and Metchnikoff, who in their own works worthily maintained the Pasteur tradition. The Pasteur school gave an enormous impetus and standing to bacteriological research, and from 1888 onwards Pasteur Institutes based on the parent Institut Pasteur Paris type were established in more than forty places within a few years. Indeed an almost continuous chain of research institutes to carry on his work were founded round the world, and much of our present-day knowledge has emanated from the efforts of the workers in them. The torch that Pasteur set alight still burns vividly after fifty years.

XI
HISTORY OF DOCTRINES OF IMMUNITY

XI

HISTORY OF DOCTRINES OF IMMUNITY

THE work of R. Koch and Pasteur respectively gave two directions to the subsequent development of bacteriology. The German school concentrated, at first at any rate, on the isolation and exhaustive study of individual bacteria both parasitic and saprophytic, and in due course this led to the foundations of many highly specialized branches of bacteriology, the magnitude of any one of which to-day is such that no single individual, however diligent, can cope with it. For the medical profession the pathogenic bacteria have claimed the greatest interest, but of as great if not greater importance are the saprophytic bacteria which are necessary for continued life on the earth.

The French school, starting with Pasteur, early concentrated on the problems of the prevention of infective disease by artificial inoculation, and the processes involved in the recovery from infection. The results, profoundly interesting, and of the greatest significance, led to the creation of that branch of science called Immunology, for, although the latter now has implications far beyond the bacteriological sphere, it was in connexion with bacterial diseases that it first took its root. By his researches on chicken cholera, anthrax, swine erysipelas, and rabies between 1880 and 1888 Pasteur first directed attention to the practical and theoretical possibilities connected with immunity. Simple as immunity appeared to be at first it was not long before it was found to be a complex and variable process. The doctrine of susceptibility or predisposition, variable like immunity, began to be developed. Natural immunity was differentiated from acquired immunity. By the end of the eighties of last century the new immunity science had begun to develop in two directions. On the one hand the Pasteur method of inoculation with attenuated virus was further developed along his original lines, and on the other hand, theories were formulated to explain the nature of the immunity process. With regard to the practical possibilities of inoculation an important advance was taken by D. E. Salmon and Theobald Smith (1886), who showed that dead virus can induce immunity against the living virulent virus. Their experiments were done on pigeons, and after inoculation with hog cholera bacilli killed by heat, they found that the pigeons could withstand with impunity multiple doses of living bacilli fatal to unprotected pigeons.

This important discovery was the basis of the methods employed on a gigantic scale to protect human beings against cholera, plague, and typhoid fever. The names of W. Haffkine, A. E. Wright, R. Pfeiffer and Kolle are associated with this advance, and A. E. Wright subsequently developed the idea of therapeutic as opposed to prophylactic inoculation along the same lines.

At an early stage investigators occupied themselves with the mechanism of immunity and two hypotheses developed side by side. The one explained the immunity as due to the occurrence or development of mysterious properties or substances in the blood-stream and fluids of the body. These substances were inimical in some way to the development of the infective agents and for this reason were described as anti-bacterial, protective, or defensive. This was the humoral theory of immunity. The other view, formulated by E. Metchnikoff, attributed the immunity to a cellular rather than a humoral activity, and for a decade or more (1880–90) the discussion on humoral versus cellular, immunity was fought out on a big scale.

That the living body fluids, and especially the blood, possess some power of inhibiting putrefaction was indicated by Timothy Lewis and D. Cunningham in 1872, and more exactly by M. Traube and Gscheidlen (1874). They found that normal blood taken direct from the vessels did not undergo putrefaction even after months. Arterial blood, drawn with aseptic precautions from rabbits 24 to 48 hours after they had been injected with bacteria-containing fluids did not become putrid. Traube and Gscheidlen observed, further, that the anti-putrefactive power of the circulating blood had limitations and they ascribed the power in general to the ozonized oxygen of thé red blood corpuscles. Watson Cheyne (1879) and Grohmann (1884) confirmed these facts. It was, however, in 1886 that J. Fodor (1843–1901), a Hungarian, drew attention to the importance of such observations, and he it was who emphasized the protective or defensive significance of the process. His work was repeated by Wyssokowitch (1886) in Flügge's Institute in Göttingen. He showed in general that bacteria introduced into the blood-stream rapidly disappeared and apparently did not leave the body by any of the channels of secretion or excretion. He opposed Fodor's idea of destruction, and considered that the disappearance was apparent rather than real. In Wyssokowitch's opinion the injected bacteria really become deposited in the internal organs such as the spleen, liver, and bone marrow. Fodor (1887, 1890) showed that there was evidence of extra-vascular destruction. He infected blood *in vitro* with anthrax bacilli and found that if kept at 38° C. a rapid diminution in the number of bacilli took place as judged by enumeration on gelatine plates. A possible flaw in this experiment was that the blood, having undergone clotting, might have entangled bacteria in the clot and thereby have given the appearance of destruction.

This criticism was set aside by G. H. F. Nuttall (1888), who employed defibrinated blood. Working in Flügge's Institute in Göttingen, he took blood from a vessel and defibrinated it with sterile quartz. Known quantities of bacteria were now added to the blood *in vitro* and plate cultures were made at intervals. Nuttall placed the blood of rabbits, mice, pigeons, sheep, and dogs, in contact with anthrax bacilli, and he clearly showed in a number of instances that there was a profound diminution in the number of bacilli. Thus, four controls which yielded an average of 15,730 anthrax colonies showed no colonies or only five colonies one hour after exposure to the action of the defibrinated blood. Nuttall also discovered that the bacteria killing (bactericidal) power of the blood was lost after a time, and that it is destroyed in the case of dogs' blood by heating to 52° C. for from 10 to 30 minutes.

The bactericidal power of rabbits' blood was rendered inert at 55° C. for 45 minutes. Heating to 48°–50° C. for 10 minutes was ineffective. The essential facts disclosed by Nuttall were rapidly confirmed by many workers and especially by F. Nissen (1889), and by Hans Buchner alone and in association with several of his co-workers. Nissen found that the bactericidal power of the blood varies according to the bacteria employed. Some bacteria were destroyed, others not. He showed also that the bactericidal property can be diminished for one species of bacteria while it remains intact for another. Behring and Nissen (1890) found that whereas the serum of normal guinea-pigs is incapable of destroying *Vibrio metchnikovi*, the serum of guinea-pigs immunized against this organism can accomplish this feat easily. They realized, however, that this is not of universal applicability as it does not occur in the case of animals immunized against *Pneumococcus*. About this time began the long series of experimental researches made by Hans Buchner (1850–1902) of Munich. He confirmed the existence of bactericidal substances in the blood and showed they also occur in the cell-free serum *in vitro*. He also showed that the bactericidal effect is manifested *in vivo*. The carotid artery of a dog was exposed and connected with a cannula. The animal was now injected with a culture of typhoid bacilli, and 50 c.c. of blood allowed to flow from the cannula to ensure that the artery was filled with blood. A small quantity of blood was again run out and tested for the presence of typhoid bacilli. The artery above was then ligated in two places and the contents of the isolated segment was examined five hours later, when it was found that a profound reduction in the number of bacilli had ensued. In the course of his prolonged researches Hans Buchner gradually elaborated the view of the existence of substances in the serum, and that these substances were inimical to bacteria. He named the substances 'alexines' (ἀλέξω, I defend) from their defensive or protective properties. This alexic theory held the ground for a number of years despite the criticisms of Metchnikoff (1890), Jetter (1891–2), and von Székely and Szana (1892). Ultimately, however, the accumulation of facts indicated that Buchner's theory was untenable, at any rate in the simple form proposed by him.

Cellular defensive mechanism. Phagocytosis

During the time that the humoral theory of immunity was being developed, the idea that an active part is played by the infected host in natural immunity was clearly fore-seen by the Russian zoologist E. Metchnikoff (1845–1916), and developed by him in a long series of papers. The central idea of his teaching was that natural immunity depends on a cellular rather than a humoral mechanism. He considered that certain cells of the body have the ability to incorporate the infective material and to destroy it by a process of intracellular digestion. The cells which possess this property were called by Metchnikoff (1884[1]) 'phagocytes', and the process 'phagocytosis'.

That cells can engulf foreign matter was of course known long before Metchnikoff's time. The amoeba obtains its nourishment in this way. Langhans (1870) showed that in haemorrhagic foci leucocytes become charged with red blood corpuscles. Birch-Hirschfeld (1872) found that cocci injected into the blood may be taken up by leucocytes. Koch (1876) also described and figured the occurrence of anthrax bacilli inside the cells of the dorsal lymph sac of the frog after particles of anthrax spleen had been implanted in that site. He saw the same phenomenon in the spleen itself of a horse dead from anthrax. Panum (1874) had actually suggested that this might be a method for the destruction of microbes. More definitely of this opinion was K. Roser (1881), but such observations passed un-heeded. It was about this time (1882, 1883) that Metchni-koff, working in Messina, made observations on starfish and concluded that certain mesodermic cells possess real digestive powers, and he extended this idea of intracellular digestion in a number of directions and among them as a cause of immunity against infective diseases. This aspect was first studied by Metchnikoff (1884[1]) in the case of the well-known crustacean *Daphnia magna* infected by a para-site, *Monospora bicuspidata*. The infected crustaceans were found in the reptile tank in the Jardin des Plantes in Paris. The cellular immunity theory was extended to bacterial infections such as anthrax (1888) and erysipelas (1887[2]). A wide dissemination of Metchnikoff's views took place through the publication of his *Leçons sur la pathologie comparée de l'inflammation* (1892), and in various articles dealing with immunity. The phagocytic theory was criticized by Baumgarten (1889[1, 2]), Weigert (1891), de Christmas (1887), and others, and the development of the humoral doctrine of immunity about this time seemed to put the phagocytic theory in the shade. It was, however, suggested that a union of the opposing views might be brought about on the supposition that bactericidal substances in the serum might be the product of certain cells. Attempts of this kind by Hankin with his theory of alexocytes dates from about 1892-3. A great mass of observations continued to be published on both theories, but it gradually emerged that neither the one view nor the other is capable of explaining all the facts of natural immunity.

Before dealing with the further development of this question it is necessary to refer to the progress of the humoral doctrine which was brought about by the discovery of antitoxin by Behring and Kitasato in 1890. It was in connexion with the immunity to tetanus and diphtheria that this great discovery took its origin. F. Loeffler (1884) discovered the diphtheria bacillus and Kitasato (1889) proved that the tetanus bacillus is the cause of lockjaw. Roux and Yersin (1889) proved that the diphtheria bacillus operates in virtue of a poison (toxin) which it elaborates, and K. Faber (1889) showed that the tetanus bacillus acts in a similar manner. Attempts were made to create an artificial immunity against diphtheria, and along this line C. Fraenkel (1890) and E. Behring (1890) did important work. By injecting three-week-old broth cultures of diphtheria bacilli which had been heated to 65°-70° C. for one hour, C. Fraenkel showed that an immunity had been established so that a subcutaneous inoculation of living diphtheria bacilli remained without effect. His results were published in the *Berliner klinische Wochen-*

schrift, 3 December 1890, and on the following day Behring and Kitasato published the first account of their discovery in the *Deutsche medicinische Wochenschrift* (4 Dec. 1890). This brief but epoch-making paper which dealt with immunity to tetanus contained the remarkable statement that the immunity of rabbits and mice which have been treated with tetanus cultures depends on the capacity of the cell-free serum to render innocuous the toxic substances elaborated by the tetanus bacillus. This capacity was so durable that it still exhibited its activity when the immune serum was transferred to the bodies of other animals. By means of immune serum Behring and Kitasato were able to protect mice which received 300 minimal lethal doses of the tetanic poison. Normal serum had no such properties. In a footnote to their paper the toxin-destroying action of the immune serum was described as 'antitoxic' and this word was employed at once and permanently.

One week after the joint publication of Behring and Kitasato on tetanus immunity Behring alone published another paper (11 Dec. 1890) on immunization against diphtheria, and he confirmed for this infection all the essential facts which in conjunction with Kitasato he had demonstrated for tetanus. The possibility of serum-therapy was opened up by these investigations. Priority claims have been made by or on behalf of Héricourt and Richet (1888), and Babes and Lepp (1889), but history must assign unchallenged to Behring and Kitasato the discovery of antitoxic immunity. In rapid succession a mass of papers appeared confirming their discovery. It is impossible to refer to these other than by the names of the authors, prominent among whom were Behring and Wernicke (1892), Brieger and Cohn (1893), Behring, Boer, and Kossel (1893), Aronson (1893), Behring and Ehrlich (1894), and Ehrlich, Kossel, and Wassermann (1894). The first person treated by diphtheria antitoxin was a child in von Bergmann's clinic in Berlin, the injection being made by Geissler on Christmas night 1891 (Wernicke, 1913).

A remarkable impulse was given to the use of antitoxin in man by the work of Roux and L. Martin (1894) in Paris, their results being communicated by Roux in an address at the International Congress of Hygiene and Demography in Budapest in that year. From the scientific point of view the highest praise must be given to Paul Ehrlich (1897, 1898), who, in most laborious researches, established the principles of the standardization of bacterial toxins and antitoxins and laid down the practical methods employed to-day. His *Werthbemessung des Diphtherieheilserums und deren theoretische Grundlagen* (1897) is a classic in the bacteriological literature. In its preparation Ehrlich carried out his experiments mainly in a small building in Steglitz (Berlin), and later (Oct. 1899) he transferred his activities to a large institute—the Kgl. Institut für experimentelle Therapie—in Frankfort a. M., to which was added (1906) the Georg-Speyer-Haus founded by Frau Franziska Speyer. From this famous research institute has emanated a great mass of scientific knowledge and its activity remains unabated to-day.

Following the discovery of Behring and Kitasato it was soon found that the ability of the animal organism to produce antitoxin is a widely spread phenomenon, and results from the inoculation of many substances classed as poisons and enzymes of animal as well as of vegetable origin. Of historic importance in this respect were the researches of Ehrlich (1891) on ricin and abrin. He showed that each of these vegetable poisons produces an antitoxin which is specific. He it was (1892) also who pointed out the differences between active and passive immunization.

The success attained in the treatment of diphtheria by specific antitoxic serum early led to the hope that similar specific antitoxins might be prepared for other infective agents such as those producing pneumonia, typhoid fever, cholera, anthrax, tuberculosis, and suppuration. It soon became evident, however, that the sera of animals treated with such agents is not similar to diphtheria antitoxin. In some instances indeed the immune serum had apparently no protective properties which could not be found in normal serum. In other cases, as for example in typhoid and cholera, the immune serum had definite specific

THEOBALD SMITH
(1859–1934)

LAZZARO SPALLANZANI
(1729–1799)

properties but manifested them in a manner differing from antitoxic sera. The researches of Nuttall, Nissen, and Behring, had shown that some thermolabile bactericidal element occurs in certain sera. Behring and Nissen found that in the case of *Vibrio metchnikovi* immunization apparently increases the bactericidal content of the serum and it was found that this also applies to infections with *Bacillus typhosus* and *Vibrio cholerae*. It was in connexion with the last-mentioned micro-organisms that R. Pfeiffer (1894) began his classical researches which showed that the destruction of the cholera vibrio takes place by a kind of dissolution which he called 'bacteriolysis' (Pfeiffer and Issaëff). Cholera vibrios were destroyed *in vitro* in the serum of animals immunized against cholera, and the same effect took place in the peritoneal cavity of the living immunized animal. By an ingenious and simple technique R. Pfeiffer showed that in the normal peritoneal cavity cholera vibrios multiply rapidly, whereas in the immunized animal they disappear. Pfeiffer removed drops of peritoneal fluid by means of capillary pipettes and found in the case of the immune animal that cholera vibrios which had been injected were almost instantly rendered motionless and swollen. They then changed into micrococcuslike bodies (spherulation) which became more and more difficult to see and ultimately they disappeared from view altogether. This process of lysis occupied about 20 minutes and was apparently independent of any cellular intervention. R. Pfeiffer showed that the serum also acts *in vitro* but its activity is abolished by dilution or by heat at 60° C. The solution or lysis of vibrios in the peritoneal cavity of an immunized animal is spoken of as 'Pfeiffer's phenomenon', and is, as Pfeiffer showed, highly specific in that vibrios closely related to *Vibrio cholerae* are not destroyed by cholera serum. He also observed that cholera immunity could be passively transmitted, for when anti-cholera serum from an immunized guinea-pig was injected into the peritoneum of a normal guinea-pig cholera vibrios subsequently introduced into the latter underwent the characteristic bacteriolysis. C. Fraenkel and Sobernheim (1894) found that cholera immune serum, even when heated to 70° C. for one hour, could protect normal guinea-pigs against lethal doses of cholera vibrios, and in this case likewise the vibrios are destroyed by the lytic process in the peritoneal cavity. Such heated serum is inert *in vitro* and it apparently becomes active *in vivo*. R. Pfeiffer regarded the active agent as a kind of pro-ferment which became converted into a ferment or enzyme in the body of the living animal. Although the observed facts reported by R. Pfeiffer were at once confirmed, some alterations were made in his interpretation as the result of added knowledge. Thus Metchnikoff (1895) showed that Pfeiffer's phenomenon takes place *in vitro* if the precaution is taken of adding some normal peritoneal exudate to the mixture of cholera vibrios and heated cholera-serum, and in the same year Jules Bordet (1895), a young Belgian, showed that normal blood-serum can be used as a substitute for peritoneal exudate. Bordet also confirmed R. Pfeiffer's observation that cholera-serum acts specifically, for, on mixing it with vibrios other than cholera no lysis took place. It was apparent, therefore, in experiments on bacteriolysis of cholera vibrios that heated immune serum cannot produce this effect, but that a mixture of normal unheated serum and heated immune serum can. On heating normal serum to 55° C., Bordet found it incapable of rendering heated immune serum active and he at once concluded that two substances or factors must be concerned in the lytic action. One of these substances is present both in normal and fresh immune serum and is thermolabile, the other is peculiar to the immune serum and is thermostable.

These classical experiments of Bordet were at once confirmed everywhere and a lytic action was found to be applicable to cells other than bacteria. Thus red blood corpuscles are destroyed by a similar mechanism—haemolysis—and this discovery was in the main the work of J. Bordet. The study of haemolysis subsequently developed to an extraordinary extent and was found to reveal facts of fundamental importance for the immunity problem.

It was long known from observations by Creite (1869)

and Landois (1875) that the serum of certain animals is destructive to the red blood corpuscles of other animals, but no great interest was taken in the fact. G. Daremberg (1891) and H. Buchner (1893) also noted that a haemolytic serum ceases to be so if heated to 55° C. In 1898 Belfanti and Carbone showed that the serum of horses into which the red blood corpuscles of rabbits had been injected acquired toxic properties for rabbits. Bordet (1898) recognized the significance of this discovery, found that it was a general principle, and proved that haemolysis is identical in type with bacteriolysis. Bordet's paper on the subject opened a new field of inquiry which was soon enlarged by numerous workers, prominent among whom may be specially mentioned Ehrlich and Morgenroth (1899–1900). The essential facts established by Bordet (1898) were that when rabbits' blood was injected into guinea-pigs the serum of the latter acquired the property of dissolving the blood corpuscles of the former *in vitro*. On adding to a suspension of rabbits' blood corpuscles in salt solution a small quantity of guinea-pig immune serum, the blood corpuscles of the rabbit rapidly clumped together (agglutination), and the haemoglobin passed into the medium leaving behind the colourless stromata or 'shadows'. Heated to 55° C. for half an hour the haemolytic property of the immune serum was found to have disappeared. If then to a mixture of rabbits' corpuscles and inactive (i.e. heated) guinea-pig immune serum Bordet added a quantity of serum from a normal guinea-pig or a normal rabbit, the phenomenon of haemolysis soon took place. The guinea-pig immune serum was found to have no effect on the corpuscles of guinea-pigs or pigeons. In fact it had a specific effect only on the corpuscles of the animal (rabbit) with which the guinea-pig had been originally injected. Bordet also showed that a haemolytic destruction of corpuscles also takes place in the peritoneal cavity of an immunized guinea-pig. If the immune serum is heated to 55° C. and is then mixed with the homologous red blood corpuscles, the latter are haemolysed in the peritoneum of a normal animal.

Taking up the subject of haemolysis at this stage, Ehrlich and Morgenroth in a series of valuable memoirs confirmed and extended the fundamental facts discovered by Bordet. In the nomenclature of Bordet the thermolabile factor in haemolysis was named 'alexine' and the thermostable element the 'substance sensibilisatrice', whereas Ehrlich and Morgenroth spoke of them as 'complement' and 'amboceptor' respectively—names which are universally used to-day. The further development of the haemolysis question is outside the scope of the present work and will not be considered further here, except in its applications to bacterial diagnosis.

Even before the subject of haemolysis had been worked out it was noted that the serum may possess properties which are neither antitoxic nor lytic. As far back as 1889 Charrin and Roger found that when *Bacillus pyocyaneus* was grown in the fluid serum of an animal which had been injected with that organism it did not grow diffusely through the medium, as it does in broth or normal serum, but in small masses which sank to the bottom of the tubes and the bacilli were found stuck together. Similar appearances were found by Metchnikoff (1891) in the cases of *Vibrio metchnikovi* and *pneumococci* grown in their respective immune sera, and he pointed out the importance of the change and the need for its further study. The clumping of bacteria is now called agglutination and has since these early observations been the subject of an extraordinary amount of study. The first systematic examination of the agglutinative process was published by Herbert Durham, who communicated results of investigations which he had carried out with Max Gruber in the Hygiene Institute in Vienna and by himself in Guy's Hospital, London. Durham's paper was read in abstract at a meeting of the Royal Society, London, on 3 January 1896, and this was followed by communications by Gruber (3 March 1896), and Gruber and Durham (31 March 1896). Even in Durham's first communication attention was directed to the possibility of the method as a diagnostic aid to disease, and this application was actually made by F. Widal of Paris (26 June 1896) and by A. S. Grünbaum (19 September 1896). Known for long as 'Widal's' or 'Gruber-Durham' reaction, it is nowadays generally spoken of simply as 'agglutination reaction'. This reaction is brought about by the addition of specific immune serum to a uniform suspension of a bacterium. Thus if a uniform suspension of actively motile typhoid bacteria is taken and to it is added a small quantity of the serum of an animal previously injected with typhoid bacteria, the motility of the bacilli is instantly arrested, and from being uniformly disseminated in the liquid medium the bacilli run into masses or clumps visible to the naked eye. From the first the agglutination reaction aroused the greatest interest and was practised everywhere in innumerable modifications as a diagnostic test, particularly for typhoid fever and as a standard test for the differentiation of allied bacteria. The literature on the subject is of great dimensions and cannot be followed in detail here.

The two aspects of the agglutination which have received the greatest attention are its specificity and its nature. With regard to the former the existence of 'group agglutination', i.e. a common agglutinating reaction acting on a group of bacteria closely allied to each other, was emphasized by Pfaundler (1899) and by Durham (1900–1). Bordet (1899), in his classical paper on the mechanism of agglutination, drew attention to the remarkable fact that agglutinins are absorbed from serum when the latter is saturated with the homologous organism. He showed that if normal horse serum is mixed with cholera vibrios and after contact the vibrios are removed by centrifugalization, the clear supernatant fluid no longer retains its agglutinating property for cholera vibrios while its agglutinating power on typhoid bacilli persists. Conversely, by saturation first with typhoid bacilli the typhoid agglutinins are removed, whereas the supernatant can still agglutinate cholera vibrios. This principle was utilized by Castellani (1902) with his absorption of agglutinin test, which is extensively employed at the present day. With regard to the mechanism of the agglutinative reaction the experiments of Bordet (1899) and

A. Joos (1901 and 1902) were of great importance, but the subsequent development cannot be dealt with here. It may merely be stated that extensive researches are still in progress on the subject. During the time that the facts respecting agglutination were being brought to light R. Kraus (1897) discovered another property in immune serum. By injecting into animals the clear filtrates from liquid cultures of cholera, typhoid, or plague bacilli, he obtained sera which produce specific precipitates when mixed *in vitro* with samples of the filtrates which had been employed for the inoculation. Thus the serum of animals injected with cholera filtrates precipitated the latter but did not precipitate the filtrates of other cultures. Kraus's discovery of precipitating sera ('precipitin') was confirmed by many workers, and Bordet (1899) showed that it also occurred with non-bacterial substances. He found, for example, that cows' milk injected into rabbits gave rise to a serum ('lactoserum') which was capable of producing a visible precipitate when added to milk. Tchistovitch (1899) showed that the same phenomenon of precipitation occurs in the case of eel-serum and its homologous antiserum. Since that time precipitins have been prepared for an enormous number of protein substances of animal and vegetable origin, and their specific characters have been utilized in medicine and in industry for the differentiation of substances otherwise indistinguishable.

In the period 1888–1900 a very large number of facts had been accumulated on the subject of immunity not only against bacteria but also against many chemical substances of protein constitution. It became evident that after the parenteric injection of such substances, the animal body develops others which act in a specific manner contrary to the substance inoculated. These contrary substances appear in the blood-serum and other fluids and have been called by the general name of 'antibodies'. The substances which evoke such antibodies have been called 'antigens'.

What is the relation of the antigen to the antibody? The most remarkable property of antibodies is their specificity.

This was first clearly enunciated by Behring (1890) in the case of antitoxin, and has been confirmed for all the other known antibodies. The specificity of antitoxin is such that it early attracted notice and gave rise to the idea that it must be produced from the toxin. Thus Hans Buchner (1893) and Metchnikoff considered that antitoxin was merely a non-poisonous modification of toxin. The more, however, the problem was studied the less likely appeared this view, which in fact was disproved by the experiments of Roux and Vaillard (1893), and Salomonsen and Madsen (1898). Their experiments made it clear that antitoxin was something created or newly produced in the body as a result of the injection of toxin. Antitoxin was also proved to be occasionally present in the serum of animals that had never been injected with toxin. The antibody was found not to appear immediately after the injection of the antigen. There is always a latent period. The discovery of this important fact we owe to Ehrlich (1891) in the case of ricin antitoxin. On feeding mice with ricin he found that for the four following days they showed no increased resistance to ricin and very little on the fifth day. By the sixth day, however, the state of immunity had been critically developed and was accompanied by the presence of antitoxin. This course in antitoxin formation has been found applicable to other antibodies such as haemolysins (Bulloch, 1901), precipitins (von Dungern, 1903), and agglutinins (Madsen and Jørgensen, 1902). Subsequent to the critical entry of the antibody in the serum fluctuations occur, the course of which was studied especially by Brieger and Ehrlich (1893), Salomonsen and Madsen (1898), Morgenroth (1899), Bulloch (1901), Forssman and Lundstrom (1902), von Dungern (1903). From their researches it was apparent that an undulatory course was pursued by antibodies during their development. This course has been character-ized by A. E. Wright as the 'ebb, flow, and reflow, and maintained high tide' of immunity.

The certain fact that the animal body generates anti-bodies early led to attempts to locate the site of formation, but divergent views resulted. R. Pfeiffer and Marx (1898) considered that cholera antibodies are produced especially in the spleen, lymph glands, and bone marrow, and this view was supported by the experiments of Deutsch (1899) and Castellani (1901).

An intravascular formation was maintained by von Dungern (1903) and by Kraus and Schiffmann (1906). Ehrlich (1891), on the other hand, was of opinion that antitoxin at any rate may be formed in any part of the body where the toxin is fixed. For example, he showed the possibility of immunizing animals against abrin by introducing the poison into the conjunctival sac, and this was supported by the later experiments of Römer (1901) and Wassermann and Citron (1905). The question cannot, however, be looked upon as settled yet. Transmission of immunity was also extensively studied, and in this respect the beautiful experiments of Ehrlich (1892) were fundamental. He injected mice with abrin, ricin, robin, or tetanus poisons and showed that males highly immunized with abrin are incapable, when paired with normal females, of transmitting their immunity to their offspring, in other words the immunizing principle is not carried in the genes of the sperma. Female mice immunized before conception bore immune young, but the immunity was of a passive type in that it disappeared in a relatively short space of time. Further, the immunity was not transmitted in the next generation and was not truly hereditary. Investigating the matter still further, Ehrlich observed that the immunity in the offspring of mice actively immunized against ricin lasted longer than the immunity of adult mice passively immunized with antitoxin. To study this in greater detail he devised an ingenious changeling experiment in which the young mice of normal and of immune mothers were transferred crosswise, i.e. the young of the non-immunized mother were put to the breast of an immunized mother and the young of the immunized mice were put to suck the milk of a normal mother. The result was conclusive. The young of the immunized mother lost their immunity while sucking the normal mother, whereas the young of the non-immunized mother put to the breast of an immunized mouse rapidly developed an immunity which must have been lactogenic. These interesting experiments have, on the whole, been confirmed with certain variations according to the experimental animal employed.

The general result of enormous numbers of experiments on immunity have clearly shown that the injection of an antigen is followed by the appearance in the humours of antibody. The antibody varies in type but is specific for the antigen. The antibody is formed by the animal, runs a well-defined course, and is associated with a corresponding immunity. This immunity does not follow the general laws of heredity.

This is a brief statement of the experimental facts, but the innate tendency of the human mind for the concrete rather than the abstract early led to attempts to formulate an explanation how such things come about. In this way arose various theories of immunity. One of the earliest of these was known as the 'exhaustion' hypothesis, and was vaguely stated by Pasteur (1880). According to this the immunity was due to the disappearance from the body of some necessary foodstuff which was used up during the first attack of the microbe. This view was rendered highly improbable by the experiments of Bitter (1888), and on theoretical grounds. Chauveau's 'retention' theory, according to which the microbe in its first invasion left behind some antiseptic-like substance, was also found not to be in accord with all the facts. Both of these hypotheses neglected the cardinal role of the active interference of the infected animal. This aspect was first clearly defined by Metchnikoff. As has been stated he considered that certain cells of the body actively destroy the infecting agents by a process which he named phagocytosis.

While Metchnikoff was collecting data in support of his theory another view was gradually built up by Hans Buchner of Munich. According to him bacteria are not destroyed directly by the body cells but by defensive substances—alexines—in the humours of the body. At first this humoral doctrine was formulated in direct opposition to the cellular doctrine. In later years the differences between the two hypotheses became toned down as it was gradually assumed that in the end the humoral substances must be cellular products. The boundaries of Metchnikoff's theory were enlarged year by year as new facts presented themselves, until the question really came to be whether the bacteria are destroyed inside the cells or outside. Attempts to reconcile the humoral and cellular hypotheses were made by E. H. Hankin (1890), who referred Buchner's alexine to a leucocytic source in his so-called 'alexocytes'. Following this came the important work of Denys and his pupils in Louvain. In particular the experiments of Denys and Havet (1894) caused Buchner to amend his views.

The Belgians named found that dogs' serum is less bactericidal than dogs' blood as a whole. Working with *Bacillus coli*, *B. subtilis*, and *Staphylococcus* they found that the leucocytes play a fundamental part, for when they are removed the plasma is found to be robbed of a considerable part of its bactericidal properties. Microscopic examination observations of the blood showed all stages of phagocytosis. Denys and Havet considered that the leucocytes act conjointly with the blood fluids in a manner varying with the animal species and also, probably, with the virus employed. Subsequently a large amount of work by other investigators seemed to show that leucocytic extracts are richer in alexine than blood-serum.

The discoveries of R. Pfeiffer, Bordet, and Ehrlich conclusively showed, however, that the phenomenon of bactericidal action was much more complicated than a simple alexic action as presumed by Buchner, for in addition to an alexine or complement a second body—amboceptor—had to be reckoned with. The relation of the phagocytes to alexic substances remained in a nebulous state for years and has not yet been cleared up, although some kind of combined action as suggested by Denys and his co-workers seemed probable. This was supported in later years by the experiments of A. E. Wright and S. R. Douglas (1903) with their opsonic theory. By the simple expedient of testing separately and in conjunction the serum and leucocytes they showed that substances—opsonins (ὀψωνιάζω, I furnish with provisions)—exist in the serum and their action is in some way to alter the microbe so that it falls an easy prey to the leucocytes. The experiments of Wright and Douglas aroused a great deal of interest but their interpretation and the accuracy of their determinations were adversely criticized by several workers. The extensive literature which gathered round the subject of opsonins was very discordant, and the interest in it seems to have died down to a considerable extent in recent years.

Towards the end of the nineteenth century it was clearly seen that the whole problem of immunity was of extreme complexity. This was apparent in the consideration of the nature of the interaction of antigen and antibody. It was known that in some way or other antibodies neutralize the effects of the homologous antigens. The exact manner in which, however, this neutralization took place was soon the subject of discussion. A great deal of the answer to the question concerns the neutralization of toxin by antitoxin, and the early view was that the toxin was merely destroyed by the antitoxin. H. Buchner (1893[2]) opposed this idea when he found that a tetanus toxin-antitoxin mixture neutral, i.e. without effect on mice, was nevertheless lethal for the guinea-pig. On injecting 23 mice with a tetanus toxin-antitoxin mixture containing 140 minimal lethal doses of toxin, 9 remained well, 11 had slight tetanus, and 3 died. Of 23 guinea-pigs inoculated with the same mixture 8 died of acute tetanus, 12 had chronic tetanus, and only 3 remained unaffected. From this he concluded that the antitoxin had not destroyed the toxin, otherwise the mixture should have been ineffective both on mice and guinea-pigs. Calmette (1895), working with a neutral mixture of snake venom and anti-venom, showed that by heating the mixture to 68° C. it became as toxic as if no anti-venom were present at all. In this connexion the experiments of Ehrlich (1897) were of fundamental importance. It is well known that solutions of ricin produce in extravascular blood a peculiar agglutination of the red blood corpuscles. This agglutinative action can be inhibited by anti-ricin.

What Ehrlich found was that the same dose of anti-ricin which *in vitro* could not prevent agglutination was also unable to prevent *in vivo* a fatal ricin intoxication. On the other hand a quantity of anti-ricin which *in vitro* prevented the agglutinating action of ricin on the red blood corpuscles also completely protected mice inoculated with ricin. Finding that there was apparently a direct union of ricin and anti-ricin *in vitro*, Ehrlich concluded that it was necessarily a chemical process, and in all his numerous subsequent publications on the subject he maintained this view. Following the law of constant proportion he considered that a definite quantity of toxin united with an unalterable quantity of antitoxin. Ehrlich's experiments on ricin-anti-ricin were repeated and others were added by many workers in the next few years. The results seemed to show that the antibody acts directly on the antigen.

The next problem was to determine the fate of the toxin in a toxin-antitoxin mixture. The work of Calmette has been referred to above, but important developments were made by Ehrlich, who was the first to introduce exact quantitative measurements as being most likely to solve the problem. He showed that the union of toxin and anti-toxin is accelerated by heat and retarded by cold. The reaction proceeds quicker in concentrated than in dilute solutions. Such facts indicated analogy with chemical processes and Ehrlich believed that the substances concerned would react in equivalent proportions. He was quickly disillusioned by finding that when different diphtheria toxins were balanced against one and the same antitoxin the quantitative results were entirely different. He now began an extremely elaborate analysis of the whole question and the results have formed the basis of the methods of standardizing antitoxin from a practical point of view. In the course of his inquiries he elaborated a complicated hypothesis which had a profound influence on the development of the whole immunity problem and led to the publication of an enormous mass of literature. One of the fundamental ideas which Ehrlich developed was that the lethal action of a toxin and its antitoxin-combining power are two separate functions. He explained this on the assumption that the toxin contained two entirely different atom groups, one of which, the 'haptophore' (ἅπτω, to seize), brings about the union with the antitoxin, whereas the other—the 'toxophore'—is, as its name implies, the carrier of the toxic action. Of the two, the haptophore is the more stable. Ehrlich considered that the toxophoric atom group can deteriorate to a non-toxic state although the haptophoric group may at the same time remain unchanged. These beliefs were not mere whims but resulted from the experiments which Ehrlich carried out. Taking a constant quantity of diphtheria antitoxin as a standard unit, Ehrlich found that, when mixed with a fresh sample of a particular diphtheria toxin, it neutralized a c.c. of the toxin and in this quantity there were β lethal doses. When the same sample of toxin was examined some months later the remarkable fact was elicited that exactly a c.c. of toxin was neutralized by the unit of antitoxin, but that this a c.c. of toxin contained no longer β lethal doses but $\beta-x$. In other words, in a given toxin the antitoxin-combining property might remain unchanged, whereas the number of lethal doses in a given quantity might become less. According to Ehrlich, the toxin could deteriorate to toxoid—a less poisonous state—while the haptophoric state remained unaltered. Starting from this experimental standpoint, Ehrlich tested in laborious work a number of different samples of diphtheria toxin and from them laid down the principles for the accurate standardization of antitoxic sera.

In addition Ehrlich (1897) put forward a general theory of immunity which, under the name of the 'side-chain' or 'receptor' theory, had for many years an enormous vogue. His view on the toxophoric and haptophoric components in a toxin have already been referred to, likewise his belief that a definite quantity of antitoxin combines with a definite quantity of toxin or its toxoid equivalent. The idea that antitoxin is a product created by the animal organism in response to toxin injection rested on a sound experimental basis.

Ehrlich now conceived the idea that a poisoning by a bacterial toxin must really be due to the specific union of the cells of some part of the animal economy with the haptophoric complex of the toxin. In his judgement it is this very union of specific complexes which must give birth to what is called antitoxin in the serum.

Following the conceptions of organic chemistry, Ehrlich pictured this union as one of atom groups—side-chains—and he was supported in his beliefs by views on the chemical structure of protoplasm which he had developed years before. In his remarkable work on the oxygen requirements of the organism Ehrlich (1885) had visualized protoplasm as consisting of a giant molecule equipped with a large number of side-chains, the function of which is the fixation of foodstuffs for the nutrition of the cell. He considered that there must be a special executive centre (*Leistungskern*) with attached side-chains of subordinate importance as far as concerns the actual living character of the protoplasm. Returning to this conception in 1897 he now supposed that it explained antitoxin production. When the side-chains of particular cells have been brought under the influence of a toxin through the intermediation of the haptophoric complex of the latter, the cell was thereby rendered physiologically defective. In order to compensate the physiological defect Ehrlich believed that the living protoplasm replied with regeneration of like side-chains, and he supported himself in this view on a general biological principle formulated by C. Weigert (1896) of excessive cell regeneration beyond the actual demand. This was a morphological idea, but Ehrlich took it in order to explain the excessive regeneration of side-chains which he had to assume in order to explain antitoxin production. In his view an excess of specific side-chains was produced, and being ultimately cast into the blood-stream constituted what had been called antitoxin. In other words antitoxin is really the cast-off receptors of certain cells for which the haptophoric complexes of a particular toxin have a specific affinity. This famous theory, partly chemical, partly biological, gave rise to an enormous amount of discussion and many of its consequences were very carefully studied. Interesting facts concerning the origin of Ehrlich's theory will be found in Heymann's (1928) paper. A full account of the theory was given by Ehrlich (1900) in his Croonian Lecture at the Royal Society, and it was chiefly in this that he introduced an extension to explain the action and mode of formation of lysins and other antibodies. He regarded the immune-body element of a lysin as essentially an amboceptor combining with the cell on the one hand and with the alexine or complement on the other. This amboceptor concept was accepted by most workers and the term is in habitual use to-day notwithstanding that it has been doubted by Bordet from the very first. Ehrlich ultimately pictured receptors as belonging to three distinct orders corresponding respectively to antitoxins, agglutinins, and precipitins and lysins respectively (Figs. 35 and 36).

Although Erhlich's theory met in general with acceptance it may be stated that Gruber and von Pirquet (1903) attacked its whole fabric and considered it was erroneous and far from representing the actual facts. Forssman (1908) also published work which seemed to be destructive of the foundations of Ehrlich's hypothesis.

The original interpretation put by Ehrlich on his quantitative experiments on toxin-antitoxin union remained the standard one for several years, but criticisms were forthcoming from several writers, to mention only Danysz (1902) and Arrhenius and Madsen (1902). Thus Danysz discovered a peculiar 'effect' since known by his name. He found in connexion with ricin and anti-ricin that a mixture of a parts of ricin with b parts of anti-ricin was less toxic when the two components were mixed simultaneously than when one half of a was added to b and the other half of a was added twenty-four hours later. This fact, which was confirmed by other workers, was considered to be adverse to Ehrlich's idea of a chemical union in fixed proportions. Arrhenius and Madsen (1902) also came to the conclusion that the neutralization of toxin and antitoxin does not proceed, as maintained by Ehrlich, in step-like form but is gradual, and the curve is similar to that in which one weak acid is neutralized by one weak base, e.g. boric acid and ammonia.

This was, however, opposed by Ehrlich (1903). Long discussions followed supporting now one side, now the

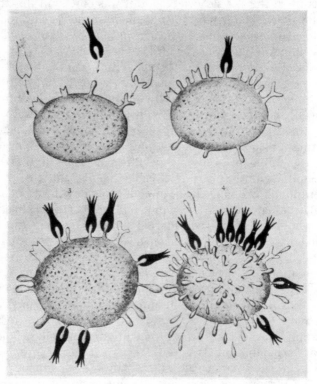

Fig. 35. Ehrlich's figures illustrating the side-chain theory. From Ehrlich (1900).

other, and by degrees a new view began to be substituted. It was considered that the union of antigen and antibody is not really a true chemical process, as had been supposed by Ehrlich and Arrhenius, but is akin to what had been called adsorption. Landsteiner and Jagić (1903), and Bordet (1903), were early upholders of this idea, which has gradually obtained wide acceptance and has led to the production of an extensive literature still growing. New theories of immunity have, however, ceased for the present to be formulated and apparently the production of this state seems to be passing into greater obscurity than was believed thirty years ago.

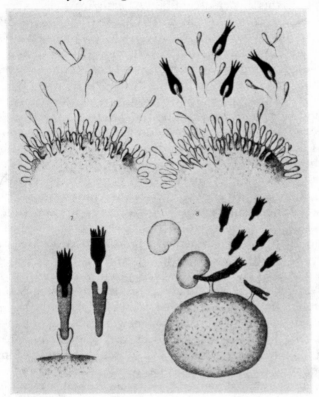

FIG. 36. Ehrlich's figures illustrating the side-chain theory. From Ehrlich (1900).

Notwithstanding, immunology as a science continues to grow and has far outstripped its original bacteriological limits. Books and journals on the subject continue to pour forth and have reached dimensions with which no single individual can expect to cope, and cannot be referred to here as historical. A word may be said on what is called anaphylaxis, which is almost a subject by itself. The original data on which this idea is based we owe to Portier and C. Richet (1902), who discovered the remarkable fact that certain poisons of animal origin evoke a condition of hypersusceptibility instead of immunity, and that as a result of this hypersusceptibility death may ensue from a dose of the poison which is otherwise ineffective. By maceration of the tentacles of certain sea anemones they obtained a poisonous fluid which, injected intravenously into dogs in a dose of 0·2 gm. per kilo, produced hypothermia, diarrhoea, and death in two or three days. Doses smaller than 0·2 gm. produced only transient symptoms. If, however, several weeks after the injection of such a dose a second sub-lethal dose was administered, violent symptoms of poisoning set in and were sometimes followed by the death of the animal in an hour but mostly in 12 to 24 hours. The condition was spoken of, perhaps unfortunately, as anaphylaxis, which was intended to convey the idea of the opposite of protection. These results were extended in a number of papers by Richet alone, and he it was who found that this experimentally produced hypersusceptibility was really a widely spread phenomenon. He recognized that all the poisons which produce this effect are protein in character, and that as a result of the first injection something is produced which in combination with the re-injected poison gives rise to a new poison—apotoxin—which is the agent that produces the deadly symptoms. The anaphylactic condition was subsequently studied on a very large scale and it has been found possible to produce it by an enormous number of different substances, animal and vegetable, toxic and non-toxic. Some bacterial poisons have been included in the class of anaphylactogens, and by many the process is believed to play an important part in the evolution of infective disease. There still remains, however, a great deal before the mystery of anaphylaxis can be estimated exactly.

The years of labour that have been devoted to the study of immunity have not been in vain and great practical results have accrued from them both with regard to the diagnosis and treatment of disease, to consider only one aspect. The modern study of immunity really dates from the experiments of Pasteur with attenuated virus. His methods of prophylaxis have been followed for many diseases and new extensions have been made to include the treatment of microbic infections.

A large field has been opened up under the name of Sero-diagnosis, and has numerous applications. The underlying basis on which the value of this method depends is the extraordinary specificity which attends immunity reactions. It is true that the specificity is not absolute, but varies qualitatively. The immunity tests depend on various characters. By means of an immune serum search may be made for an unknown antigen and vice versa.

The possibility of such a practical test was first made known by R. Pfeiffer (1894), who found that the bacteriolytic destruction of cholera vibrios in the peritoneum was so specific that this method could be employed for the differentiation of cholera vibrios from vibrios otherwise indistinguishable. Bordet (1895) utilized this principle and applied it *in vitro* instead of *in vivo*. A great extension of the sero-diagnostic method took place with the discovery of agglutination, and Durham early foreshadowed its practical application for the differentiation of allied bacteria by agglutinating serum. Widal went a step farther by employing known bacteria (e.g. typhoid bacillus) to determine the presence or absence of specific agglutination in the serum of the patient in connexion with the diagnosis of enteric fever. Previously unknown fevers inside the enteric group have been discovered by sero-diagnosis. It was in this way that Schottmüller (1900) separated the paratyphoid fevers. The same principle has been applied on a wide scale in other cases, and this method is of ever growing importance. The application of specific precipitating sera in forensic medicine, hygiene, and the industries is thoroughly established and practised as a routine.

Of particular medical importance is the so-called complement-fixation test, the principle of which was discovered by Bordet and Gengou (1901) as follows. If to a quantity of unheated normal guinea-pig-serum is added a quantity of sensitized red blood corpuscles (i.e. corpuscles charged with specific amboceptor) haemolysis occurs. They made, however, the observation that if cholera vibrios are added to such a mixture they remain intact instead of undergoing spherulation and lysis, as might perhaps have been expected. The same occurs if the cholera vibrios and blood corpuscles are introduced in the reverse order. On testing heated plague-serum mixed with plague bacilli and normal serum, Bordet and Gengou found that the subsequent addition of sensitized red blood corpuscles did not result in haemolysis. The explanation they gave was that the sensitizing substance in the plague-serum fixes the complement in the presence of plague bacilli so that it is prevented from acting on the sensitized blood corpuscles. Other experiments of a like kind led the two Belgian workers to the belief that the phenomenon described had widespread applicability and that it is possible to diagnose the existence of a 'sensitizer' in the serum by the presence or absence of haemolysis of sensitized corpuscles. Gengou (1902) showed further that 'sensitizers' (amboceptors) are developed in the blood-serum of animals which have been injected with milk, and that such sensitizers are also capable of fixing complement. The same reaction also occurred in the sera of animals injected with egg albumen, fibrinogen, or other substances.

These facts were extended by C. Moreschi (1905), who showed that complement fixation occurs in the presence of normal serum mixed with the serum of an animal injected with normal serum, and he demonstrated that extraordinarily small quantities of serum ($\frac{1}{100000}$ c.c.) can be detected by this method of diagnosis. M. Neisser and Sachs (1905) recommended the method for the diagnosis of blood in medico-legal work. In 1906 Wassermann, Neisser, and Bruck published their historic account in which they described the discovery of antibodies to syphilis antigen in the serum of syphilitic monkeys. In the same year Wassermann and Plaut (1906), by demonstrating syphilitic antibodies in the cerebro-spinal fluids of general paralytics, proved this disease to be syphilis, and Wassermann, Neisser, Bruck, and Schucht (1906) demonstrated similar antibodies in the blood-serum of syphilitics. Since that time the 'Wassermann reaction' has been practised to an enormous extent in the diagnosis of syphilis and is regarded as a test of deadly accuracy. The complement fixation test has been permanently accepted also in the case of the diagnosis of glanders. But here we may stop as it brings us well into the present century. The later ramifications of the immunity problem, the development of the doctrines of chemotherapy, and the general advance of bacteriology are still too unsettled to find a place in a work on history.

From the small beginnings in 1675, when the great Dutch microscopist Antony van Leeuwenhoek first saw with his lenses minute living creatures in rain water, down to the end of the nineteenth century, a new science has developed and has reached immense dimensions. I have attempted to trace how this knowledge has been built up, often by laborious efforts, by tortuous routes, and sometimes through trackless wastes. But one thing is certain: the science of bacteriology is built on broad and sound foundations. If I have succeeded in infusing interest in this subject among the younger students of to-day I am amply rewarded for my studies among the bacteriological works of the last two centuries.

BIBLIOGRAPHY

CHAPTER I

ADAMS, FRANCIS, trans. of Hippocrates, Lond., Sydenham Soc., 1849, i. 352; also trans. of Paulus Aegineta, Sydenham Soc., 1844, i. 284.

APPIAN, *Romanorum historiarum: De bello Mithridatico*, chap. 76; *De rebus Illyricis*, chap. 4; *De rebus Punicis*, chap. 73.

BENEDICTUS, ALEX., *De observatione in Pestilentia*, Venetiis, 1493.

BERNHEIM, 'Contagion', *Dict. encycl. d. sc. méd.*, Paris, 1877, xx. 1–60.

BOGHURST, W., *Loimographia, an account of the great plague of London in the year 1665*, edited by J. F. Payne, Lond., 1894.

CHOULANT, J. L., *Handbuch der Bücherkunde für die ältere Medicin*, 1828.

DAREMBERG, C. V., *Histoire des sciences médicales*, Paris, 1870, 1303 pp.

DIODORUS SICULUS, *Bibliothecæ historicæ*, lib. v. 82; xii. 45, 58; xiii. 86; xiv. 71.

DION CASSIUS, *Historia romana*, lib. lxxii, cap. 15 (in Teubner's edition, p. 207).

DIONYSIUS OF HALICARNASSUS, *Antiquitatum Romanorum* lib. ix. 42, 68; lib. x. 53.

FOSSEL, V., 'Geschichte der epidemischen Krankheiten', *Handb. d. Geschichte der Med.*, hrsg. v. Neuburger u. Pagel, 1903; ii. 737–901.

FRACASTORIUS, H., *De sympathia et antipathia rerum*, liber unus, *De contagione et contagiosis morbis et curatione*, lib. iii, Venetiis, 1546. French and Latin of *De contagione* in Meunier's transl., Paris, 1893. German transl. by V. Fossel, 1910 (Klassiker der Medizin series). English transl. by Wilmer Cave Wright, 1930 (New York: G. P. Putnam's sons).

GANSZYNIEC, 'Apollon als Heilgott', *Arch. f. Geschichte d. Med.*, 1923, xv. 33–42.

GOODALL, E. W., 'Fracastor as an Epidemiologist', *Proc. Roy. Soc. Med.*, 1937, xxx. 341–50.

GREENWOOD, M., 'Galen as an Epidemiologist', *Proc. Roy. Soc. Med.*, 1921, xiv, Sect. Hist. of Med. 3.

GRUNER, C. G., *Morborum antiquitates*, Vratislaviae, 1774.

HAESER, H., *Lehrbuch der Geschichte der Medicin und der epidemischen Krankheiten*, Jena, 1882, Bd. III.

HIRSCH, A., *Handbuch der historisch-geographischen Pathologie*, Erlangen, 1881–3, 2 pts.; English trans. New Sydenham Soc., Lond., 1883–6, 3 vols.

LE CLERC, D., *Histoire de la médecine, etc.*, Amsterdam, 1702.

LIVY, *History*, lib. iii. 6; iv. 30; v. 6, 12; xxv. 26.

LÖFFLER, F., *Vorlesungen über die geschichtliche Entwickelung der Lehre von den Bacterien*, Leipzig, 1887, 252 pp., 3 plates.

LUCRETIUS, *De rerum natura*, lib. vi.

MARX, C. F. H., *Origines contagii*, Caroliruhae et Badae, 1824, 153 pp. *Additamenta ad origines contagii*, 1826, 51 pp.

OMODEI, A., 'Cognito agli storici e filosofi antichi', *Ann. Univ. di med. compilati dal Sig. Dottore Annibale Omodei*, Milano, 1822, xxii. 170–97.

OZANAM, J. A. F., *Histoire médicale, génerale et particulière des maladies épidémiques, contagieuses et épizootiques qui ont régné en Europe depuis les temps les plus reculés, jusqu'à nos jours*, 2ᵉ éd., Paris, 1835, 4 vols.

PUSCHMANN, 'Die Geschichte der Lehre von der Ansteckung', *Wien. med. Wchnschr.*, 1895, xlv. 1425, 1471, 1509, 1551, 1587, 1623, 1655, 1695.

SEIDEL, 'Die Lehre von der Kontagion bei den Arabern', *Arch. f. Geschichte d. Med.*, 1913, vi. 81–93.

SINGER, C., and DOROTHEA, 'The scientific position of Girolamo Fracastoro with special reference to the source, character, and influence of his theory of infection', *Ann. Med. Hist.*, 1917, i. 1–34.

SPRENGEL, K. P. J., *Versuche einer pragmatischen Geschichte der Arzneikunde*, 1792–1803, 5 vols.

THUCYDIDES, *History of the Peloponnesian War*, book ii, chap. 48.

VIRGIL, *Georgics*, lib. iii; *Eclogue* i.

YEATS, G. D., 'Some observations on the opinions of the ancients respecting Contagion', *Quart. J. Lit. Sc. and Arts*, Lond., 1819, vi. 124–34.

CHAPTER II

ADAMS, G. (the elder), *Micrographia illustrata or the Knowledge of the microscope explained . . . to which is added a translation of Mr. Joblott's observations on the animalcula, etc.*, 4°, Lond., 1746, 2nd ed. 1747, 3rd ed. 1771.

ADAMS, G. (the younger), *Essays on the microscope . . . a description of three hundred and seventy-nine animalcula . . . with atlas of 31 plates*, 4°, Lond., 1787, 2nd ed. 1798.

ALMQUIST, ERNST, 'Linné und die Mikroorganismen', *Ztschr. f. Hyg. u. Infektionskrankh.*, 1909, lxiii. 151–76.

ANDRY DE BOISREGARD, N., *De la génération des vers dans le corps de l'homme . . . avec trois lettres . . . sur le sujet de vers; les deux premières par N. Hartsoeker, et l'autre par G. Baglivi*, 12°, Paris, 1700; 3rd ed. 1741, 2 vols. Engl. trans., Lond., 1701.

BAKER, HENRY, 'An account of Mr. Leeuwenhoek's microscopes', *Phil. Trans.*, 1740, xli, pt. 2, no. 458, pp. 503–19.

—— *The microscope made easy, etc.*, Lond., 1742, 3rd ed. 1744, 4th ed. 1754, 5th ed. 1769.

—— *Employment for the microscope in two parts . . .*, 8°, Lond., 1753, 2nd ed. 1764.

BEIJERINCK, M. W., 'De infusies en de ontdekking der bakterien', *Jaarboek voor 1913 van de Koninklijke Akademie van Weten- schappen gevestigd te Amsterdam*; also in Beijerinck's *Verzamelde Geschriften*, Delft, 1922, 5. deel, 119–40.

BONOMO, G. C., *Osservazioni intorno a' pellicelli del corpo umano*, Firénze, 1687, 1 pl.; English abstr. *Phil. Trans.*, Lond., 1704, xxiii. 1296–9, 2 figs.

BÜTSCHLI, O., 'Protozoa', in H. G. Bronn's *Klassen und Ordnungen des Thier-Reichs*, Leipz., 1887–9, Bd. I, Abt. III, Infusoria, pp. 1100–1227.

COLE, F. J., *The History of Protozoology*, Lond., 1926, 64 pp.

COLUMELLA, L. J. M., *De re rustica*, Ex officina Roberti Stephani, Parisiis, 1543, lib. i, cap. v (De acqua).

CORTI, BONAVENTURA, *Osservazioni microscopiche sulla Tremella e sulla circolazione dell fluido in una pianta acquajuola (appellata Cara)*, Lucca, 1774.

DOBELL, C., 'A protozoological bicentenary. Antony van Leeuwen- hoek (1632–1723) and Louis Joblot (1645–1723)', *Parasitology*, 1923, xv. 308–19.

——— *Antony van Leeuwenhoek and his 'little animals'*, Lond., 1932, 435 pp.

EICHHORN, J. C., *Beyträge zur Naturgeschichte der kleinsten Wasser- Thiere*, &c., 4°, Danzig [1775]; *Zugabe . . . mit zwey neuentdecten Wasser-Thieren*, 1783.

ETTMÜLLER, M., Observatio medica. De sironibus. *Acta eruditorum*, Lipsiae, 1682, 317, 2 figs.

FOLKES, M., 'Some account of Mr. Leeuwenhoek's curious micro- scopes lately presented to the Royal Society', *Phil. Trans.*, 1722, 1723, xxxii, no. 380, pp. 446–53.

VON GLEICHEN, W. F. (called Russworm), *Abhandlung über die Saamen und Infusionsthierchen und über die Erzeugung; nebst microskopischen Beobachtungen des Saamens der Thiere und verschiedener Infu- sionen*, 4°, Nürnberg, 1778, 171 pp., 32 plates.

HARTING, P., *Gedenkenboek van het den 8sten September 1875 gevierde 200-jarig Herinneringsfeest der ontdekking van de mikroskopische wezens, door Antony van Leeuwenhoek*, 8°, 's Gravenhage and Rotterdam, 1876, 146 pp.

——— *Het mikroskoop, deszelfs gebruik, geschiedenis en tegenwoordige toestand*, 8°, Utrecht and Tiel, 1848–54, 4 vols.

——— *Das Mikroskop. Theorie, Gebrauch, Geschichte und gegenwärtiger Zustand desselben . . . Deutsche Original-Ausgabe vom Verfasser . . . vervollständigt*, Aus dem Holländischen übertragen von F. W. Theile, 8°, Braunschweig, 1859, 2 Aufl. 1866, 3 vols.

HAUPTMANN, AUGUST, *Uhralter Wolckensteinischer warmer Badt- und Wasser-Schatz; zu unser lieben Frawen auf dem Sande genand, etc.*, 2 pts., Part I, Leipz., 1657; Part II, Francof. 1650; continuous pagination, 252 pp.

HILL, JOHN, *Essays in natural history and philosophy containing a series of discoveries by the assistance of microscopes*, Lond., 1752, 415 pp.

―― *An history of animals containing descriptions of the birds, beasts, and insects of the several parts of the world and including accounts of the several classes of animalcules visible only by the assistance of microscopes*, Lond., 1752, p. 1.

HOOKE, R., *Micrographia; or some physiological descriptions of minute bodies made by magnifying glasses, with observations and inquiries thereupon*, fol., Lond., 1665.

JOBLOT, L., *Descriptions et usages de plusieurs nouveaux microscopes tant simples que composez; avęc de nouvelles observations faites sur une multitude innombrable d'insects et d'autres animaux de diverses espèces, qui naissent dans de liqueurs préparées et dans celles qui ne le sont point*, 4°, Paris, 1718, 1 vol. in 2 parts.

―― *Observations d'histoire naturelle, faites avec le microscope, sur un grand nombre d'insectes et sur les animalcules qui se trouvent dans les liqueurs préparées et dans celles qui ne le sont pas*, etc., Paris, 1754, 2 vols. in one; vol. ii dated 1755.

KIRCHER, A., *Scrutinium physico-medicum contagiosae Luis, quae pestis dicitur, quo origo, causae, signa, prognostica Pestis . . . una cum appropriatis remediorum antidotis nova doctrina in lucem eruuntur*, etc., Romae, 1658; another edition, cum praefatione D. C. Langii, Lipsiae, 1659.

―― *Ars magna lucis et umbrae in decem libros digesta*, etc., fol. Romae, 1646, Pars 2, Caput 8, Pragmatia II, p. 834; editio altera, Amstelodami, 1671.

KONARSKI, WLODIMIR, 'Un savant Barrisien précurseur de M. Pasteur: Louis Joblot [1645–1723]', *Mém. Soc. let. sc. et ars de Bar-le-Duc.*, 1895, 3ᵉ s., iv. 205–333, 4 plates.

LANCISI, G. M., *De noxiis paludum effluviis eorumque remediis*, Libri duo, Coloniae Allobrogum, 1718, 354 pp., chaps. 16 and 18.

LANGE, C. J., *Pathologia animata seu animadversiones in pathologiam spagiricam* . . ., 4°, Francof. a. M., 1688, 698 pp.

LEDERMÜLLER, MARTIN F., *Mikroskopische Gemüths- und Augen-Ergötzung*, Beyreuth, 1760–5, 3 vols.

VAN LEEUWENHOEK, A., 'Observations communicated to the publisher by Mr. Antony van Leewenhoeck in a Dutch letter of the 9th of Oct. 1676 here Englished, concerning little animals by him observed in rain- well- sea- and snow-water; as also in water wherein pepper had lain infused', *Phil. Trans.*, 1677, 13th year, no. 133, pp. 821–31.

―― Letter 39 (according to his own enumeration) dated 17 Sep. 1683 (MS. Roy. Soc.) completely published (in Dutch) in a tract entitled *Ondervindingen en Beschouwingen*, etc., *door Antoni van Leeuwenhoek*, 4°, Leyden, 1684. Bound up in *Werken*, Dutch edit., vol. i, pp. 1–19 (6th pagination) with figures but fig. G

omitted. Latin trans. (complete) first issued in *Arcana naturae
detecta*, 4⁰. Lugd. Bat. 1695, pp. 41–53 with best figs. including
fig. G. In both Latin and Dutch editions the letter is wrongly
dated 12 Sep. 1683. English abstr. [1st version] in *Phil. Trans.*,
no. 159, vol. xiv, 20 May 1684, pp. 568 [*bis*]–574 (568 should be
598) with figs. (Fig. G very poorly shown). English abstr.
[2nd version] in *Phil. Trans.*, no. 197, vol. xvii, Feb. 1693, pp. 646–
9, with figs., G missing. Partial English trans. in Hoole's *Select
Works*, 1798, vol. i, pp. 118–19 (fig. incomplete).

VAN LEEUWENHOEK, A., *Ontledingen en Ontdekkingen . . . Vervat in
verscheyde brieven geschreven aan de . . . Koninglijke . . . Societeit
tot London*, 12 pt., 4⁰, Leyden, 1685.

—— *Arcana naturae ope et beneficio exquisitissimorum microscopiorum
detecta*, etc., sm. 4⁰, Delphis Batavorum, 1695.

—— *Opera omnia seu Arcana Naturae, ope . . . microscopiorum
detecta*, quatuor tomis distincta, Editio novissima . . . emendatior,
4 tom., 4⁰, Lugduni Batavorum, Delphis, 1715, 1719, 1722.

—— *The select works of A. van Leeuwenhoek containing his microscopical [sic] discoveries . . .* translated from the Dutch and Latin
editions . . . by S. Hoole, 2 vols., 4⁰, Lond., 1798; 1807.

LESSER, F. C., *Insecto-theologia oder . . . Versuch wie ein Mensch durch . . .
Betrachtung deren sonst wenig geachteten Insecten zu . . . Erkänntniss
. . . der Allmacht . . . des grossen Gottes gelangen könne*, Franckf. u.
Leipz., 1738.

LINNÉ, C., *Amoenitates academicae*, Holmiae, 1763, vi. 4.

—— *Systema naturae per regna tria naturae*, ed. xii. reformata,
Holmiae, 1767, tom. i, pars 2, p. 1326.

M.A.C.D., *Système d'un medecin anglois sur la cause de toutes les
especes de maladies avec les surprenantes configurations des differentes
especes de petits insectes, qu'on voit par le moyen d'un bon microscope
dans le sang et dans les urines de differens malades, et meme de tous
ceux qui doivent le devenir*, Paris, 1726, 34 pp., 87 illust.

——*Suite du système d'un medecin anglois sur la guerison des maladies . . .*,
Paris, 1727, 22 pp.

MARTEN, BENJ., *A new theory of consumptions; more especially of a
phthisis or consumption of the lungs*, Lond., 1720, 2nd edit., 1722,
154 pp.

MÜLLER, O. F., *Vermium terrestrium et fluviatilium seu animalium
infusoriorum, helminthicorum et testaceorum non marinorum succincta
historia*, 2 vols. 4⁰, Hav. et Lips., 1773–4.

—— *Animalcula infusoria fluviatilia et marina quae detexit, systematice
descripsit et ad vivum delineari curavit Otho Fridericus Müller*, 4⁰,
Hauniae, 1786, 367 pp., 50 pls.

NYANDER, JOHANNES C., *Resp. Exanthemata viva . . . Praes. C. Linnaeo*,
etc., 4⁰, Upsaliae, [1757].

PAULLINI, C. F., *Cynographia curiosa seu canis descriptio*, 4⁰, Norimb.,
1685, 258 pp.

PLENCIZ, M. A., *Opera medico-physica in quatuor tractatus digesta quorum primus contagii morborum*, . . . Vindobonae, 1762, Tract I, Sectio I.

POWER, H., *Experimental philosophy in three books containing new experiments, microscopical, mercurial, magnetical*, Lond., 1664, 193 pp.

RAMSTRÖM, C. L., *Generatio ambigena* . . . *Praes. C. Linnaeo*, 4°, Upsaliae [1759].

SINGER, C., 'Benjamin Marten a neglected predecessor of Louis Pasteur', *Janus*, Haarlem, 1911, vi. 81–98.

—— *The development of the doctrine of contagium vivum, 1500–1750*, Lond., priv. print., 1913, 15 pp.

SPALLANZANI, L., *Saggio di osservazioni microscopiche concernenti il sistema della generazione dei Signori di Needham e Buffon*, Modena, 1765.

TREMBLEY, A., *Mémoires pour servir à l'histoire d'un genre de polypes d'eau douce à bras en forme de cornes*, 4°, Leyden, 1744.

VALLISNERI, A., *Opere fisico-mediche stampate e manoscritte* . . . *raccolte da Antonio Vallisneri*. . . . Venezia, 1733, iii. 218.

VARRO, MARCUS TERENTIUS, *M.T.V. rerum rusticorum libri iii*, edit. Teubner, 1912, lib. i, cap. xii. 2, p. 27.

WILLNAU, C., *Ledermüller und v. Gleichen, zwei deutsche Mikroskopisten der Zopfzeit*, Leipz., 1926, 24 pp.

WRISBERG, H. A., *Observationum de animalculis infusoriis satura*, Gottingae, 1765, 110 pp., 2 pls.

CHAPTER III

ANDRAL and GAVARRET, 'Recherches sur le développement du penicillium glaucum sous l'influence de l'acidification dans les liquides albumineux normaux et pathologiques', *Ann. chimie et phys.*, 1843, 3ᵉ s., viii. 385–401, 1 pl.

ANONYMOUS, 'Das enträthselte Geheimniss der geistigen Gährung', *Ann. d. Pharmacie*, Heidelb., 1839, xxix. 100–4.

APPERT, *L'art de conserver pendant plusieurs années toutes les substances animales et végétales*, Paris, 1810, 116 pp.; 4ᵉ édit., 1831; Engl. trans., Lond., 1811.

BALLING, C. J. N., *Die Gährungschemie wissenschaftlich begründet*, etc., 1845, 3 vols., 3rd edit. 1865, 4 vols.

BERZELIUS, 'Einige Ideen über eine bei der Bildung organischer Verbindungen in der lebenden Natur wirksame, aber bisher nicht bemerkte Kraft', *Jahresb. über die Fortschr. der physischen Wissenschaften*, Tübingen, 1836, xv. 237–45.

—— 'Weingährung', *Jahresb. über die Fortschr. der physischen Wissenschaften*, Tübingen, 1839, xviii. 400–3.

BLONDEAU, C., 'Des fermentations', *J. de pharm. et de chim.*, 1847, xii. 244, 336.

BOERHAAVE, H., *Elementa chemiae*, Lipsiae, 1732, 2 vols.; Engl. trans. by Peter Shaw, Lond., 1741, 2nd ed.

BREFELD, O., 'Untersuchungen über Alkoholgährung', *Verhandl. d. phys.-med. Gesellsch.*, Würzburg, 1874, N.F., v. 163–78.

BUCHNER, E., BUCHNER, H., and HAHN, M., *Die Zymasegärung. Untersuchungen über den Inhalt der Hefezellen und die biologische Seite des Gärungsproblems*, München u. Berl., 1903, 416 pp.

CAGNIARD-LATOUR, *L'Institut*, 1836, nos. 164 (June 29), p. 209; 165 (July 6), p. 215; 166 (July 15), p. 224; 167 (July 20), p. 237; 185 (Nov. 23), p. 389; 1837, no. 199 (Feb. 18), p. 73.

—— 'Mémoire sur la fermentation vineuse', *Compt. rend. Acad. d. sc.*, 1837, iv. 905–6.

—— 'Mémoire sur la fermentation vineuse', *Ann. chimie et phys.*, 1838, lxviii. 206–22.

CHEVREUL, E., (no title), *J. des savants*, année 1856, 94–105.

COLIN, 'Mémoire sur la fermentation du sucre', *Ann. chim. et phys.*, 1825, xxviii. 128–42.

DELBRÜCK, M., and SCHROHE, A., *Hefe, Gärung und Fäulnis—eine Sammlung der grundlegenden Arbeiten von Schwann, Cagniard-Latour und Kützing sowie von Aufsätzen zur Geschichte der Theorie der Gärung und der Technologie der Gärungsgewerbe*, hrsg. v. Delbrück, M., and Schrohe, A., Berlin, P. Parey, 1904, 232 pp.

DELBRÜCK, *Illustriertes Brauerei-Lexikon*, 2. Aufl. hrsg. v. F. Hayduck, Berl., 1925, 3 vols.

DESMAZIÈRES, J.-B.-H.-J., 'Recherches microscopiques sur le genre Mycoderma', *Recueil des travaux de la soc. de sc. de l'agriculture et des arts de Lille*, année 1825, Lille, 1826, pp. 297–323, 1 pl.; also *Ann. d. sc. nat.*, 1827, x. 42–67.

DRYANDER, J., *Catalogus bibliothecae historico-naturalis Josephi Banks*, Lond., 1800, v. 229.

DUCLAUX, E., *Pasteur: histoire d'un esprit*, 8°, Paris, 1896, 400 pp.; English trans. by Erwin F. Smith and Florence Hedges, Philad. and Lond., 1920, 383 pp.

FABBRONI, ADAM, *Dell'arte di fare il vino*, Firenze, 1787, 264 pp., 1 pl.; French trans. by F. R. Baud, An x (1801), 220 pp.

FOURCROY, 'Notice d'un mémoire du Cit. Fabroni sur les fermentations vineuse, putride, aceteuse et sur l'etherification', *Ann. de chimie*, An vii (1799), xxxi. 299–328.

FRANKLAND, PERCY, and MRS. PERCY FRANKLAND, *Pasteur*, Lond., Cassell & Co., 1898, 224 pp.

FREDERICQ, 'Theodor Schwann', obit. notice in *Annuaire de l'acad. roy. des sciences, des lettres et des beaux-arts de Belgique*, 1885, Année li, p. 205.

FREMY, E., *Sur la génération des ferments*, Paris, 1875, 217 pp.

GAY-LUSSAC, 'Extrait d'un mémoire sur la fermentation', *Ann. de chimie*, 1810, lxxvi. 245–59.

HALLIER, E., *Gährungserscheinungen. Untersuchungen über Gährung,*

Fäulniss und Verwesung mit Berücksichtigung der Miasmen und Contagien sowie der Desinfection, Leipz., 1867, 116 pp., 1 pl.

HANSEN, E. C., *Practical Studies in Fermentation*, Lond. and N.Y., 1896, 279 pp.

HARDEN, A., *Alcoholic Fermentation*, Lond., 3rd ed. 1923, 194 pp.; 4th ed. 1932.

HUEPPE, F., 'Untersuchungen über die Zersetzungen der Milch durch Mikroorganismen', *Mitth. a. d. Kaiserl. Gesundheitsamte*, 1884, ii. 309–71.

INGENKAMP, C., *Die geschichtliche Entwicklung unserer Kenntniss von Fäulniss und Gährung*, Inaug. Diss. Bonn., 1885; also *Ztschr. f. klin. Med.*, 1886, x. 59–107.

KÜTZING, F., 'Microscopische Untersuchungen über die Hefe und Essigmutter nebst mehreren andern dazu gehörigen vegetabilischen Gebilden', *J. f. prakt. Chemie*, 1837, xi. 385–409, 2 pls.

LAVOISIER, A. L., *Traité élémentaire de chymie*, Paris, 1789, 2 tom., 2nd edit., 1793; Engl. trans. by Kerr, Edinb., 1790; 4th ed. 1799.

LEEUWENHOEK, *Arcana naturae detecta*, Delph. Batav., 1695, p. 1.

LIEBIG, J. VON, 'Über die Erscheinungen der Gährung, Fäulniss und Verwesung und ihre Ursachen', *Ann. d. Physik u. Chemie*, 1839, 2. R. xviii. 106–50.

—— *Organic Chemistry in its application to agriculture and physiology*, edit. from MSS. of author by Lyon Playfair, Lond., 1840, 387 pp.

—— *Chemische Briefe*, Leipz. u. Heidelb., 1878, Brief 21.

—— 'Über die Gährung und die Quelle der Muskelkraft', *Sitzungsb. d. K. bayer. Akad. d. Wissensch. zu München*, 1869, ii. 323, 393 French trans. *Ann. chim. et phys.*, 1871, 4e s., xxiii. 5, 149.

LIEBIG and KOPP, *Annual report of the progress of Chemistry*, 1849, i. 356.

LÜDERSDORFF, F. W., 'Über die Natur der Hefe', *Ann. d. Physik u. Chemie*, 1846, lxvii. 408–11.

MAYER, Adolf, *Lehrbuch der Gährungs-Chemie*, Heidelb., 1874, 166 pp.

Mémoires de l'Institut national des sciences et arts, An XI (1803), Tome iv, p. 2.

MITSCHERLICH, E., 'Über die chemische Zersetzung und Verbindung vermittelst Contact-Substanzen', *Ber. über die zur Bekanntmach. geeigneten Verhandl. der Königl. Preuss. Akad. der Wissensch. zu Berlin*, 1841, pp. 379–96; also *Ann. d. Chem. und Pharm.*, 1842, xliv. 186–206.

MULDER, G. J., *Liebig's question to Mulder, tested by morality and science*, transl. by P. F. H. Fromberg, 8°, Lond., 1846.

MÜLLER, JOHANNES, no title, *Mitth. a. d. Verhandl. d. Gesellsch. naturforschenden Freunde zu Berlin*, 1837, Jahrg. ii, p. 9.

v. NÄGELI, C. W., *Theorie der Gährung. Ein Beitrag zur Molekular-Physiologie*, München, 1879, 156 pp.

OLMSTED, J. M. D., 'Claude Bernard's posthumously published attack on Pasteur and Pasteur's defense', *Ann. med. history*, 1937, ix. 114–24.

Pasteur, L., 'Recherches sur le dimorphisme', *Ann. chimie et phys.*, 1848, xxiii. 267–94.

—— 'Mémoire sur la relation qui peut exister entre la forme cristalline et la composition chimique et sur la cause de la polarisation rotatoire', *Compt. rend. Acad. d. sc.*, 1848, xxvi. 1848.

—— 'Mémoire sur la fermentation appelée lactique', *Compt. rend. Acad. d. sc.*, 1857, xlv. 913–16.

—— 'Mémoire sur la fermentation alcoolique', *Compt. rend. Acad. d. sc.*, 1857, xlv. 1032–6.

—— 'Mémoire sur la fermentation appelée lactique', *Ann. chimie et phys.*, 1858, 3ᵉ ser., lii. 404–18.

—— 'Mémoire sur la fermentation de l'acide tartrique', *Compt. rend. Acad. d. sc.*, 1858, xlvi. 615–18.

—— 'Nouveaux faits concernant l'histoire de la fermentation alcoolique', *Compt. rend. Acad. d. sc.*, 1858, xlvii. 1011–13.

—— 'Nouveaux faits pour servir à l'histoire de la levure lactique', *Compt. rend. Acad. d. sc.*, 1859, xlviii. 337–8.

—— 'Mémoire sur la fermentation alcoolique', *Ann. chimie et phys.*, 1860, 3ᵉ ser. lviii. 323–426, 9 figs.

—— 'Note relative au Penicillium glaucum et à la dissymétrie moléculaire des produits organiques naturels', *Compt. rend. Acad. d. sc.*, 1860, li. 298–9.

—— 'Expériences et vues nouvelles sur la nature des fermentations', *Compt. rend. Acad. d. sc.*, 1861, lii. 1260–4.

—— 'Animalcules infusoires vivant sans gaz oxygène libre et déterminant des fermentations', *Compt. rend. Acad. d. sc.*, 1861, lii. 344–7.

—— 'Études sur les mycodermes. Rôle de ces plantes dans la fermentation acétique', *Compt. rend. Acad. d. sc.*, 1862, liv. 265–72.

—— 'Recherches sur la putréfaction', *Compt. rend. Acad. d. sc.*, 1863, lvi. 1189–94.

—— 'Mémoire sur la fermentation acétique', *Ann. scientif. de l'école normale supérieure*, 1864, i. 113–58, 5 figs.

—— *Études sur le vin, ses maladies, causes qui les provoquent; procédés nouveaux pour le conserver et pour le vieillir*, Paris, 1866, 264 pp., 43 figs.; 2nd edit. 1873, 344 pp.

—— *Études sur le vinaigre, sa fabrication, ses maladies, moyens de les prévenir; nouvelles observations sur la conservation des vins par la chaleur*, Paris, 1868, 119 pp., 7 illus.

—— 'Nouvelles observations sur la nature de la fermentation alcoolique', *Compt. rend. Acad. d. sc.*, 1875, lxxx. 452–7.

—— 'Examen critique d'un écrit posthume de Claude Bernard sur la fermentation alcoolique' *Compt. rend. Acad. d. sc.*, 1878, lxxxvii. 813–19.

—— *Études sur la bière, ses maladies, causes qui les provoquent, procédé pour la rendre inaltérable avec une théorie nouvelle de la fermentation*, Paris, 1876, 387 pp., 12 pls.; English trans. by Frank Faulkner and D. Constable Robb, Lond., 1879, 418 pp.

PERSOON, C. H., *Mycologia europaea seu completa omnium fungorum in variis Europaeae regionibus detectorum enumeratio*, Erlangae, 1822, Sectio prima, p. 96.

QUEVENNE, T. A., 'Sur la levure et la fermentation vineuse', *J. de pharm.*, 1838, xxiv. 265, 329; German trans. in *J. f. prakt. Chemie*, 1838, xiv. 328, 458.

ROCHE, PH., *Les précurseurs de Pasteur, histoire des fermentations*, Paris, 1904–5, Thèse No. 434.

SCHLOSSBERGER, J., 'Über die Natur der Hefe mit Rücksicht auf die Gährungserscheinungen', *Ann. d. Chem. u. Pharm.*, 1844, li. 193–212.

SCHWANN, TH., 'Vorläufige Mittheilung betreffend Versuche über die Weingährung und Fäulniss', *Ann. d. Physik u. Chemie*, 1837, xli. 184.

—— *Mikroskopische Untersuchungen über die Übereinstimmung in der Struktur und dem Wachsthum der Thiere und Pflanzen*, Berl., 1839, p. 234; English trans. Lond., Sydenham Soc., 1847, p. 197.

SPIESS, G. A., *J. B. van Helmont's System der Medicin verglichen mit den bedeutenderen Systemen älterer und neuerer Zeit*, Frankf. a. M., 1840.

STAHL, G. E., *Zymotechnia fundamentalis oder allgemeine Grund-Erkänntniss der Gährungs-Kunst*. Stettin u. Leipz., 1748, 304 pp.

THÉNARD, 'Sur la fermentation vineuse', *Ann. de chimie*, An xi (1802–3), xlvi. 294–320.

TRAUBE, MORITZ, *Theorie der Fermentwirkungen*, Berl., 1858, 119 pp.

—— *Gesammelte Abhandlungen*, Berl., 1899, 583 pp.

TURPIN, 'Mémoire sur la cause et les effets de la fermentation alcoolique et acéteuse', *Compt. rend. Acad. d. sc.*, 1838, vii. 369–402; also with 9 plates in *Mém. de l'Acad. roy. des sciences de l'Inst. de France*, 1840, xvii. 93–180.

WILLIS, THOMAS, *Diatribae duae medico-philosophicae quarum prior agit de fermentatione sive de motu intestino particularum in quovis corpore . . .*, 12°, Londini, 1659, 339 pp.; Engl. trans. by S. Pordage, fol., Lond., 1681, 178 pp.

[WÖHLER, F.], *Briefwechsel zwischen J. Berzelius und F. Wöhler*, Leipz., 1901, Bd. II., S. 72.

CHAPTER IV

APPERT, *L'art de conserver pendant plusieurs années toutes les substances animales et végétales*, Paris, 1810, 116 pp.; 2nd ed. 1811; 4th ed. 1831; Engl. trans., Lond., 1811.

ARISTOTLE, *Historia animalium*, lib. v. 1, 19, 31; vi. 15, 16; trans. by D'Arcy Thompson, Oxford, 1910.

BAKER, HENRY, *The microscope made easy*, Lond., 1744, 3rd ed., Part II.

—— *Employment for the microscope*, Lond., 1764; 2nd ed., Part II.

BARCLAY, J., *An inquiry into the opinions ancient and modern concerning life and organization*, Edinb., 1822, 542 pp.

BASTIAN, H. C., *The beginnings of life; being some account of the nature, modes of origin and transformations of lower organisms*, Lond. (Macmillan), 1872, 2 vols. Vol. i, 475 pp.; Vol. 2, 640 pp.

—— *Studies in heterogenesis*, Lond., 1903, 354 pp.

—— 'On the great importance from the point of view of medical science of the proof that bacteria and their allies are capable of arising de novo', *Lancet*, 1903, ii. 1220–4.

—— 'The evidence for the heterogenetic origin of bacteria', *Lancet*, 1904, i. 1151.

—— *The evolution of life*, Lond. [1907], 319 pp.

—— *The nature and origin of living matter*, Lond., 1905, 344 pp.; revised and abbreviated ed., Lond., 1910, 144 pp.

—— *The origin of life; being an account of experiments with certain superheated saline solutions in hermetically sealed vessels*, Lond., 1911, 76 pp.

BENNETT, J. HUGHES, 'The atmospheric germ theory', *Edinb. M. J.*, 1867–8, xiii. 810–34.

BERNARD, CLAUDE, *Compt. rend. Acad. d. sc.*, 1859, xlviii. 33.

—— *Leçons sur les propriétés physiologiques et les altérations pathologiques des liquides de l'organisme*, Paris, 1859, tome i. 488–91.

BONIFAS, *De la génération spontanée*, Paris, 1858.

BONNET, C., *Considérations sur les corps organisés*, Amsterdam, 1762, 2 vols.; 2nd ed. 1768.

BORY DE SAINT VINCENT, J. B. M. G., articles 'Microscopiques' and 'Psychodiaires' in *Encyclopédie méthodique, Histoire naturelle des zoophytes*, Paris, 1824, tome ii. 515, 657.

BRESCHET, G., article 'Déviation organique' in *Dict. de méd.*, 1823, tome vi (in table), p. 530.

BROWNE, T., *Pseudodoxia epidemica*, 1646, in Sir Thomas Browne's *Works*, ed. by Simon Wilkin, Lond., 1836, vol. ii.

BUONANNI, F., *Observationes circa Viventia quae in rebus non-viventibus reperiuntur cum micrographia curiosa; sive rerum minutissimarum observationes quae ope microscopii recognitae ad vivum exprimuntur*, Romae, 1691.

BURDACH, K. F., *Die Physiologie als Erfahrungswissenschaft*, Leipz., 1826, i.

CABANIS, P. J. G., *Rapport du physique et du moral de l'homme*, 12°, Paris, 1824, 3 vols.

CAUVET, 'Exposé des principales expériences faites au sujet des générations dites spontanées', *Rec. de mém. de méd. . . . mil.*, Paris, 1862, 3ᵉ sér., vii. 162, 261, 356, 443, 512.

CAVALERI, G. M., 'Di alcuni esperimenti intorno alla questione della generazione spontanea degli infusori', *Rendic., R. Ist. Lomb. di sc. e lett.*, Milano, 1865, ii. 331–45.

CAZENEUVE, P., 'La génération spontanée d'après les livres d'Henry Baker et de Joblot', *Rev. scient.*, Paris, 1894, 4ᵉ sér., i. 161–6.

CHAMBERLAND, CH., 'Recherches sur l'origine et le développement des organismes microscopiques', *Ann. scientif. de l'école normale supérieure*, 1878, 2ᵉ sér., vii, Suppl. 94 pp.

CHEYNE, W. WATSON, *Antiseptic Surgery, its principles, practice, history and results*, Lond., 1882, 598 pp.

COHN, F., 'Untersuchungen über Bacterien', *Beiträge zur Biologie der Pflanzen*, 1877, ii, Heft 2, 249–76.

CRACE-CALVERT, F., 'Action of heat on protoplasmic life', *Proc. Roy. Soc.*, 1871, xix. 472–6.

—— 'On protoplasmic life', *Proc. Roy. Soc.*, 1871, xix. 468–72.

CRIVELLI, G. B., and MAGGI, 'Eterogenia. Ancora di alcune esperienze con infusioni organiche, chiusa a fuoco in palloncini di vetro e scaldate a 150° centigradi', *Rendic., R. Ist. Lomb. di sc. e lett.*, Milano, 1873, 2ᵉ ser., vi. 23–6.

CUNNINGHAM, D. D., *Microscopical examinations of air*, Calcutta, 1873; also as 'Appendix A' in the *9th Annual Rep. San. Commissioner with the Gov. of India* (for 1872), Calcutta, 1873.

DELBRÜCK, M., and SCHROHE, A., *Hefe, Gärung und Fäulnis. Eine Sammlung der grundlegenden Arbeiten von Schwann, Cagniard-Latour und Kützing, sowie von Aufsätzen zur Geschichte der Theorie der Gärung und der Technologie der Gärungsgewerbe*, hrsg. v. M. Delbrück und A. Schrohe, Berl., 1904, 232 pp.

DONNÉ, A., 'Expériences sur l'altération spontanée des œufs', *Compt. rend. Acad. d. sc.*, 1863, lvii. 448.

—— 'De la génération spontanée des moisissures végétales et des animalcules', *Compt. rend. Acad. d. sc.*, 1866, lxiii. 301–5.

—— 'Sur la génération spontanée des animalcules infusoires', *Compt. rend. Acad. d. sc.*, 1866, lxiii. 1072–3.

—— 'Expériences nouvelles sur les générations spontanées', *Compt. rend. Acad. d. sc.*, 1872, lxxv. 521–3.

DÖPPING, O., and STRUVE, H., 'Versuche über Fäulniss und Gährung', *J. f. prakt. Chemie*, 1847, xli. 255–77.

DUCLAUX, E., *Pasteur: histoire d'un esprit*, 8°, Paris, 1896, 400 pp.; English trans. 1920, 383 pp.

DUJARDIN, F., *Histoire naturelle des zoophytes*, Paris, 1841, 684 pp., 20 pls.

DUMAS, *Compt. rend. Acad. d. sc.*, 1859, xlviii. 35.

EHRENBERG, C. G., *Die Infusionsthierchen als vollkommene Organismen. Ein Blick in das tiefere organische Leben der Natur*, folio, Leipz., 1838, text 547 pp.; atlas 64 plates.

FERCHAULT DE RÉAMUR, R. A., *Mémoires pour servir à l'histoire des insectes*, 1734–42, 4°, 6 tom.

FLOURENS, M. J. P., *Buffon: histoire de ses travaux et ses idées*, Paris, 1844.

—— (remarks on communication of Joly and Musset), *Compt. rend. Acad. d. sc.*, 1863, lxvii. 845.

FRAY, J. B., *Essai sur l'origine des corps organisés et inorganisés et sur quelques phénomènes de physiologie animale et végétale*, Paris, 1817.

FREDERICQ, L., 'Théodore Schwann', *Annuaire de l'Acad. roy. des sc.* *etc. de Belgique*, 1885, li. 205.

GAULTIER DE CLAUBRY, 'Note relative aux générations spontanées des végétaux et des animaux', *Compt. rend. Acad. d. sc.*, 1859, xlviii. 334–6.

VON GLEICHEN, W. F. (called Russworm), *Abhandlung über die Saamen und Infusionsthierchen und über die Erzeugung: nebst mikroskopischen Beobachtungen des Saamens der Thiere und verschiedener Infusionen*, 8°, Nürnberg, 1778, 177 pp., 32 pls.

GRUITHUISEN, F. VON PAULA, *Organozoonomie, oder ueber das niedrige Lebensverhältniss als Propadeutik zur Anthropologie; mit einem Anhange*, München, 1811, 239 pp.

—— *Beyträge zur Physiognosie und Eautognosie für Freunde der Naturforschung auf dem Erfahrungswege: von den Jahren 1809–11*, München, 1812, 447 pp., 4 pls.

GSCHEIDLEN, R., 'Ueber die Abiogenesis Huizingas', *Arch. f. d. ges. Physiol.*, 1874, ix. 163–73.

HELMHOLTZ, 'Ueber das Wesen der Fäulniss und Gährung', *Arch. f. Anat. Physiol. u. wissensch. Med.*, 1843, 453–62.

VAN HELMONT, J. B., *Ortus medicinae, id est initia physicae inaudita*, ed. of F. M. van Helmont, Amsterodami, 1652. Tract. *Imago fermenti impregnat massam semine*, chap. 9, p. 92, in Chandler's transl. *Oriatrike or Physick refined*, Lond., 1662, p. 113, par. 9.

[HILL, J.], *Lucina sine concubitu*, Lond. 1758, 48 pp.

—— *Essays in natural history and philosophy: discoveries by the microscope*, Lond., 1752.

HOFFMANN, HERMANN, 'Mykologische Studien über die Gährung', *Bot. Ztg.*, 1860, Jahrg. XVIII, 41, 49.

—— 'Neue Beobachtungen über Bacterien mit Rücksicht auf generatio spontanea', *Bot. Ztg.*, 1863, Jahrg. XXI, 304, 315.

HOOKE, R., *Micrographia: or some physiological descriptions of minute bodies made by magnifying glasses*, Lond., 1665.

HUIZINGA, 'Zur Abiogenesis-Frage', *Arch. f. d. ges. Physiol.*, 1873, vii. 549–74.

—— 'Weiteres zur Abiogenesisfrage', *Arch. f. d. ges. Physiol.*, 1874, viii. 180–9.

HUPPERT, H., 'Ueber die Urzeugung; vom physiologisch-chemischen Standpunkte aus nach den neueren Untersuchungen bearbeitet', *Schmidts Jahrb.*, 1866, cxxix. 1–29.

INGENKAMP, C., 'Die geschichtliche Entwicklung unserer Kenntniss von Fäulniss und Gährung', *Ztschr. f. klin. Med.*, 1886, x. 59–107, 1 pl.

JOBLOT, L., *Descriptions et usages de plusieurs nouveaux microscopes*, 4°, Paris, 1718, 1 vol. in 2 parts.

JOLY, N., 'Examen critique du mémoire de M. Pasteur relatif aux générations spontanées', *Mém. de l'Acad. imp. des sciences*, etc., Toulouse, 1863, 6ᵉ sér., i. 215–41.

JOLY, N., *Conférence publique sur l'hétérogénie ou génération spontanée*, Paris, 1864.

JOLY and MUSSET, 'Étude microscopique de l'air', *Compt. rend. Acad. d. sc.*, 1860, l. 647.

—— —— 'Quelques nouvelles expériences en faveur de l'hétérogénie', *Compt. rend. Acad. d. sc.*, 1862, lv. 488.

—— —— 'Études physiologiques sur l'hétérogénie', *Compt. rend. Acad. d. sc.*, 1862, lv. 490.

—— —— 'Réponse de MM. N. Joly et Ch. Musset aux observations critiques de M. Pasteur relatives aux expériences exécutées par eux dans les glaciers de la Maladetta', *Compt. rend. Acad. d. sc.*, 1863, lvii. 845.

—— —— 'Nouvelles expériences tendant à infirmer l'hypothèse de la panspermie localisée', *Compt. rend. Acad. d. sc.*, 1864, lviii. 1152.

KOCH, R., and WOLFFHÜGEL, G., 'Untersuchungen über die Desinfection mit heisser Luft', *Mitth. a. d. kaiserl. Gesundheitsamte*, 1881, i. 301–21.

KOCH, GAFFKY, and LOEFFLER, 'Versuche über die Verwendbarkeit heisser Wasserdämpfe zu Desinfectionszwecken', *Mitth. a. d. kaiserl. Gesundheitsamte*, 1881, i. 322–40.

KOCH, R., 'Ueber Desinfection', *Mitth. a. d. kaiserl. Gesundheitsamte*, 1881, i. 234–82.

KÜTZING, F., 'Microscopische Untersuchungen über die Hefe und Essigmutter nebst mehreren andern dazu gehörigen vegetabilischen Gebilden', *J. f. prakt. Chemie*, 1837, ii. 385–409.

LACAZE-DUTHIERS, 'Lettre sur la question des générations spontanées addressée à M. Milne-Edwards', *Compt. rend. Acad. d. sc.*, 1859, xlviii. 118–20.

MONET DE LAMARCK, J. B. P., *Philosophie zoologique: ou exposition des considérations relatives à l'histoire naturelle des animaux*, Paris, 1809.

[LELARGE DE LIGNAC, J. A.], *Lettres à un Ameriquain sur l'histoire naturelle générale et particulière de M. de Buffon*, Hambourg, 1751.

LISTER, J., 'On lactic fermentation', *Trans. Path. Soc.*, Lond. 1878, xxix. 425–67.

LOEWEL, H., 'Note additionelle au 3ᵉ mémoire sur la sursaturation des dissolutions salines', *Ann. chimie et phys.*, 1853, xxxvii. 179–80.

MAGGI, L., 'Sulle distinzioni introdotte nella generazione spontanea', *Rendic., R. Ist. Lomb. di sc. e lett.*, Milano, 1874, 2ᵉ ser., vii. 488–93.

MANTEGAZZA, P., 'Richerche sulla generazione degli infusorii e descrizione di alcune nuove specie', *Giorn. dell' I. R. Istituto Lombardo di scienze lettere ed arti e biblioteca italiano*, Milano, 1852, iii. 467–90, 1 pl.

—— 'Sulla generazione spontanea', *Gaz. med. ital. lomb.*, Milano, 1864, 5ᵉ ser., iii. 73.

MEUNIER, V., 'Expérience relative à la question des générations spontanées', *Compt. rend. Acad. d. sc.*, 1865, lxi. 377.
—— 'Nouvelle expérience relative à la question des générations
spontanées', *Compt. rend. Acad. d. sc.*, 1865, lxi. 449.
—— 'Expériences sur le développement de la vie dans les ballons à
cols recourbés', *Compt. rend. Acad. d. sc.*, 1865, lxi. 1060.
MILNE-EDWARDS, H., 'Remarques sur la valeur des faits qui sont con-
sidérés par quelques naturalistes comme étant propres à prouver
l'existence de la génération spontanée des animaux', *Compt. rend.
Acad. d. sc.*, Paris, 1859, xlviii. 23–9; also *Ann. des sc. nat.*
(*zoologie*), 1858, ix. 353–61.
—— *Leçons sur la physiologie et l'anatomie comparée de l'homme et des
animaux*, 1863, tome viii, leçon 71, 236–98.
MIQUEL, P., *Les organismes vivants de l'atmosphère*, Paris, 1883,
310 pp.
MOREAU DE MAUPERTUIS, P. L., *Vénus physique*, 12°, Le Haye, 1746.
MÜLLER, O. F., *Animalcula infusoria fluviatilia et marina, quae detexit,
systematice descripsit et ad vivum delineari curavit O. F. M. opus
cura O. Fabricii*, 4°, Hanniae, 1786.
MUSSET, CH., *Nouvelles recherches expérimentales sur l'hétérogénie ou
génération spontanée*, Toulouse, 1862, 44 pp., 1 pl.
NÄGELI, C., *Gattungen einzelliger Algen physiologisch und systematisch
bearbeitet*, Zürich, 1849, 139 pp., 8 plates.
NEEDHAM, TURBERVILLE, 'A summary of some late observations upon
the generation, composition, and decomposition of animal and
vegetable substances', *Phil. Trans.*, Lond. 1749, no. 490, p. 615.
—— *Nouvelles observations microscopiques, avec des découvertes intéres-
santes sur la composition et la décomposition des corps organisés*,
Paris, 1750, 524 pp.
—— *Idée sommaire ou vue générale du système physique et métaphysique
de Monsieur Needham sur la génération des corps organisés*, Bruxelles,
1776, 16 pp.
OEHL, E., and CANTONI, 'Richercho sullo sviluppo degli infusorii
considerato in sè stesso e in relazione colla loro genesi', *Ann.
Univ. di med.*, Milano, 1866, cxcvi. 32.
PASTEUR, L., 'Expériences relatives aux générations spontanées',
Compt. rend. Acad. d. sc., 1860, l. 303, 675.
—— 'De l'origine des ferments: nouvelles expériences relatives aux
générations dites spontanées', *Compt. rend. Acad. d. sc.*, 1860, l.
849–54.
—— 'Nouvelles expériences relatives aux générations dites sponta-
nées', *Compt. rend. Acad. d. sc.*, 1860, li. 348–52.
—— 'Mémoire sur les corpuscules organisés qui existent dans l'atmo-
sphère. Examen de la doctrine des générations spontanées', *Ann.
d. sc. naturelles*, 1861, xvi. 5–98; also *Ann. chimie et phys.*, 1862,
lxiv. 5–110.
—— 'Note en réponse à des observations critiques presentées à

l'Académie par MM. Pouchet, Joly et Musset dans la séance du 21 septembre dernier', *Compt. rend. Acad. d. sc.*, 1863, lvii. 724.

PASTEUR, L., 'Observations verbales à la suite de la communication de M. Davaine', *Compt. rend. Acad. d. sc.*, 1865, lxi. 526.

—— 'Note sur l'altération de l'urine à propos d'une communication du Dr. Bastian de Londres', *Compt. rend. Acad. d. sc.*, 1876, lxxxiii. 176–80.

—— 'Sur l'altération de l'urine. Réponse a M. le Dr. Bastian', *Compt. rend. Acad. d. sc.*, 1876, lxxxiii. 377–8.

PAYEN, *Compt. rend. Acad. d. sc.*, 1859, xlviii. 29–30.

PAYEN, DE QUATREFAGES, BERNARD, CL., and DUMAS, 'Observations sur la question des générations spontanées', *Ann. des sc. nat. (zoologie)*, 1858, ix. 360–6.

PENNETIER, G., *L'origine de la vie*; préface par F.-A. Pouchet, 16°, Paris, 1868, 303 pp.

POUCHET, F. A., 'Note sur des proto-organismes végétaux et animaux, nés spontanément dans l'air artificiel et dans le gaz oxygène', *Compt. rend. Acad. d. sc.*, 1858, xlvii. 979–82; also *Ann. des sc. nat. (zoologie)*, 1858, ix. 347–50.

—— 'Remarques sur les objections relatives aux proto-organismes recontrés dans l'oxygène et l'air artificiel', *Compt. rend. Acad. d. sc.*, 1859, xlviii. 148–58.

—— 'Étude des corpuscules en suspension dans l'atmosphère', *Compt. rend. Acad. d. sc.*, 1859, xlviii. 546–51.

—— *Hétérogénie ou traité de la génération spontanée, basé sur de nouvelles expériences*, Paris (Baillière), 1859, 672 pp., 3 pls.

—— 'Genèse de proto-organismes dans l'air calciné et à l'aide de corps putrescibles portés à la température de 150 degrés', *Compt. rend. Acad. d. sc.*, 1860, l. 1014–18.

—— 'Moyen de rassembler dans un espace infiniment petit tous les corpuscules normalement invisibles contenus dans un volume d'air déterminé', *Compt. rend. Acad. d. sc.*, 1860, l. 748–9.

—— Recherches sur les corps introduits par l'air dans les organes respiratoires des animaux', *Compt. rend. Acad. d. sc.*, 1860, l. 1121–7.

—— 'Observations faites sur l'air de la cime du Mont Blanc à 14,800 pieds d'altitude', *Compt. rend. Acad. d. sc.*, 1863, lvii. 765.

—— *Nouvelles expériences sur la génération spontanée et la résistance vitale*, Paris (Masson), 1864, 256 pp.

POUCHET and HOUZEAU, 'Expériences sur les générations spontanées. Développement de certains proto-organismes dans l'air artificiel', *Compt. rend. Acad. d. sc.*, 1858, xlvii. 982–4; also *Ann. des sc. nat. (zoologie)*, 1858, ix. 350–2.

POUCHET, F. A., JOLY, N., and MUSSET, CH., 'Expériences sur l'hétérogénie exécutées dans l'intérieur des glaciers de la Maladetta (Pyrénées d'Espagne)', *Compt. rend. Acad. d. sc.*, 1863, lvii. 558–61.

PUSCHKAREW, B. M., 'Ueber die Verbreitung der Süsswasserprotozoen durch die Luft', *Arch. f. Protistenk.*, 1913, xxviii. 323.

QUATREFAGES, *Compt. rend. Acad. d. sc.*, 1859, xlviii. 30–3.

'Rapport sur les expériences relatives à la génération spontanée', *Compt. rend. Acad. d. sc.*, 1865, lx. 384–97.

REDI, F., *Esperienze intorno alla generazione degl'insetti*, 4°, Firenze, 1668 (5th impression 1688); Latin trans., Amstelodami, 1671; Engl. trans. by Mab Bigelow, Chicago, 1909, 160 pp.

ROBERTS, W., 'Studies on biogenesis', *Phil. Trans.*, Lond., 1874, clxix. 457–77.

—— 'The doctrine of contagium vivum and its applications to medicine', *Brit. M. J.*, 1877, ii. 168.

ROSS, A., *Arcana microcosmi or the hid secrets of man's body discovered in an anatomical duel between Aristotle and Galen concerning the parts thereof, etc.*, 12°, Lond., 1652, 267 pp.

SAMUELSON, PAUL, 'Ueber Abiogenesis', *Arch. f. d. ges. Physol.*, 1874, viii. 277–88.

SCHRÖDER, H., and VON DUSCH, TH., 'Ueber Filtration der Luft in Beziehungen auf Fäulniss und Gährung', *Ann. d. Chem. u. Pharm.*, 1854, lxxxix. 232–43.

SCHRÖDER, H., 'Ueber Filtration der Luft in Beziehung auf Fäulniss, Gährung und Krystallisation', *Ann. d. Chem. u. Pharm.*, 1859, cix. 35–52.

—— 'Ueber Filtration der Luft in Beziehung auf Gährung, Fäulniss und Krystallisation', *Ann. d. Chem. u. Pharm.*, 1861, cxvii. 273–95.

SCHULZE, FRANZ, 'Vorläufige Mittheilung der Resultate einer experimentellen Beobachtung über Generatio aequivoca', *Ann. d. Physik u. Chemie*, Leipz., 1836, xxxix. 487–9, 1 fig.

SCHWANN, TH., *Versamml. der Naturforscher und Aerzte zu Jena*, 26 Sep. 1836; *vide* Isis, von Oken, Jahrg. 1837, Heft 5, column 524.

—— 'Vorläufige Mittheilung betreffend Versuche über die Weingährung und Fäulniss', *Ann. d. Physik u. Chemie*, 1837, xli. 184.

—— *Mikroskopische Untersuchungen über die Uebereinstimmung in der Struktur und dem Wachsthum der Thiere und Pflanzen*, 8°, Berl., 1839; Engl. trans., Lond., Sydenham Soc., 1847.

SPALLANZANI, L., *Saggio di osservazioni microscopiche concernenti il sistema della generazione dei Signori di Needham e Buffon*, Modena, 1765; reprinted in *Nuova raccolta d'opuscoli, etc.*, Venezia, 1767, xv. 208–323, and in *Opere di Lazzaro Spallanzani*, Milano, 1826, tome v. 257–360; French trans. by Regley, entitled *Nouvelles recherches sur les découvertes microscopiques*, Lond. and Paris, 1769, with notes by Needham.

—— *Lazari Spallanzani in Regio Ticinensi Gymnasio publici naturalis historiae professoris Prolusio*, Mutinae, 1770.

—— *Osservazioni e sperienze intorno agli animalucci delle infusioni*; in *Opuscoli di fisica animale, e vegetabile dell' Abate Spallanzani*, Modena, 1776, i. 3–221; partial trans. in English by Dalyell in

Tracts on the nature of animals and vegetables, Edinb., 1799, 1–69; French trans. by Senebier, Genève, 1787.

STRAUS, I., 'De la génération spontanée', *Arch. de méd. exp. et d'anat. path.*, 1889, i. 139, 329.

TERREIL, A., 'Observations sur les générations dites spontanées', *Compt. rend. Acad. d. sc.*, 1861, lii. 851–2.

TREVIRANUS, G. R., *Biologie oder Philosophie der lebenden Natur für Naturforscher und Aerzte*, Göttingen, 1803, Bd. II. 264–406.

TYNDALL, J., 'Observations on the optical deportment of the atmosphere in reference to the phenomena of putrefaction and infection', *Brit. M. J.*, 1876, i. 121.

—— 'The optical deportment of the atmosphere in relation to the phenomena of putrefaction and infection', *Phil. Trans.* (1876), Lond., 1877, clxvi. 27–74.

—— 'Further researches on the deportment and vital persistence of putrefactive and infective organisms from a physical point of view', *Phil. Trans.* (1877), Lond. 1878, clxvii. 149–206.

—— Letter to Huxley, 14 Feb. 1877, *Proc. Roy. Soc.*, 1877, xxv. 569–70.

—— 'On Schulze's mode of intercepting the germinal matter of the air', *Proc. Roy. Soc.*, Lond., 1878, xxvii. 99–100.

—— *Floating matter of the air in relation to putrefaction and infection*, Lond., 1881, 338 pp.

VALLISNERI, A., *Dialoghi del Signor Dottore A. Valsinieri [sic] sopra la curiosa origine di molti insetti*, Venezia, 1700.

—— *Considerazioni ed esperienze intorno alla generazione de vermi ordinari del corpo umano*, Padova, 1710, 160 pp.

—— 'Della curiosa origine degli sviluppi e de costumi ammirabili di molti insetti', in *Opere fisico-mediche*, Venice, 1733, tomo i.

VOLTAIRE, *Dictionnaire philosophique* (article 'Dieu'), 1764; *Histoire de Jenni* (chap. iv); *L'homme aux quarante écus*, 1768 (chap. vi); *La défense de mon oncle*, 1769 (chap. xix); *Singularités de la Nature*, 1769 (chap. xx).

WYMAN, JEFFRIES, 'Experiments on the formation of infusoria in boiled solutions of organic matter enclosed in hermetically sealed vessels and supplied with pure air', *Amer. J. of Science and Arts*, 1862, 2nd ser., xxxiv. 79–87.

—— 'Observations and experiments on living organisms in heated water', *Amer. J. of Science and Arts*, 1867, xliv. 152–69.

QUESTION OF BACTERIA IN FLUIDS AND TISSUES OF HEALTHY ANIMALS AND PLANTS

AMAKO, T., 'Untersuchungen über das Conradi'sche Ölbad und den Bacteriengehalt der Organe gesunder Tiere', *Ztschr. f. Hyg. u. Infektionskr.*, 1910, lxvi. 166–76.

BERNHEIM, H., 'Die parasitären Bacterien der Cerealien', *München. med. Wchnschr.*, 1888, xxxv. 743, 767.

BIEROTTE, E., and MACHIDA, 'Untersuchungen über Keimgehalt normaler Organe', *München. med. Wchnschr.*, 1910, lvii. 636–7.

BILLROTH, TH., *Untersuchungen über die Vegetationsformen von Coccobacteria septica, etc.*, Berl., 1874, 244 pp., 5 plates.

VAN DEN BROEK, J. H., 'Onderzoek aangaaende het agens, waardoor in versch. druivensap, benevens in druivensap, waarin door koking alle kiemen of cellen verniegtigd zijn, de alkoholische gisting intreedt', *Anteekeningen van det Verhandelde in de sectie voor Natuuren-Geneeskunde van het Provinciaal Utrechtsche Genootschap van Kunsten en Wetenschappen*, 1858–9, Utrecht, 1859, pp. 30–62.

—— 'Onderzoek aangaande de alkoholische gisting van druivensap', *Anteekeningen van het Verhandelde in de sectie voor Natuur- en-Geneeskunde van het Provinciaal Utrechtsche Genootschap van Kunsten en Wetenschappen*, Utrecht, 1859, pp. 42–53.

—— 'Untersuchungen über die geistige Gährung des Traubensaftes und über die Fäulniss thierischer Substanzen in frischem Zustande: Versuche über das in der Atmosphäre enthaltene Agens, welches diese beiden Zersetzungen einleitet', *Ann. d. Chem. u. Pharm.*, 1860, cxv. 75–86.

CARRIÈRE, G., and VANVERTS, J., 'Études sur les lésions produites par la ligature expérimentale des vaisseaux de la rate', *Arch. de méd. expér. et d'anat. path.*, 1899, xi. 498–520.

CAZENEUVE, P., and LIVON, CH., 'Nouvelles recherches sur la fermentation ammoniacale de l'urine et la génération spontanée', *Compt. rend. Acad. d. sc.*, 1877, lxxxv. 571–4.

CHAUVEAU, A., 'Nécrobiose et gangrène. Étude expérimentale sur les phénomènes de mortification qui se passent dans l'organisme animal vivant', *Compt. rend. Acad. d. sc.*, 1873, lxxvi, 1092–6.

CHIENE, J., and EWART, J. C., 'Do bacteria or their germs exist in the organs of healthy living animals?' *J. Anat. and Physiol.*, 1878, xii. 448–53.

CONRADI, H., 'Ueber den Keimgehalt normaler Organe', *München. med. Wchnschr.*, 1909, lvi. 1318–20; also *Centralbl. f. Bakteriol.*, 1909, Ref. xliv, Beiheft, p. 139.

FERNBACH, A., 'De l'absence des microbes dans les tissus végétaux', *Ann. de l'Inst. Past.*, 1888, ii. 567–70.

VON FODOR, J., 'Bacterien im Blute lebender Thiere', *Arch. f. Hyg.*, 1886, iv. 129–48.

FORD, W. W., 'On the bacteriology of normal organs', *J. Hyg.*, Camb., 1901, i. 277–84.

GALIPPE, V., 'Note sur la présence de micro-organismes dans les tissus végétaux', *Compt. rend. Soc. de biol.*, Paris, 1887, 8e sér., iv. 410, 557.

—— 'Présence de micro-organismes dans les végétaux', *Compt. rend. Soc. de biol.*, 1890, 9e sér., ii. 85–7.

GAYON, U., 'Sur l'altération spontanée des œufs', *Compt. rend. Acad. d. sc.*, 1873, lxxvi. 232.

GAYON, U., 'Sur les altérations spontanées des œufs', *Compt. rend. Acad. d. sc.*, 1873, lxxvii. 213.

HAUSER, G., 'Ueber das Vorkommen von Mikroorganismen im lebenden Gewebe gesunder Thiere', *Arch. f. exp. Path. u. Pharmacol.*, 1886, xx. 162–202.

HENSEN, 'Bemerkungen zu dem Aufsatz "Ueber Abstammung und Entwickelung von Bacterium termo",' *Arch. f. mikroskop. Anat.*, 1867, iii. 342–4.

KOCH, ROBERT, *Untersuchungen über die Aetiologie der Wundinfectionskrankheiten*, 8°, Leipz., 1878, 80 pp., 5 pls.; Engl. trans., London, New Sydenham Soc., 1880.

KOUKOL-YASNOPOLSKY, W., 'Ueber die Fermentation der Leber und Bildung von Indol', *Arch. f. d. ges. Physiol.*, 1876, xii. 78–86.

LISTER, J., 'A further contribution to the natural history of bacteria and the germ theory of putrefactive changes', *Quart. J. Microscop. Sc.*, 1873, N.S., xiii. 380–408, 3 pls.

—— 'A contribution to the germ theory of putrefaction and other fermentable changes and to the natural history of torulae and bacteria', *Trans. Roy. Soc.*, Edinb., 1875, xxvii. 313–44.

—— 'On the nature of fermentation', *Quart. J. Microscop. Sc.*, 1878, xviii. 177–94.

LÜDERS, JOHANNA, 'Ueber Abstammung und Entwickelung des Bacterium termo Duj. = Vibrio lineola Ehrb.', *Arch. f. mikroskop. Anat.*, 1867, iii. 317–41, 1 pl.

MORGAN, H. DE R., 'The presence of bacteria in the organs of healthy animals', *Lancet*, 1904, ii. 21.

MOTT, F. W., and HORSLEY, V., 'On the existence of bacteria or their antecedents in healthy tissues', *J. Physiol.*, Camb., 1880–2, iii. 188–94, 1 pl.

NENCKI, M., and GIACOSA, P., 'Giebt es Bacterien oder deren Keime in den Organen gesunder lebender Thiere', *J. f. prakt. Chemie*, 1879, N.F. xx. 34–44.

PASCHUTIN, V., 'Einige Versuche über Fäulniss und Fäulnissorganismen', *Arch. f. path. Anat.*, 1874, lix. 490–510.

PASTEUR, L., 'Examen du rôle attribué au gaz oxygène atmosphérique dans la destruction des matières animales et végétales après la mort', *Compt. rend. Acad. d. sc.*, 1863, lvi. 734–40.

—— 'Observations verbales présentées après la lecture de la note de M. Donné', *Compt. rend Acad. d. sc.*, 1866, lxiii. 305–8.

—— *Études sur la bière, ses maladies, causes qui les provoquent, procédé pour la rendre inaltérable, avec une théorie nouvelle de la fermentation*, Paris, 1876, 387 pp.

RINDFLEISCH. 'Untersuchungen über niedere Organismen', *Arch. f. path. Anat.*, 1872, liv. 396–407.

ROBERTS, W., 'Studies on biogenesis', *Phil. Trans.*, Lond., 1874, clxiv. 457–77.

ROSENBACH, J., 'Ueber einige fundamentale Fragen in der Lehre von den chirurgischen Infectionskrankheiten', *Deutsche Zeitschr. f. Chir.*, 1880, xiii. 344–64.

SANDERSON, J. B., 'Further report of researches concerning the intimate pathology of contagion', *Rep. Med. Off. Privy Council*, 1870, Lond., 1871, xiii. 48–69; 1874, N.S. iii. 11.

SCHWANN, TH., 'Vorläufige Mittheilung betreffend Versuche über die Weingährung und Fäulniss', *Ann. d. Physik u. Chemie*, 1837, xli. 184.

SELTER, HUGO, 'Bakterien im gesunden Körpergewebe und deren Eintrittspforten', *Ztschr. f. Hyg.*, 1906, liv. 363–84.

SERVEL, 'Sur la naissance et l'évolution des bactéries dans les tissus organiques mis à l'abri du contact de l'air', *Compt. rend. Acad. d. sc.*, 1874, lxxix. 1270.

TIEGEL, E., 'Ueber Coccobacteria septica (Bilbroth) im gesunden Wirbelthierkörper', *Arch. f. path. Anat.*, 1874, lx. 453–70.

WATSON CHEYNE, W., 'On the relation of organisms to antiseptic dressings', *Trans. Path. Soc.*, Lond., 1879, xxx. 557–82.

—— *Antiseptic surgery, its principles, practice, history and results*, Lond., 1882, 616 pp.

WRZOSEK, A., 'Experimentelle Beiträge zur Lehre von dem latenten Mikrobismus', *Virchow's Arch. f. path. Anat.*, 1904, clxxviii. 82–111.

ZAHN, F. W., 'Untersuchungen über das Vorkommen von Fäulnisskeimen im Blut gesunder Thiere', *Arch. f. path. Anat.*, 1884, xcv. 401–7.

ZWEIFEL, P., 'Untersuchungen über die wissenschaftliche Grundlage der Antisepsis und die Entstehung des septischen Gifts', *Ztschr. f. physiol. Chemie*, 1882, vi. 386–421.

—— 'Gibt es im gesunden, lebenden Organismus Fäulnisskeime?' *Tagebl. d. Versamml. deutsch. Naturf. u. Aerzte*, Strassb., 1885, lvii. 303–5.

CHAPTER V

VON BERGMANN, E., *Das putride Gift und die putride Intoxication*, 1. Abt., 1. Lief., 8°, Dorpat, 1868, 63 pp.

VON BERGMANN, E., and SCHMIEDEBERG, O., 'Ueber das schwefelsäure Sepsin (das Gift faulender Substanzen)', *Centralbl. f. d. med. Wissensch.*, 1868, vi. 497–8.

VON BERGMANN, E., 'Zur Lehre von der putriden Intoxication', *Deutsche Ztschr. f. Chir.*, 1872, i. 373–98.

BLUMBERG, C., 'Experimenteller Beitrag zur Kenntniss der putriden Intoxication', *Arch. f. path. Anat.*, etc., 1885, c. 377–415.

BRIEGER, L., 'Einige Beziehungen der Fäulnissprodukte zu Krankheiten', *Ztschr. f. klin. Med.*, 1881, iii. 465–90.

—— 'Ueber die Bedeutung der Fäulnissalcaloide', *Verhandl. d. Cong. f. innere Med.*, Wiesb., 1883, ii. 277.

—— *Ueber Ptomaine*, Berl., 1885, 80 pp.

—— *Weitere Untersuchungen über Ptomaine*, Berl., 1885, 83 pp.

—— *Untersuchungen über Ptomaine*, Dritter Theil, Berl., 1886, 119 pp.

BROWN, A. M., *A treatise on the animal alkaloids, cadaveric and vital, or the ptomaines and leucomaines chemically, physiologically and pathologically considered in relation to scientific medicine*, Lond., 1887, 2nd ed., 1889.

CHISHOLM, C., 'An essay towards an inquiry how far the effluvia from dead animal bodies passing through the natural process of putrefaction are efficient in the production of malignant pestilential fever . . .', *Edinb. M. and S. J.*, 1810, vi. 389–420.

CLEMENTI, G., and THIN, G., 'Untersuchungen über die putride Infection', *Med. Jahrb.*, Wien, 1873, 292–303.

DANCE, 'De la phlébite utérine et de la phlébite en général considérées principalement sous le rapport de leurs causes et de leurs complications', *Arch. gén. de méd.*, 1828, xviii. 473–525; 1829, xix. 1–52, 163–202.

DUPUY, 'Injection de matière putride dans la veine jugulaire d'un cheval', *Arch. gén. de méd.*, 1826, xi. 297–8.

EVANS, W. J., *A clinical treatise on the endemic fevers of the West Indies*, Lond., 1837, 309 pp.

FAUST, E. S., 'Ueber das Fäulnisgift Sepsin', *Arch. f. exper. Path. u. Pharmakol.*, 1904, li. 248–69.

FORNET, W., and HEUBNER, W., 'Versuche über die Entstehung des Sepsins', *Arch. f. exper. Path. u. Pharmakol.*, 1908, Suppl.-Bd. 176–180, 1 pl.

GASPARD, B., 'Mémoire physiologique sur les maladies purulentes et putrides, sur le vaccin, etc.', *J. de physiol. expér.*, Paris, 1822, ii. 1–45.

—— 'Second mémoire physiologique et médical sur les maladies putrides', *J. de physiol. expér. et path.*, 1824, iv. 1–69.

GAUTIER, A., 'Les alcaloïdes dérivés des matières protéiques sous l'influence de la vie des ferments et des tissus', *J. de l'anat. et de physiol. norm. et path.*, 1881, xvii. 333–63.

—— 'Sur la découverte des alcaloïdes dérivés des matières protéiques animales', *Compt. rend. Acad. d. sc.*, 1882, xciv. 1119–22.

—— *Les toxines microbiennes et animales*, Paris, 1896, 617 pp.

GUSSENBAUER, C., *Sephthämie, Pyohämie und Pyo-sephthämie. Deutsche Chirurgie*, hrsg. v. Billroth u. Luecke, Stuttg., 1882, Lief. 4, 293 pp.

HEMMER, M., *Experimentelle Studien über die Wirkung faulender Stoffe auf den thierischen Organismus: gekrönte Preisschrift*, München, 1866, 170 pp.

HILLER, A., 'Ueber putrides Gift', *Centralbl. f. Chir.*, 1876, iii. 145, 177.

—— *Die Lehre von der Fäulniss auf physiologischer Grundlage*, Berl., 1879, 547 pp.

HUEPPE, 'Ueber Beziehung der Fäulniss zu den Infectionskrankheiten', *Med. chir. Centralbl.*, Wien, 1888, xxiii. 25, 73, 85, 97, 109, 157.

KEHRER, F. A., 'Ueber das putride Gift', *Arch. f. exper. Path. u. Pharmakol.*, 1874, ii. 33–61, 4 tab.

LEURET, F., *Sur l'altération du sang*, Paris, 1826, 19 pp.

MAGENDIE, 'Remarques sur la notice précédente avec quelques expériences sur les effets des substances en putréfaction', *J. de physiol. expér. et path.*, 1823, iii. 81–8.

MEYER, J., 'Impfversuche mit dem Blute und den Ausleerungen Cholerakranker', *Arch. f. path. Anat.*, etc., 1852, iv. 29–54.

MORGAGNI, J. B., *De sedibus et causis morborum per anatomen indagatis*, Venetiis, 1761, lib. iv, epist. li, par. 16–40.

NENCKI, M., 'Zur Geschichte der basischen Fäulnissprodukte', *J. f. prakt. Chem.*, 1882, N.F., xxvi. 47–52.

PANUM, P. L., *Bidrag til Laeren om den saakaldte putride eller septiske Infection*. Bibliothek for Laeger, Kjøbenh., 1856, 4. Raekke, Bind viii, 253–85, Abstr. in Schmidts *Jahrb.* 1859, ci. 213–17.

—— 'Experimentelle Beiträge zur Lehre von der Embolie', *Arch. f. path. Anat.*, etc., 1862, xxv. 308, 433.

—— 'Das putride Gift, die Bakterien, die putride Infection oder Intoxication und die Septicämie', *Arch. f. path. Anat.*, etc., 1874, lx. 301–52.

—— 'Studier over Göring och Förrådnelse mit sårligt Hensyn til de mikroskopiske Organismers Andel i Fermentvirkningerne', *Nord. med. Ark.*, Stockholm, 1876, viii, no. 25, 1–35; 1878, x. no. 4, 1–53.

PASTEUR, L., 'Recherches sur la putréfaction', *Compt. rend. Acad. d. sc.*, 1863, lvi. 1189–94.

PIORRY, P.-A., *Traité de diagnostic et de séméiologie*, Bruxelles, 1837, 2ᵉ éd., 638 pp.

PRINGLE, J., 'Experiments and observations upon septic and antiseptic substances', in his *Obser. on the Diseases of the Army*, etc., 8°, Lond., 1752, 365–431; 2nd ed. 1753, 307–403; 3rd ed. 1761, 314–411; 4th ed. 1764; 5th ed. 1765; 7th ed. 1775.

RAVITSCH, J., *Zur Lehre von der putriden Infection und deren Beziehung zum sogennanten Milzbrande. Experimentelle und microscopische Untersuchungen*, Berl., 1872, 118 pp.

ROSENBACH, J., 'Gibt es verschiedene Arten der Fäulniss?' *Deutsche Ztschr. f. Chir.*, 1881–2, xvi. 342–68.

ROSENBERGER, J. A., *Ueber das Wesen des septischen Giftes*, Leipz., 1882; also *Festschr. zur dritten Saecularfeier der Alma Julia Maximiliana*, Würzb., 1882, Bd. i. 233–55.

SALOMONSEN, C. J., *Studier over Blodets Forraadnelse*, Kjøbenh., 1877, 171 pp., 3 pls.

—— 'Bidrag til Lären om den putride Forgiftning', *Nord. med. Ark.* Stockholm, 1881, xiii, No. 9, 1–42, 1 pl.

SAMUEL, S., 'Ueber die Wirkung des Fäulnissprocesses auf den lebenden Organismus', *Arch. f. exper. Path. u. Pharmakol.*, 1873, i. 317–55.

SANDERSON, J. B., 'Report of a further investigation of the properties of the septic ferment', *Rep. Med. Off. Privy Council*, Lond., 1876, N.S., no. viii, 11–22.

—— 'On the preparation and properties of the septic extract of muscle', *Practitioner*, Lond., 1877, xix. 19–36.

SCHWANN, TH., 'Vorläufige Mittheilung betreffend Versuche über die Weingährung und Fäulniss', *Ann. d. Physik u. Chemie*, 1837, xli. 184.

SCHWENINGER, F., 'Ueber die Wirkung faulender Substanzen auf den lebenden thierischen Organismus', *Aerztl. Int.-Bl.*, München, 1866, xiii. 590, 612, 623, 654, 668; also reprint, München, 1866, 31 pp.

SELMI, F., 'Sulla esistenza di principii alcaloidi naturali nei visceri freschi e putrefatti onde il perito chimico può essere condotto a conclusioni eronee nella ricerca degli alcaloidi venefici', *Mem. Accad. d. Sc. d. Ist. di Bologna*, 1872, 3ᵉ ser., ii. 81–6.

SELMI, F. [*et al.*], 'Sugli alcaloidi dei cadaveri', *Bull. d. sc. med. di Bologna*, 1876, 5ᵉ ser., xxii. 256–69.

SEMMER, E., 'Putride Intoxication und septische Infection, metastatische Abscesse und Pyämie', *Arch. f. path. Anat.*, etc., 1881, lxxxiii. 99–116.

STICH, A., 'Die acute Wirkung putrider Stoffe im Blute', *Ann. d. Charité-Krankenhaus . . . zu Berlin*, 1853, iii., 2. Heft, 192–250.

THIERSCH, C., *Infectionsversuche an Thieren mit dem Inhalte des Choleradarmes*, München, 1855–6, 86 pp.

TROUSSEAU and DUPUY, 'Expériences et observations sur les altérations du sang considerées comme causes ou comme complications des maladies locales', *Arch. gén. de méd.*, 1826, xi. 373–95.

VAUGHAN, V. C., and NOVY, F. A., *Ptomaines and leucomaines or the putrefactive and physiological alkaloids*, Philad., 1888, 314 pp.; 2nd ed. (with altered title) 1891, 391 pp.

VIRCHOW, R., 'Thrombose und Embolie. Gefässentzündung und septische Infection', in his *Gesammelte Abhandl. z. wissensch. Med.*, Frankf. a. M., 1856, pp. 219–732.

WEBER, O., 'Experimentelle Studien über Pyämie, Septicämie und Fieber', *Deutsche Klinik*, 1864, xvi. 461, 473, 485, 493; 1865, xvii. 13, 21, 33, 41, 53, 61, 69.

ZUELZER, W., 'Studien über die putride Intoxikation', *Arch. f. exper. Path. u. Pharmakol.*, 1878, viii. 133–9.

ZUELZER, W., and SONNENSCHEIN, 'Ueber das Vorkommen eines Alkaloids in putriden Flüssigkeiten', *Berl. klin. Wchnschr.*, 1869, vi. 121–2.

CHAPTER VI

D'ARCET, F., *Recherches sur les abcès multiples et sur les accidents qu'amène la présence du pus dans le système vasculaire, suivies de quelques remarques sur les altérations du sang*, Paris, thèse, 1842.

ARNOTT, J. A., 'Pathological enquiry into the secondary effects of inflammation of the veins', *Med.-Chir. Tr.*, Lond., 1829, xv. 1–131.

BEHIER, 'Discussion sur la septicémie', *Bull. Acad. de méd.*, Paris, 1873, 2ᵉ sér., ii. 147–87.

BÉRARD, P. H., Article 'Pus, Pyogénie', *Dict. de méd.*, Paris, 1842, 2ᵉ éd., xxvi. 411–504.

BIRCH-HIRSCHFELD, 'Untersuchungen über Pyämie', *Arch. d. Heilk.*, 1873, xiv. 193–240, 1 pl.

BONNET, 'Mémoire sur la composition et l'absorption du pus', *Gaz. méd. de Paris*, 1837, 2ᵉ sér., v. 593–602.

BOULEY, 'Sur la septicémie', *Bull. Acad. de méd.*, Paris, 1872, 2ᵉ sér., i. 922–33.

BOYER, A., 'Mémoire sur les résorptions purulentes', *Gaz. méd. de Paris*, 1834, 2ᵉ sér., ii. 193–6.

DE CASTELNAU, H., and DUCREST, F. M., 'Rechercher les cas dans lesquels on observe les abcès multiples et comparer ces cas sous leurs différents rapports', *Mém. Acad. roy. de méd.*, Paris, 1846, xii. 1–151.

CHAUVEL, J., Article 'Septicémie', in *Dict. encyclop. d. sc. méd.*, Paris, 1880, 3ᵉ sér., viii. 700–76; 1881, ix. 1–228.

CHEYNE, W. W., *Antiseptic surgery, its principles, practice, history, and results*, Lond., 1882, 616 pp.

—— 'Lectures on suppuration and septic diseases', *Brit. M. J.*, 1888, i. 404, 452, 524.

DE CHRISTMAS, J., 'Recherches expérimentales sur la suppuration', *Ann. de l'Inst. Pasteur*, 1888, ii. 469–78.

CLEMENTI, G., and THIN, G., 'Untersuchungen über die putride Infection', *Med. Jahrb.*, Wien, 1873, 292–303.

COLIN, [G.], 'Nouvelles recherches sur l'action des matières putrides et sur la septicémie', *Bull. Acad. de méd.*, Paris, 1873, 2ᵉ sér., ii. 1175, 1208, 1240.

—— 'De la diversité des effets produits par les matières septiques, suivant leur degré d'altération', *Bull. Acad. de méd.*, Paris, 1878, 2ᵉ sér., vii. 1139–58.

COUNCILMAN, W. T., 'Zur Aetiologie der Eiterung', *Arch. f. path. Anat.*, etc., 1883, xcii. 217–20.

COZE, L., and FELTZ, V., *Recherches cliniques et expérimentales sur les maladies infectieuses, étude spécialement au point de vue de l'état du sang et de la présence des ferments*, Paris, 1872, 324 pp., 6 pls.

CRUVEILHIER, J., *Traité d'anatomie pathologique générale*, Paris, 1862, tome iv, 455.

DANCE, 'De la phlébite utérine et de la phlébite en général considérées principalement sous le rapport de leurs causes et de leurs complications', *Arch. gén. de méd.*, 1828, xviii. 473; 1829, xix. 5, 161.

DAVAINE, 'Recherches sur quelques questions relatives à la septicémie', *Bull. Acad. de méd.*, Paris, 1872, 2ᵉ sér., i. 907, 976.

DAVAINE, 'Réponse à M. Colin sur ses communications relatives à la septicémie', *Bull. Acad. de méd.*, Paris, 1873, 2ᵉ sér., ii. 1272–1281.

—— 'Recherches sur quelques-unes des conditions qui favorisent ou qui empêchent le développement de la septicémie', *Bull. Acad. de méd.*, Paris, 1879, 2ᵉ sér., viii. 121–38.

'Discussion sur l'infection purulente', *Bull. Acad. de méd.*, Paris, 1869, xxxiv. 345, 379; 1871, xxxvi. 150, 182, 202, 227, 249, 272, 282, 307, 334, 376, 428, 450, 486, 620, 664, 702, 736, 768, 786, 805, 835. Résumé of above discussion in *Gaz. d. hôp. Par.*, 1871, xliv. 509, 521, 533.

'Discussion sur la septicémie', *Bull. Acad. de méd.*, Paris, 1873, 2ᵉ sér., ii. 55, 84, 124, 147, 300, 423, 451, 460, 487, 515, 1058.

DONNÉ, A., *Recherches . . . sur les globules du sang, du pus, du mucus, et sur ceux des humeurs de l'œil*, Paris, 1831.

—— 'Mémoire sur les caractères distinctifs du pus et les moyens de reconnaître la présence de ce liquide dans les différens fluides auxquels il se trouve mélangé, particulièrement dans le sang; suivi d'expériences nouvelles relatives à l'action du pus sur le sang', *Arch. gén. de méd.*, 1836, 2ᵉ sér., xi. 443–70.

DREYER, U., 'Ueber die zunehmende Virulenz des septischen Giftes (Davaine)', *Arch. f. exper. Path. u. Pharmakol.*, 1874, ii. 149–82.

DUNCAN, J. MATTHEWS, 'An address on the treatment of puerperal fever', *Lancet*, 1880, ii. 683, 721.

FELTZ, V., 'Recherches expérimentales sur la pathogénie des infarctus et le processus inflammatoire dans la septicémie', *J. de l'anat. et physiol.*, etc., Paris, 1873, ix. 355–68.

—— 'Sur la septicémie expérimentale', *Compt. rend. Acad. d. sc.*, 1874, lxxix. 1268–70.

—— 'Expériences démontrant que la septicité du sang putréfié ne tient pas à un ferment soluble', *Compt. rend. Acad. d. sc.*, 1877, lxxxiv. 789–91.

—— 'Expériences démontrant que la septicité du sang putréfié tient aux ferments figurés', *Compt. rend. Acad. d. sc.*, 1877, lxxxiv. 953–5.

FRANK, G., 'Über Sepsis (Synonyme: Septische Erkrankungen, Pyämie und Septikämie)', *Ergebn. d. allg. Path. u. path. Anat.*, 1902, Wiesb., 1904, viii. 472–508.

GAFFKY, G., 'Experimentell erzeugte Septicämie mit Rücksicht auf progressive Virulenz und accomodative Züchtung', *Mitth. a. d. k. Gesundheitsamte*, 1881, i. 80–133.

GRAWITZ, P., 'Beitrag zur Theorie der Eiterung', *Arch. f. path. Anat.*, etc., 1889, cxvi. 116–53.

GRAWITZ, P., and DE BARY, W., 'Ueber die Ursachen der subcutanen Entzündung und Eiterung', *Arch. f. path. Anat.*, etc., 1887, cviii. 67–102.

GUSSENBAUER, C., *Sephthämie, Pyohämie und Pyo-sephthämie*, Deutsche

Chirurgie, hrsg. v. Billroth u. Luecke, Stuttg., 1882, Lief. 4, 293 pp.

HAUSER, G., *Ueber Fäulnissbakterien und deren Beziehungen zur Septicämie, ein Beitrag zur Morphologie der Spaltpilze*, 8°, Leipz., 1885.

HODGSON, J., *A treatise on the diseases of arteries and veins containing the pathology and treatment of aneurisms and wounded arteries*, Lond., 1815, 603 pp.

HUETER, C., 'Die septikämischen und pyämischen Fieber', *Handb. d. allg. u. spec. Chir.*, 8°, Erlang., 1869, i, 2 Abt., A, Abschn. ii, no. 1, 1–127.

HUNTER, J., 'Observations on the inflammation of the internal coats of veins', *Tr. Soc. Improve. M. and Chir. Knowl. 1783–92*, Lond., 1793, i. 18–29.

JANOWSKI, W., 'Ueber die Ursachen der acuten Eiterungen', *Beitr. z. path. Anat. u. z. allg. Path.*, 1889, vi. 227–76.

KLEBS, E., 'Die Ursache der infectiösen Wundkrankheiten', *Correspondenzbl. f. schweiz. Aerzte*, 1871, i. 241–6.

—— *Beiträge zur pathologischen Anatomie der Schusswunden. Nach Beobachtungen in den Kriegslazarethen in Carlsruhe, 1870 und 1871*, Leipz., 1872, 132 pp., 10 pls.

KLEMPERER, G., 'Ueber die Beziehung der Mikroorganismen zur Eiterung', *Ztschr. f. klin. Med.*, 1886, x. 158–92.

KOCH, R., *Untersuchungen über die Aetiologie der Wundinfectionskrankheiten*, 8°, Leipz., 1878, 80 pp., 5 pls., Engl. trans., New Sydenham Soc., Lond., 1880.

KOCHER, T., 'Zur Aetiologie der acuten Entzündungen', *Arch. f. klin. Chir.*, 1879, xxiii. 101–16.

KREIBOHM and ROSENBACH, 'Experimentelle Beiträge zur Frage: Kann Eiterung ohne Mitbetheiligung von Mikroorganismen durch todte Stoffe entstehen?' *Arch. f. klin. Chir.*, 1888, xxxvii. 737–44.

LABORDE, J. V., 'Recherches sur la septicémie expérimentale à l'aide d'un procédé nouveau de transmission de la maladie et qui permet l'étude sur l'organisme même de l'action des divers agents réputés antiseptiques', *Compt. rend. Soc. de biol.*, 1874, Paris, 1875, 6ᵉ sér., i. 49–58.

LENHARTZ, H., *Die septischen Erkrankungen*, Wien, 1903, 553 pp., 13 pls.

LEUBE, W., 'Zur Diagnose der "spontanen" Septicopyämie', *Deutsch. Arch. f. klin. Med.*, 1878, xxii. 235–78.

MOXON, W., and GOODHART, J. F., 'Observations on the presence of bacteria in the blood and inflammatory products of septic fever and on the "cultivation" of septicaemia', *Guy's Hosp. Rep.*, Lond., 1875, 3rd ser., xx. 229–60.

OGSTON, A., 'Ueber Abscesse', *Arch. f. klin. Chir.*, 1880, xxv. 588–600, 1 pl.

—— 'Report upon micro-organisms in surgical diseases', *Brit. M. J.*, 1881, i. 369–74 (Corrigenda on p. 453).

OGSTON, A., 'Micrococcus poisoning', *J. Anat. and Physiol.*, 1882, xvi. 526–67; 1883, xvii. 24–58.

ONIMUS, 'Contribution à l'étude de la septicémie', *Gaz. hebd. de méd.*, Paris, 1873, 2e sér., x. 410–16.

ORTHMANN, E. G., 'Ueber die Ursache der Eiterbildung', *Arch. f. path. Anat.* etc., 1882, xc. 549–59.

PASSET, 'Ueber Mikroorganismen der eiterigen Zellgewebsentzündung des Menschen', *Fortschr. d. Med.*, 1885, iii. 33, 68, 1 pl.

PIORRY, P. A., *Traité de diagnostic et de séméiologie*, Bruxelles, 1837, 2e éd., 638 pp.

'Report of the Committee appointed by the Pathological Society of London to investigate the nature and causes of those infective diseases known as Pyaemia, Septicaemia, and purulent infection', *Tr. Path. Soc.*, Lond., 1879, xxx. 1–118.

RIBES, F., 'Exposé succinct des recherches faites sur la phlébite', *Rev. méd. franç. et étrang.*, Paris, 1825, iii. 5–41.

ROSENBACH, F. J., *Mikro-Organismen bei den Wund-Infections-Krankheiten des Menschen*, Wiesb., 1884, 122 pp., 5 pls.

ROSENBERGER, J. A., 'Experimentelle Studien über Septikämie', *Centralbl. f. d. med. Wissensch.*, 1882, xx. 65, 515.

ROSER, W., 'Die specifische Natur der Pyämie', *Arch. d. Heilk.*, 1860, i. 39–50.

—— 'Zur Pyämie-Frage', *Arch. d. Heilk.*, 1862, ii. 368; 1863, iv. 92.

—— 'Zur Verständigung über den Pyämiebegriff', *Arch. d. Heilk.*, 1867, viii. 15–24.

SANDERSON, J. B., 'Preparations showing the results of certain experimental inquiries relating to the nature of the infective agent in pyaemia', *Tr. Path. Soc.*, Lond., 1871–2, xxiii. 303–8.

SAVORY, W. S., 'On pyaemia', *Lancet*, 1867, i. 74, 108, 139, 201.

SCHEURLEN, E., 'Die Entstehung und Erzeugung der Eiterung durch chemische Reizmittel', *Arch. f. klin. Chir.*, 1885, xxxii. 501–10.

—— 'Weitere Untersuchungen über die Entstehung der Eiterung; ihr Verhältniss zu den Ptomainen und zur Blutgerinnung', *Arch. f. klin. Chir.*, 1887. xxxvi. 925–33.

SCHÜLLER, M., 'Experimentelle Beiträge zum Studium der septischen Infection', *Deutsche Ztschr. f. Chir.*, 1876, vi. 113–90.

SÉDILLOT, C. E., *De l'infection purulente ou pyoémie*, Paris, 1849, 518 pp.

[SEMMELWEIS], 'Höchst wichtige Erfahrungen über die Aetiologie der in Gebäranstalten epidemischen Puerperalfieber', *Ztschr. d. k.-k. Gesellsch. d. Aerzte zu Wien*, 1847, Jahrg. IV, 242; 1849, Jahrg. V, Bd. i. 64.

SEMMER, E., 'Putride Intoxication und septische Infection, metastatische Abscesse und Pyämie', *Arch. f. path. Anat.*, etc., 1881, lxxxiii. 99–116.

STEINHAUS, J., *Die Aetiologie der acuten Eiterungen. Litterarisch-kritische, experimentelle und klinische Studien*, Leipz., 1889, 184 pp.

STRAUS, I., 'Du rôle des micro-organismes dans la production de la suppuration', *Compt. rend. Soc. de biol.*, 1884, 7ᵉ sér., v. 651–7.

THIN, G., and CLEMENTI, G., 'Studii sperimentali sulla infezione putrida', *Morgagni*, Napoli, 1873, xv. 673–99; also trans., *Edinb. M. J.*, 1873, xix. 40–53.

USKOFF, N., 'Giebt es Eiterung unabhängig von niederen Organismen?' *Arch. f. path. Anat.*, etc., 1881, lxxxvi. 150–9.

VELPEAU, A., *Sur quelques propositions de médecine*, Paris, 1823. Thèse.

—— 'De l'infection purulente, aperçu historique sur cette maladie; observation remarquable', *Gaz. d. hôp. Par.*, 1842, xv. 254, 285.

VIRCHOW, R., 'Embolie und Infection', in his *Ges. Abhandl. z. wissenschaftl. Med.*, Frank.-a.-M., 1856, 636 pp.

VULPIAN [*et al.*], 'Contributions à l'étude de la septicémie', *Compt. rend. Soc. de biol.*, 1872, Paris, 1874, 5ᵉ sér., iv, pt. 2, 49–59.

WEIGERT, C., 'Bismarckbraun als Färbemittel', *Arch. f. mikr. Anat.*, 1878, xv. 258–60.

WOLFF, 'Zur Bakterienlehre bei accidentellen Wundkrankheiten', *Arch. f. path. Anat.*, etc., 1880, lxxxi. 193, 385.

WUNDERLICH, C. A., 'Ueber spontane und primäre Pyämie', *Arch. f. physiol. Heilk.*, Stuttg., 1857, N.F., i. 89–121.

CHAPTER VII

ACERBI, F. ENRICO, *Dottrina teorico-pratica del morbo petecchiale con nuove ricerche intorno l'origine, l'indole, le cagioni predisponenti ed effettrici, la cura e la preservazione del morbo medesimo in particolare e degli altri contagi in generale*, Milano, 1822, 483 pp., 8 tav.

AUDOUIN, 'Recherches anatomiques et physiologiques sur la maladie contagieuse qui attaque les vers à soie et qu'on désigne sous le nom de muscardine', *Compt. rend. Acad. d. sc.*, 1836, iii. 82–9.

BALSAMO-CRIVELLI, G., 'Sovra la nuova specie di Mucedinea del genre Botrytis che se svolge sovra i bachi da seta e le crisalidi morte da calcino . . .', *Bibl. Italiana*, Milano, 1835, lxxix. 125–9.

—— 'Sopra l'origine e lo sviluppo della Botrytis bassiana e sopra una specie di mucorino anch' esso parassito', *Bibl. Ital.*, Milano, 1838, xc. 367–70.

BALY, W., and GULL, W. W., *Report on the nature and import of certain microscopic bodies found in the intestinal discharges of cholera*, Lond., 1849, 28 pp., 19 figs.

BASSI, A., *Del mal del segno calcinaccio o moscardino, malattia che affligge i bachi da seta e sul modo di liberarne le bigattaje anche le più infestate*, Lodi, 1835, Parti I, Teorica; 1836, Parti II, Pratica; 2nd ed., Milano, 1837.

—— *Memoria . . . in addizione alla di lui opera sul calcino*, Milano, 1837, 2nd ed.

BASSI, A., *Sui contagi in generale e specialmente su quelli che affligono l'umana specie*, Lodi, 1844.

—— *Discorsi sulla natura e cura della pellagra*, Milano, 1846.

—— *Dei parassiti generatori dei contagi e rispettivi rimedi*, Lodi, 1851.

—— *Della natura dei morbi ossia mali contagiosi e del modo di prevenirli e curarli*, Lodi, 1853.

—— *Opere di Agostino Bassi*, Pavia, 1925, 673 pp.

BAYLE, G. L., *Recherches sur la phthisie pulmonaire*, Paris, 1810.

BENNETT, J. H., 'On the parasitic vegetable structures found growing in living animals', *Trans. Roy. Soc.*, Edinb., 1844, xv. 277–94.

BERG, F. T., 'Torsk i mikroskopiskt anatomiskt hänseende', *Hygiea*, Stockh., 1841, iii. 541–50.

BICHAT, X., *Traité des membranes en général et de diverses membranes en particulier*, nouv. édit. par Husson, Paris, An XI (1802), 300 pp.

BOISSIER DE SAUVAGES, F., *Nouvelles classes de maladies, qui dans un ordre semblable à celui des botanistes comprennent les genres et les espèces de toutes les maladies avec leurs signes et leurs indications*, 12°, Avignon, [1731].

—— *Nosologia methodica, sistens morborum classes juxta Sydenhami mentem et botanicorum ordinem*, Editio ultima auctior, 4°, Amstelod., 1768.

BRETONNEAU, P., *Des inflammations spéciales du tissu muqueux et en particulier de la diphthérite ou inflammation pelliculaire connue sous le nom de croup, d'angine maligne, d'angine gangréneuse*, etc., Paris, 1826, 540 pp., 3 pls.

—— 'Notice sur la contagion de la dothinentérie', *Arch. gén. de méd.*, 1829, xxi. 57–78.

—— *Traités de la dothinentérie et de la spécificité publiés pour la première fois d'après les manuscrits originaux avec un avant-propos et des notes de L. Dubreuil-Chambardel*, Paris (Vigot), 1922, 368 pp.

BRITTAN, F., 'Report of a series of microscopical investigations on the pathology of cholera', *Lond. M. Gaz.*, 1849, N.S., ix. 530–42.

BROUSSAIS, F. J. V., *Examen de la doctrine médicale généralement adoptée et des systèmes modernes de nosologie*, 8°, Paris, 1816.

—— *Histoire des phlegmasies ou inflammations chroniques*, 3rd ed., Paris, 1822, 3 tom.

BUDD, W., *Malignant cholera, its mode of propagation and its prevention*, Lond., 1849, 30 pp.

BUEHLMANN, F., 'Ueber eine eigenthümliche auf den Zähnen des Menschen vorkommende Substanz', *Arch. f. Anat., Physiol. u. wissenschaftl. Med.*, 1840, 442–5.

CLEMENS, TH., 'Physiologische Reflexionen und Untersuchungen über Miasma und Contagium', *Arch. f. physiol. Heilk.*, Stuttg., 1853, xii. 281–308; 1854, xiii. 39–60.

COQUERELLE, J., *Bretonneau (1778–1862). La doctrine spécifique, ses origines et son évolution*, Paris, 1892–3. Thèse no. 31.

CULLEN, W., *Synopsis nosologiae methodicae*, 8°, Edinb., 1769, 303 pp.

DAVAINE, 'Sur des animalcules infusoires trouvés dans les selles de malades atteints du choléra et d'autres affections', *Compt. rend. Soc. de biol.*, 1854, 2ᵉ sér., i. 129.

DECHAMBRE, A., 'Spécificité', article in *Dict. encyclop. d. sc. méd.*, Paris, 1881, 3ᵉ sér., x. 800–7.

DONNÉ, AL., *Recherches microscopiques sur la nature des mucus et la matière des divers écoulemens des organes génito-urinaires chez l'homme et chez la femme: description des nouveaux animalcules découverts dans quelques-uns de ces fluides: observations sur un nouveau mode de traitement de la blenorrhagie*, Paris, 1837, 70 pp., 1 pl.

—— *Cours de microscopie complémentaire des études médicales: anatomie microscopique et physiologique des fluides de l'économie*, Paris, 1844, Atlas 1845.

EICHSTEDT, 'Pilzbildung in der Pityriasis Versicolor', *Notiz. a. d. Geb. d. Nat. u. Heilk.*, Weimar, 1846, xxxix. 270.

FABER, K., *Nosography in modern internal medicine*, New York, 1923, 222 pp.

FILDES, P., and McINTOSH, J., 'The aetiology of influenza', *Brit. J. Exper. Path.*, 1920, i. 119–26.

FISCHER, I., 'Marc Anton Plenciz', *Wiener med. Wchnschr.*, 1927, lxxvii. 735–6.

GOODSIR, J., 'History of a case in which a fluid periodically ejected from the stomach contained vegetable organisms of an undescribed form', *Edinb. M. and S. J.*, 1842, lvii. 430–43.

GRUBY, 'Recherches anatomiques sur une plante cryptogame qui constitue le vrai muguet des enfants', *Compt. rend. Acad. d. sc.*, 1842, xiv. 634–6.

—— 'Recherches sur la nature, le siège et le développement du porrigo decalvans ou phytalopécie', *Compt. rend. Acad. d. sc.*, 1843, xvii. 301–3.

—— 'Recherches sur les cryptogames qui constituent la maladie contagieuse du cuir chevelu décrit sous le nom de teigne tondante (Mahon) Herpes tonsurans (Cazenave)', *Compt. rend. Acad. d. sc.*, 1844, xviii. 583–5.

HENLE, *Pathologische Untersuchungen*, Berl., 1840, 274 pp. Reprint of section 'Von den Miasmen und Contagionen . . .', Einleitung von F. Marchand, Leipz., 1910, 38 pp

HOLLAND, H., *Medical Notes and Reflections*, Lond., 1839, chap. xxxiv, p. 560.

KLENCKE, H., *Untersuchungen und Erfahrungen im Gebiete der Anatomie, Physiologie, Mikrologie und wissenschaftlichen Medicin*, Leipz., 1843, Bd. I, p. 123.

KOCH, R., *Untersuchungen über die Aetiologie der Wundinfectionskrankheiten*, Leipz., 1878, p. 27.

—— 'Die Aetiologie der Tuberkulose', *Mitth. a. d. k. Gesundheitsamte*, 1884, ii. 3.

316   BIBLIOGRAPHY

LAENNEC, R. T. H., *De l'auscultation médiate* . . ., Paris, 1819, tome i, chap. 2, pp. 19–40.

V. LINNÉ, C., *Genera morborum in auditorum usum*, Upsaliae, 1763, 32 pp.

LOUIS, P. C. A., *Recherches anatomiques . . . sur la maladie connue sous le noms de fièvre typhoide, putride . . . gastro-entérite* . . ., Paris, 1829, 2 vols.

MALMSTEN, P. H., 'Trichophyton tonsurans', *Hygiea*, Stockh., 1845, vii. 325, 483, 1 pl.; trans., *Arch. f. Anat. u. Physiol.*, 1848, p. 1, Pl. 1, Figs. 1–3.

MARTEN, BENJ., *A new theory of consumptions; more especially of a phthisis or consumption of the lungs*, Lond., 1720, 2nd ed., 1722, 154 pp.

MAYER, 'Beobachtung von Cysten mit Fadenpilzen aus dem äussern Gehörgange eines Mädchens', *Arch. f. Anat., Physiol. u. wissensch. Med.*, 1844, 404–8.

MITCHELL, J. K., *On the cryptogamous origin of malarious and epidemic fevers*, Philad., 1849, 137 pp.

MONTAGNE, 'Expériences et observations sur le champignon entomoctone ou histoire botanique de la muscardine', *Compt. rend. Acad. d. sc.*, 1836, iii. 166–70.

MÜLLER, J., and RETZIUS, A., 'Ueber parasitische Bildungen', *Arch. f. Anat., Physiol. u. wissensch. Med.*, 1842, 193–212.

NEALE, ADAM, *Researches to establish the truth of the Linnaean doctrine of animate contagions; wherein the origin, causes, mode of diffusion, and cure of epidemic diseases, spasmodic cholera, dysentery, plague, small-pox, hooping-cough, etc., are illustrated by facts from the natural history of mankind of animals and of vegetables and from the phenomena of the atmosphere*, Lond., 1831, 258 pp.

PACINI, F., 'Osservazioni microscopiche e deduzioni pathologiche sul cholera asiatico', *Gaz. med. ital. fed. toscana*, Firenze, 1854, 2ᵉ sér., iv. 397, 405; reprint 1854.

—— *Sulla causa specifica del colera asiatico, il suo processo pathologico e la indicazione curative che ne resulta*, Firenze, 1865, 62 pp.; French trans. by Janssens, Bruxelles, 1865, 42 pp.

PINEL, P., *Nosographie philosophique, ou la méthode de l'analyse appliquée à la médecine*, 2ᵉ éd., Paris, An XI (1802), 3 vols.; 6ᵉ éd., 1818.

PLENCIZ, M. A., *Opera medico physica in quatuor tractatus digesta quorum primus contagii morborum una cum additamento de lue bovina. Anno 1761 epidemice grassante sistit* . . ., Vindobonae, 1762.

POUCHET, 'Infusoires dans les déjections des cholériques', *Compt. rend. Acad. d. sc.*, 1849, xxviii. 555–6.

REMAK, 'Zur Kenntniss von der pflanzlichen Natur der Porrigo lupinosa', *Med. Ztg. Berl.*, 1840, ix, no. 16, pp. 73–4.

RIQUIER, G. C., *Agostino Bassi e la sua opera*, Pavia, 1924, 63 pp.

ROBIN, CH., *Histoire naturelle des végétaux parasites qui croissent sur l'homme et sur les animaux vivants*, Paris, 1853, 702 pp., Atlas, 15 pls.

ROLLESTON, J. D., 'Bretonneau: his life and work', *Proc. Roy. Soc. Med.*, Sect. History of Med., 1924, xviii. 1–12.

SAGAR, J. B. M., *Systema morborum symptomaticum secundum classes, ordines Genera et species cum characteribus, differentiis et therapeiis*, Viennae, 1776.

SCHÖNLEIN, J. L., 'Zur Pathogenie der Impetigines', *Arch. f. Anat. u. Physiol.*, 1839, p. 82 (Plate III, fig. 5).

—— *Allgemeine und specielle Pathologie und Therapie*, St. Gallen, 1839, 4th ed., Theil I, p. 5.

SLUYTER, TH., *De vegetabilibus organismi animalis parasitis ac de novo epiphyto in pityriasi versicolore obvio*, Berl., [1847], 32 pp., 1 pl.

SNOW, JOHN, 'On the pathology and mode of Communication of ·Cholera', *Lond. Med. Gaz.*, 1849, N.S. ix. 745, 923.

—— *On the mode of communication of Cholera*, Lond., 1849, 31 pp.; 2nd ed., 1855, 162 pp.

SWAYNE, J. G., 'An account of certain organic cells peculiar to the evacuations of cholera', *Lancet*, 1849, ii. 368, 398, 410.

TROUSSEAU, A., 'De la maladie à laquelle M. Bretonneau . . . a donné le nom de dothinentérie ou dothinentérite', *Arch. gén. de méd.*, Paris, 1826, x. 67, 169.

—— 'Spécificité', Leçon 31 in his *Clinique médicale de l'Hôtel-Dieu de Paris*, 2ᵉ éd., 1865, tome i, 464–81.

VIRCHOW, R., 'Sarcine', *Arch. f. path. Anat.*, etc., 1847, i. 264–71.

VITTADINI, C., 'Della natura del calcino o mal del segno', *Mem. dell. I.R. Istituto Lombardo*, 1852, iii. 447–512, 2 pls.

WAGNER, R., 'Fragmente zur Physiologie der Zeugung vorzüglich zur mikroskopischen Analyse des Spermas', *Abhandl. d. math.-phys. Kl. d. K. Bayer. Akad. d. Wissensch.* (1831–6), München, 1837, ii. 381–417.

CHAPTER VIII

BEALE, LIONEL S., *Disease germs, their supposed nature*, Lond., 1870[1], 82 pp., 4 pls.

—— *Disease germs: their real nature*, London, 1870[2], 176 pp., 20 pls.

—— *Disease germs, their nature and origin*, 2nd ed., Lond., 1872, 472 pp., 28 pls.

BÉCHAMP, A., 'Sur les microzymas géologiques', *Compt. rend. Acad. d. sc.*, 1881, xcii. 1291.

—— 'Du rôle et de l'origine de certains microzymas', *Compt. rend. Acad. d. sc.*, 1881, xcii. 1344–7.

—— 'Les microzymas au point de vue physiologique et pathologique', *Trans. Internat. M. Congr.*, 7 sess. London, 1881, i. 352–74.

—— 'Les microzymas sont-ils des organismes vivants? Exposition d'une théorie expérimentale de l'antisepticité', *Bull. Acad. de méd.*, Paris, 1882, 2ᵉ sér., xi. 497–547.

—— 'Les microzymas et les zymases', *Arch. de physiol. norm. et path.*, Paris, 1882, 2ᵉ sér., x. 28–62.

BÉCHAMP, A., *Les microzymas dans leurs rapports avec l'hétérogénie, l'histogénie, la physiologie et la pathologie. Examen de la panspermie atmosphérique continue ou discontinue, morbifère ou non morbifère*, Paris (Baillière), 1883, 992 pp., 5 pls.

—— 'Sur la théorie du microzyma et le système microbien', *Gaz. méd. de Par.*, 1886, 7ᵉ sér., iii. 353, 365, 385, 409, 445, 509, 577, 601, 613.

—— *Le Sang et son troisième élément anatomique*, Paris, 1899, 248 pp.; Engl. trans., Lond., 1912, 424 pp.

BERGEY'S *Manual of determinative bacteriology, a key for the identification of organisms of the class Schizomycetes*, arranged by a committee of the Society of American Bacteriologists, Baltimore, 1923, 442 pp.

BIEDERT, PH., 'Beitrag zur Frage nach der Constanz der Spaltpilze (Kokkobacillus zymogenus und Bacterium termo)', *Arch. f. path. Anat.*, 1885, c. 439–59.

BILLROTH, TH., *Untersuchungen über die Vegetationsformen von Coccobacteria septica und den Antheil, welchen sie an der Entstehung und Verbreitung der accidentellen Wundkrankheiten haben*, Berlin, 1874, 244 pp., 5 pls.

BILLROTH, TH., and EHRLICH, F., 'Untersuchungen über Coccobacteria septica', *Arch. f. klin. Chir.*, 1877, xx. 403–33, 1 pl.

BRAUELL, 'Versuche und Untersuchungen betreffend den Milzbrand des Menschen und der Thiere', *Arch. f. path. Anat.*, &c., 1857, xi. 132–44.

—— 'Weitere Mittheilungen über Milzbrand und Milzbrandblut', *Arch. f. path. Anat.*, etc., 1858, xiv. 432–66.

BUCHANAN, R. E., 'Studies in the nomenclature and classification of bacteria', *J. Bact.*, Balt., 1916, i. 591–6; 1917, ii. 347–50; 1918, iii. 27, 301.

BUCHNER, HANS, *Die Naegeli'sche Theorie der Infectionskrankheiten in ihren Beziehungen zur medicinischen Erfahrung*, Leipz., 1877, 112 pp.

—— 'Ueber die experimentelle Erzeugung des Milzbrandcontagiums aus den Heupilzen', in Naegeli, *Untersuchungen über niedere Pilze*, München and Leipz., 1882, 140–77.

—— 'Kritisches und Experimentelles über die Frage der Constanz der pathogenen Spaltpilze', in Naegeli's *Untersuchungen über niedere Pilze*, München and Leipz., 1882, 231–85.

—— 'Kurze Uebersicht über die Entwickelung der Bacterienforschung seit Naegeli's Eingreifen in dieselbe', *München. med. Wchnschr.*, 1891, xxxviii. 435, 454.

BUHL, L., 'Einiges über Diphtherie', *Ztschr. f. Biol.*, München, 1867, iii. 341.

CHAUVEAU, A., 'Nature du virus vaccin. Détermination expérimentale des éléments qui constituent le principe actif de la sérosité vaccinale virulente', *Compt. rend. Acad. d. sc.*, 1868[1], lxvi. 289–93.

CHAUVEAU, A., 'Nature du virus vaccin. Nouvelle démonstration de l'inactivité du plasma de la sérosité vaccinale virulente', *Compt. rend. Acad. d. sc.*, 1868², lxvi. 317–20.

—— 'Nature des virus. Détermination expérimentale des éléments qui constituent le principe virulent dans le pus varioleux et le pus morveux', *Compt. rend. Acad. d. sc.*, 1868³, lxvi. 359–63.

—— 'Théorie de la contagion médiate ou miasmatique appelée encore infection. De la méthode à suivre pour la détermination des conditions qui rendent les milieux infectieux', *Compt. rend. Acad. d. sc.*, 1868⁴, lxvii. 696–700.

—— 'Théorie de la contagion médiate ou miasmatique appelée encore infection. Détermination expérimentale des conditions qui donnent aux sujets contagifères la propriété d'infecter les milieux', *Compt. rend. Acad. d. sc.*, 1868⁵, lxvii. 746–53.

—— 'Théorie de la contagion médiate ou miasmatique appelée encore infection. Des voies par lesquelles s'opère l'infection des sujets sains exposés à la contagion', *Compt. rend. Acad. d. sc.*, 1868⁶, lxvii. 898–903.

—— 'Isolement des corpuscules solides qui constituent les agents spécifiques des humeurs virulentes; démonstration directe de l'activité de ces corpuscules', *Compt. rend. Acad. d. sc.*, 1869, lxviii. 828–9.

CIENKOWSKI, L., *Zur Morphologie der Bacterien*, Mém. de l'Acad. impér. d. sc. de St. Pétersbourg, 1877, 7ᵉ sér., xxv, no. 2, 18 pp., 2 pls.

COHN, F., 'Untersuchungen über die Entwickelungsgeschichte der mikroskopischen Algen und Pilze', *Novorum actorum Acad. Caes. Leop.-Carol. Nat. curiosorum*, Vratislaviae et Bonnae, 1854, xxiv. 101–256, 6 pls.

—— *Ueber Bacterien, die kleinsten lebenden Wesen.* Samml. gemeinverständl. wissenschaftl. Vorträge, hrsg. v. R. Virchow u. Fr. v. Holtzendorff, Berl., 1866, no. 165.

—— 'Ueber den Brunnenfaden (Crenothrix polyspora) mit Bemerkungen über die mikroskopische Analyse des Brunnenwassers', *Beitr. z. Biol. d. Pflanzen*, 1870, Bd. I, Heft. 1, 108–131, 1 pl.

—— 'Organismen in der Pockenlymphe', *Arch. f. path. Anat.*, etc., 1872¹, lv. 229–38.

—— 'Ueber Pilze und Contagien', *Jahresber. d. schlesischen Gesellsch. f. vaterländ. Cultur* (1871), Breslau, 1872², xlix. 191–4.

—— 'Untersuchungen über Bacterien', *Beitr. z. Biol. d. Pflanzen*, Breslau, 1872³, Bd. I, Heft 2, 126–224, 1 pl.

—— 'Untersuchungen über Bacterien. II', *Beitr. z. Biol. d. Pflanzen*, 1875, Bd. I, Heft 3, 141–207, 2 pls.

—— 'Untersuchungen über Bacterien. IV. Beiträge zur Biologie der Bacillen', *Beitr. z. Biol. d. Pflanzen*, 1876, Bd. II, Heft 2, 248–76.

—— 'Ueber die morphologische Einheit der Spaltpilze und über Naegeli's Anpassungs-theorie', *Deutsche med. Wchnschr.*, 1879, v. 73–6.

DAVAINE, C., 'Recherches sur les infusoires du sang dans la maladie connue sous le nom de sang de rate', *Compt. rend. Acad. d. sc.*, 1863[1], lvii. 220, 351, 386.

—— 'Nouvelles recherches sur la maladie du sang de rate considérée au point de vue de sa nature', *Compt. rend. Soc. de biol.*, 1863[2], Paris, 1864, 3e sér., v, pt. 2, 193–202.

—— 'Nouvelles recherches sur les infusoires du sang dans la maladie connue sous le nom de sang de rate', *Compt. rend. Soc. de biol.*, 1863[3], Paris, 1864, 3e sér., v. 149–52.

—— 'Recherches sur une maladie septique de la vache regardée comme de nature charbonneuse', *Compt. rend. Acad. d. sc.*, 1865, lxi. 368–70.

—— 'Expériences sur une maladie septique de la vache regardée à tort comme de nature charbonneuse', *Compt. rend. Soc. de biol.*, 1865, Paris, 1866, 4e sér., ii. 152–3.

—— 'Expériences relatives à la durée de l'incubation des maladies charbonneuses et à la quantité de virus nécessaire à la transmission de la maladie', *Bull. Acad. de méd.*, 1868, xxxiii. 816–21.

—— 'Études sur la contagion du charbon chez les animaux domestiques', *Bull. Acad. de méd.*, 1870[1], xxxv. 215–35.

—— 'Études sur la genèse et la propagation du charbon', *Bull. Acad. de méd.*, 1870[2], xxxv. 471–98.

—— (sur la découverte des bactéridies), *Bull. Acad. de méd.*, Paris, 1875, 2e sér., iv. 581–4.

—— *L'Œuvre de C.-J. Davaine*, Paris, 1889, 864 pp.

DE BARY, A., 'Bericht über die in den Cholera-Ausleerungen vorgefundenen Pilze', *Virchow-Hirsch Jahresber. über die Leistungen und Fortschritte in der ges. Med.*, 1867, Jahrg. II, Bd. II, Abt. I, 240–52; also *Bot. Ztg.*, 1868, Jahrg. XXVI, 686, 713, 736, 761, 787.

—— *Vergleichende Morphologie und Biologie der Pilze, Mycetozoen und Bacterien*, Leipz., 1884, 558 pp.; Engl. trans., Oxford, 1887, 525 pp.

DELAFOND, *Recueil de méd. vét.*, Paris, 1860, 4e sér., vii. 726–52.

DE-TONI, J. B., and TREVISAN, V., *Sylloge Schizomycetum*, Patavii, 1889; reprint from Sacardo's *Sylloge Fungorum*, viii. 923–1090.

DUJARDIN, F., *Histoire naturelle des zoophytes. Infusoires, comprenant la physiologie et la classification de ces animaux et la manière de les étudier à l'aide du microscope*, Paris, 1841, 684 pp., 22 pls.

DUNBAR, W. P., *Zur Frage der Stellung der Bakterien, Hefen und Schimmelpilze in System: die Entstehung von Bakterien, Hefen und Schimmelpilzen aus Algenzellen*, München u. Berlin, 1907, 60 pp., 4 pls.

EBERTH, C. J., *Zur Kenntnis der bacteritischen Mykosen*, 4°, Leipz., 1872, 28 pp., 1 pl.

—— 'Ueber diphtheritische Endocarditis', *Arch. f. path. Anat.*, etc., 1873, lvii. 228–37; also *Untersuch. a. d. pathol. Instit. zu Zürich*, 1873, Heft 1, 95–9.

EBERTH, C. J., 'Mycotische Endocarditis', *Arch. f. path. Anat.*, etc., 1875, lxv. 352–8.

—— 'Mycotische Endocarditis', *Arch. f. path. Anat.*, etc., 1878, lxxii. 103–7.

EHRENBERG, C. G., *Die Infusionsthierchen als vollkommene Organismen: ein Blick in das tiefere organische Leben der Natur*, Leipz., 1838, text 547 pp., Atlas, 1838, 64 pls.

EIDAM, E., 'Untersuchungen über Bacterien. III. Beiträge zur Biologie der Bacterien', *Beitr. z. Biol. d. Pflanzen*, 1875, Bd. I, Heft 3, 208–24.

FISCHER, ALFRED, 'Untersuchungen über Bakterien', *Jahrb. f. wissenschaftl. Botanik*, Berl., 1895, xxvii. 1–163, 5 pls.

FLÜGGE, C., 'Sind die von Dr. Zopf in seinem Handbuch über die Spaltpilze gelehrten Anschauungen vereinbar mit den Ergebnissen der neueren Forschungen über Infectionskrankheiten?' *Deutsche med. Wchnschr.*, 1884, x. 741–3.

GERBER and BIRCH-HIRSCHFELD, 'Ueber einen Fall von Endocarditis ulcerosa und das Vorkommen von Bakterien bei dieser Krankheit', *Arch. d. Heilk.*, 1876, xvii. 208–31.

GILKINET,. ALFRED, *Mémoire sur le pléomorphisme des champignons*, Mémoires couronnés . . . par l'Acad. roy. d. sc. . . . de Belgique, Bruxelles, 1875 (collection in 8°), tome xxvi, pp. 1–121.

HALLIER, E., 'Ueber einen pflanzlichen Parasiten auf dem Epithelium bei Diphtheritis', *Bot. Ztg.*, 1865[1], xxiii. 144–6.

—— 'Ueber Leptothrix buccalis', *Bot. Ztg.*, 1865[2], xxiii. 181–3.

—— *Das Cholera-Contagium. Botanische Untersuchungen Aerzten und Naturforschern mitgetheilt*, Leipz., 1867[1], 400 pp., 1 pl.

—— *Gährungserscheinungen. Untersuchungen über Gährung, Fäulniss und Verwesung mit Berücksichtigung der Miasmen und Contagien sowie der Desinfection*, Leipz., 1867[2], 116 pp., 1 pl.

—— *Parasitologische Untersuchungen bezüglich auf die pflanzlichen Organismen bei Masern, Hungertyphus, Darmtyphus, Blattern, Kuhpocken, Schafpocken, Cholera nostras*, etc., Leipz., 1868[1], 80 pp., 2 pls.

—— 'Researches into the nature of vegetable parasitic organisms', *Med. Times and Gazette*, Lond., 1868[2], ii. 222–3.

—— 'Ueber die Parasiten der Ruhr', *Ztschr. f. Parasitenkunde*, Jena, 1869[1], i. 71–5.

—— 'Die Parasiten der Infectionskrankheiten', *Ztschr. f. Parasitenkunde*, Jena, 1869[2], i. 117, 291; 1870, ii. 67, 113; 1872, iii. 7, 157; 1873, iv. 56.

—— 'Beweis dass der Micrococcus der Infectionskrankheiten keimfähig und von höheren Pilzformen abhängig ist und Widerlegung der leichtsinnigen Angriffe des Herrn Collegen Bary zu Halle', *Ztschr. f. Parasitenkunde*, Jena, 1870, ii. 1–20.

—— 'Beweis dass der Cryptococcus keimfähig und von höheren Pilzformen abhängig ist und Widerlegung der Ansichten der

Bary'schen Schule über die Bierhefe', *Ztschr. f. Parasitenkunde*, Jena, 1872, iii. 217–44.

HALLIER, E., *Die Parasiten der Infectionskrankheiten bei Menschen, Thieren und Pflanzen*, Buch I. Die Plastiden der niederen Pflanzen, 1878.

HEIBERG, H., 'Ein Fall von Endocarditis ulcerosa puerperalis mit Pilzbildung im Herzen (Mycosis endocardii)', *Arch. f. path. Anat.*, etc., 1872, lvi. 407–15.

HUEPPE, F., *Die Formen der Bakterien und ihre Beziehungen zu den Gattungen und Arten*, Wiesbaden, 1886, 152 pp.

HUETER, C., 'Pilzsporen in den Geweben und im Blut bei Gangraena diphtheritica', *Centralbl. f. d. med. Wissensch.*, 1868, vi. 177.

HUME, E. DOUGLAS, *Béchamp or Pasteur? A lost chapter in the history of biology*, Chicago—London, 1923, 280 pp.

HUXLEY, 'On the relations of Penicillium, Torula, and Bacterium', *Quart. J. Microscop. Sc.*, 1870, N.S., x. 355–62.

JENSEN, ORLA, 'Die Hauptlinien des natürlichen Bacteriensystems', *Centralbl. f. Bakteriol.*, 2 Abt., 1908, xxii. 97, 305.

—— 'Vorschlag zu einer neuen bakteriologischen Nomenclatur', *Centralbl. f. Bakteriol.*, 2 Abt., 1909, xxiv. 477–80.

KELLY, H. A., 'Jules Lemaire the first to recognize the true nature of wound infection and inflammation, and the first to use carbolic acid in medicine and surgery', *J. Am. M. Ass.*, 1901, xxxvi. 1083–8.

KLEBS, E., 'Beiträge zur Kenntniss der Micrococcen', *Arch. f. exp. Path. u. Pharmakol.*, 1873, i. 31–64.

—— 'Beiträge zur Kenntniss der pathogenen Schistomyceten', *Arch. f. exp. Path. u. Pharmakol.*, 1875, iv. 107, 207, 409.

KLOB, J. M., *Pathologisch-anatomische Studien über das Wesen der Cholera-Processes*, Leipz., 1867, 82 pp., 1 pl.

KÜTZING, F., 'Microscopische Untersuchungen über die Hefe und Essigmutter nebst mehreren andern dazu gehörigen vegetabilischen Gebilden', *J. f. prakt. Chemie*, 1837, xi. 385–409, 2 pls.

LANKESTER, E. RAY, 'On a peach-coloured bacterium—Bacterium rubescens, N.S.', *Quart. J. Microscop. Sc.*, 1873, N.S., xiii. 408–25.

—— 'Further observations on a peach- or red-coloured bacterium—Bacterium rubescens', *Quart. J. Microscop. Sc.*, 1876[1], N.S., xvi. 27–40, 1 pl.

—— 'Note on Bacterium rubescens and Clathrocystis roseopersicina', *Quart. J. Microscop. Sc.*, 1876[2], N.S., xvi. 278–83.

—— 'The pleomorphism of the Schizophyta', *Quart. J. Microscop. Sc.*, 1885–6, N.S., xxvi. 499–505.

LEHMANN, K. B., and NEUMANN, R. O., *Atlas und Grundriss der Bakteriologie und Lehrbuch der speciellen bakteriologischen Diagnostik*, München, 1896, 7. Aufl. 1926–7.

LEMAIRE, J., *Du Coaltar saponiné, désinfectant énergique, arrêtant les fermentations*, Paris, 1860, 92 pp.

LEMAIRE, J., *De l'acide phénique, de son action sur les végétaux, les animaux, les ferments, les venins, les virus, les miasmes et de ses applications à l'industrie, à l'hygiène, aux sciences anatomiques et à la thérapeutique*, Paris, 1865, 2ᶜ éd., 754 pp.

LEPLAT and JAILLARD, 'De l'action des bactéries sur l'économie animale', *Compt. rend. Acad. d. sc.*, 1864, lix. 250–2.

—— —— 'Note au sujet d'expériences prouvant que le charbon de la vache, inoculé aux lapins, les tue avec les phénomènes du sang de rate sans que leur sang contienne aucune trace de bactéridies', *Compt. rend. Acad. d. sc.*, 1865¹, lxi. 298–301.

—— —— 'Nouvelles expériences pour démontrer que les bactéridies ne sont pas le cause de sang de rate', *Compt. rend. Acad. d. sc.*, 1865², lxi. 436–40.

LETZERICH, L., 'Beiträge zur Kenntniss der Diphtheritis', *Arch. f. path. Anat.*, etc., 1869, xlv. 327.

—— 'Die Entwickelung des Diphtheriepilzes', *Arch. f. path. Anat.*, etc., 1873, lviii. 303.

LEWIS, T. R., *The microscopic organisms found in the blood of man and animals and their relation to disease*, 4°, Calcutta, 1879, 90 pp. 3 pls.

LEYDEN, E., and JAFFE, M., 'Ueber putride (fötide) Sputa nebst einigen Bemerkungen über Lungenbrand und putride Bronchitis', *Deutsch. Arch. f. klin. Med.*, 1867, ii. 488–519.

LISTER, J., 'A further contribution to the natural history of bacteria and the germ theory of fermentative changes', *Quart. J. Microscop. Sc.*, 1873¹, N.S., xiii. 380–408, 3 pls.

—— 'A contribution to the germ theory of putrefaction and other fermentative changes and to the natural history of torulae and bacteria', *Tr. Roy. Soc. Edinb.* (1872–6), 1876, xxvii. 313–44, 5 pls.

LÖHNIS, F., *Studies upon the life-cycles of the bacteria*, Mem. of the National Acad. of Sc. Washington, 1921, xvi, mem. 2.

LÜDERS, JOH[ANNA], 'Ueber Abstammung und Entwickelung des Bacterium termo Duj. = Vibrio lineola Ehrb.', *Arch. f. mikr. Anat.*, 1867, iii. 317–41, 1 pl.

METCHNIKOFF, E., 'Contributions à l'étude du pléomorphisme des bactériens', *Ann. de l'Inst. Pasteur*, 1888, iii. 61–8, 1 pl.

MIGULA, W., *System der Bakterien. Handbuch der Morphologie, Entwickelungsgeschichte und Systematik der Bakterien*, Jena, Bd. I, 1897, 368 pp.; B. II, 1900, 1,068 pp.

MÜLLER, O. F., *Vermium terrestrium et fluviatilium seu animalium infusoriorum, helminthicorum et testaceorum non marinorum succincta historia*, Havniae et Lipsiae, 1773, vol. i; 1774, vol. ii.

—— *Animalcula infusoria et marina systematice descripsit et ad vivum delineari curavit O.F.M. opus cura O. Fabricii*, 4°, Hauniae, 1786, 367 pp., 50 pls.

MÜLLER, REINER, '80 Jahre Seuchenbakteriologie. Die Seuchenbakteriologen vor Robert Koch: Pollender 1849, Brauell 1856, Delafond 1856, Davaine 1863. Ein 50-Jahr-Gedenken des Milzbrandforschers Pollender', *Zentralbl. f. Bakteriol.*, etc., 1 Abt. 1929, cxv. orig. 1–17 (portrait of Pollender).

VON NÄGELI, C., *Gattungen einzelliger Algen physiologisch und systematisch bearbeitet*, 4°, Zürich, 1849, 139 pp., 8 pls.

—— 'Ueber die neue Krankheit der Seidenraupe und verwandte Organismen', *Amtl. Ber. über die 33. Versamml. deutsch. Naturf. u. Aerzte zu Bonn, Sep. 1857*, Bonn, 1859, xxxiii. 133; also *Bot. Ztg.*, 1857, xv. 760.

—— *Die niederen Pilze in ihren Beziehungen zu den Infectionskrankheiten und der Gesundheitspflege*, München, 1877, 285 pp.

—— *Untersuchungen über niedere Pilze, aus dem pflanzenphysiologischen Institut in München*, München u. Leipz., 1882, 285 pp.

—— 'Zur Umwandlung der Spaltpilzformen', in Naegeli, *Untersuch. über niedere Pilze*, 1882, 129–39.

NASSILOFF, 'Ueber die Diphtheritis', *Arch. f. path. Anat.*, etc., 1870, l. 550.

OBERMEIER, O., 'Vorkommen feinster, eine Eigenbewegung zeigender Fäden im Blute von Recurrenskranken', *Centralbl. f. d. med. Wissensch.*, 1873, xi. 145–7.

OERTEL, M., 'Studien über Diphtherie', *Aerztl. Intell.-Blatt*, München, 1868, xv. 407.

PASTEUR, L., 'Animalcules infusoires vivant sans gaz oxygène libre et déterminant des fermentations', *Compt. rend. Acad. d. sc.*, 1861, lii. 344–7.

—— 'Observations verbales à la suite de la communication de M. Davaine', *Compt. rend. Acad. d. sc.*, 1865, lxi, 526.

—— *Études sur la maladie des vers à soie, moyen pratique assuré de la combattre et d'en prévenir le retour*, Paris, 1870, 2 vols.

PERTY, M., *Zur Kenntniss kleinster Lebensformen nach Bau, Funktionen, Systematik mit Specialverzeichniss der in der Schweiz beobachteten*, Bern, 1852, 218 pp., 17 Tafeln.

PIERRE, I., 'Étude sur la maladie des animaux d'espèces ovine et bovine connue sous le nom de sang de rate', *Compt. rend. Acad. d. sc.*, 1864, lix. 689–93.

POLLENDER, 'Mikroskopische und microchemische Untersuchung des Milzbrandblutes sowie über Wesen und Kur des Milzbrandes', *Vrtljschr. f. gerichtl. u. öff. Med.*, Berl., 1855, viii. 103–14.

PRAŻMOWSKI, ADAM, *Untersuchungen über die Entwicklungsgeschichte und Fermentwirkung einiger Bacterien-Arten*, Leipz., 1880, 55 pp.

RAYER, 'Inoculation du sang de rate', *Compt. rend. Soc. de biol.*, 1850, ii. 141.

VON RECKLINGHAUSEN, 'Ueber Pilzmetastasen', *Sitzungsber. d. physical.-med. Gesellsch. zu Würzburg, Sitz. vom 10 Juni 1871*, p. xii.

RICHTER, H. E., 'Die neueren Kenntnisse von den krankmachenden Schmarotzerpilzen nebst phytophysiologischen Vorbegriffen',

Schmidts *Jahrbücher der in- und ausländischen ges. Med.*, 1867, cxxxv. 81–98; 1868, cxl. 101–28.

ROBIN, C., 'Remarques sur les fermentations bactériennes', *J. de l'anat. et de la physiol.*, 1879, xv. 465–91.

SALISBURY, J. H., 'Remarks on fungi with an account of experiments showing the influence of fungi of wheat straw on the human system; and some observations which point to them as the probable source of "camp measles" and perhaps of measles generally', *Amer. J. Med. Sc.*, 1862[1], N.S., xliv. 17–28, 1 pl.

—— 'Inoculating the human system with straw fungi to protect it against the contagion of measles; with some additional observations relating to the influence of fungoid growth in producing disease and in the fermentation and putrefaction of organic bodies', *Amer. J. Med. Sc.*, 1862[2], N.S., xliv. 387–94.

—— 'On the cause of intermittent and remittent fevers with investigations which tend to prove that these affections are caused by certain species of Palmellae', *Amer. J. Med. Sc.*, 1866, N.S., li. 51–75.

—— 'A brief description of what appears to be two newly discovered skin diseases; one originating in the cat and the other in the dog. Both cryptogamic and contagious and both capable of being transmitted from the animal to the human body', *Amer. J. Med. Sc.*, 1867, N.S., liii. 379–83.

—— 'Description of two new algoid vegetations, one of which appears to be the specific cause of syphilis and the other of gonorrhoea', *Amer. J. Med. Sc.*, 1868, N.S., lv. 17–25.

—— 'Vegetations found in the blood of patients suffering with erysipelas', *Zeitschr. f. Parasitenk.*, Jena, 1873[1], iv. 1–5.

—— 'Infusorial catarrh and asthma. Discovery of the cause of one form of hay fever, hay asthma, catarrhal fever', *Zeitschr. f. Parasitenk.*, Jena, 1873[2], iv. 6–11.

SALOMONSEN, C. J., *Fra Bakteriologiens Kamptid*. Reprint from *Tilskueren*, 1921, 319–39.

SANDERSON, J. B., 'Introductory report on the ultimate nature of contagion', *Rep. Med. Off. Privy Council, 1869*, Lond., 1870, xii. 229–56.

—— 'Further report of researches concerning the intimate pathology of contagion', *Rep. Med. Off. Privy Council, 1870*, Lond., 1871, xiii. 48–69.

SCHROETER, J., 'Ueber einige durch Bacterien gebildete Pigmente', *Beitr. z. Biol. d. Pflanzen*, 1872, Bd. I, Heft 2, 109–26.

—— article on 'Pilze' in *Kryptogamen-Flora von Schlesien . . .* hrsg. von Prof. Dr. Ferdinand Cohn, Breslau, 1889, Bd. III, erste Hälfte, pp. 136–74.

SÉDILLOT, C., 'De l'influence des découvertes de M. Pasteur sur les progrès de la chirurgie', *Compt. rend. Acad. d. sc.*, 1878, lxxxvi. 634–40.

SIGNOL, 'Présence des bactéries dans le sang', *Compt. rend. Acad. d. sc.*, 1863, lvii. 348–51.

TALAMON, C., 'Note sur le microbe de la diphthérie', *Progrès méd.*, 1881, ix. 122, 498; also *Bull. Soc. d'anat.*, Paris, 1881, liv. 44, 68.

THOMÉ, O. W., 'Cylindrotaenium cholerae asiaticae: ein neuer in den Cholera-Ausleerungen gefundener Pilz', *Arch. f. path. Anat.*, etc., 1867, xxxviii. 221–45.

VAN TIEGHEM, PH., and MONNIER, G., 'Sur le polymorphisme du mucor mucedo', *Compt. rend. Acad. d. sc.*, 1872, lxxiv. 997–1001.

TIGRI, A., 'Sur la présence d'infusoires du genre Bacterium dans le sang humain', *Compt. rend. Acad. d. sc.*, 1863, lvii. 633.

—— 'Sulla causa specifica ed essenziale della difteria delle fauci e del bronchi constitua da forme crittogamiche', *Gior. med. di Roma*, 1869, v. 31.

TOMMASI and HUETER, 'Ueber Diphtheritis', *Centralbl. f. d. med. Wissensch.*, 1868, vi. 531.

TULASNE, L.-R., 'Note sur l'appareil reproducteur dans des lichens et les champignons', *Compt. rend. Acad. d. sc.*, Paris, 1851, xxxii. 427, 470.

TURPIN, 'Mémoire sur la cause et les effets de la fermentation alcoolique et acéteuse', *Compt. rend. Acad. d. sc.*, 1838, vii. 369–402.

WALDEYER, 'Ueber die pathologische Bedeutung der Bacterien, Vibrionen, etc.', *Jahresber. d. schlesischen Gesellsch. f. vaterländ. Cultur* (1871), Breslau, 1872, xlix. 205.

WARD, H. MARSHALL, 'On the characters or marks employed for classifying Schizomycetes', *Ann. Botany*, Lond., 1892, vi. 103–44.

WARMING, EUG., 'Om nogle ved Danmarks Kyster levende Bakterier', *Vidensk. Meddelelser fra den naturhistoriske Forening i Kjøbenhavn*, 1875, nos. 20–8, pp. 307–420, 4 pls.

WASSERZUG, E., 'Variations de forme chez les bactéries', *Ann. de l'Inst. Pasteur*, 1888, ii. 75–83.

WEIGERT, C., *Ueber pockenähñliche Gebilde in parenchymatösen Organen und deren Beziehung zu Bacteriencolonien*, Breslau, 1875; also in Weigert's *Gesammelte Abhandlungen*, 1906, ii. 61–89.

—— 'Ueber eine Mykose bei einem neugeborenen Kinde', *Jahresber. d. schlesischen Gesellsch. f. vaterländ. Cultur* (1875), Breslau, 1876, liii. 229–30.

WINGE, 'Fall af endocardit hos en man som på Rikshospitalet vårdades för pyemie', *Förh. Svensk. Läk.-Sällsk. Sammank.*, Stockh., 1870, pp. 172–5.

WINOGRADSKY, S., 'Ueber Schwefelbacterien', *Bot. Ztg.*, 1887, xlv. 489, 513, 529, 545, 569, 585, 606.

—— 'Ueber Eisenbacterien', *Bot. Ztg.*, 1888[1], xlvi. 261–70.

—— 'Sur le pléomorphisme des bactéries', *Ann. de l'Inst. Pasteur*, 1888[2], iii. 249–64.

WINSLOW, C.-E. A., and BROADHURST, J., *et al.*, 'The families and genera of the bacteria; preliminary report of the Committee of the Society of American Bacteriologists on characterization and classification of bacterial types', *J. Bact.*, 1917, ii. 505–66.

WOOD, H. C., 'An examination into the truth of the asserted production of general diseases by organized entities', *Amer. J. Med. Sc.*, 1868, N.S., lvi. 333–52.

ZOPF, W., *Entwickelungsgeschichtliche Untersuchung über Crenothrix polyspora, die Ursache der Berliner Wassercalamität*, Berlin, 1879.

—— *Zur Morphologie der Spaltpflanzen, Spaltpilze und Spaltalgen*, Leipz., 1882, 74 pp.

—— *Die Spaltpilze, nach dem neuesten Standpunkte bearbeitet*, Breslau, 1885, 3. Aufl., 127 pp.

CHAPTER IX

BARBER, M. A., 'The pipette method in the isolation of single microorganisms and in the inoculation of substances into living cells', *Philippine J. of Sc.*, 1914, ix. 307–60, 2 pls.

BEIJERINCK, M. W., 'De Infusies en de Ontdekking der Bakterien', *Jaarb. d. Koninkl. Akad. v. Wetensch.*, Amsterdam, 1913, 1–28; also in *Verzamelde Geschriften van M. W. Beijerinck*, Delft, 1922, V. Deel. 119–40.

BERT, P., [Nouvelles recherches sur le sang de rate], *Compt. rend. Soc. de biol.*, 1876, Paris, 1877, 6ᵉ. sér., iii. 380–1.

—— 'De l'emploi de l'oxygène à haute tension comme procédé d'investigation physiologique des venins et des virus', *Compt. rend. Acad. d. sc.*, 1877[1], lxxxiv. 1130–3.

—— 'Sur le sang dont la virulence résiste à l'action de l'oxygène comprimé et à celle de l'alcool', *Compt. rend. Acad. d. sc.*, 1877[2], lxxxv. 293–5.

—— 'Sur la nature du charbon', *Compt. rend. Soc. de biol.*, 1877[3], Paris, 1879, 6ᵉ sér., iv. 317–20.

BILLROTH, TH., and EHRLICH, F., 'Untersuchungen über Coccobacteria septica', *Arch. f. klin. Chir.*, 1877, xx. 403–33, 1 pl.

BÖHMER, F., 'Zur pathologischen Anatomie der Meningitis cerebromedullaris epidemica', *Aerztl. Intellig.-Blatt*, 1865, xii. 539.

BREFELD, O., *Botanische Untersuchungen über Schimmelpilze*, Heft I, Mucor mucedo, Chaetocladium Jones ii, Piptocephalis Freseniana: Zygomyceten, Leipz., 1872, 64 pp., 6 pls.

—— 'Untersuchungen über die Entwickelung der Empusa muscae und Empusa radicans und die durch sie verursachten Epidemien der Stubenfliegen und Raupen', *Abhandl. d. naturf. Gesellsch. zu Halle*, Halle, 1873, xii. 1–50, 4 pls.

—— 'Untersuchungen über die Alkoholgährung', *Verhandl. d. phys.-med. Gesellsch. Würzburg*, 1874, N.F., v. 163–78.

—— *Botanische Untersuchungen über Schimmelpilze*, Heft II, Die Entwickelungsgeschichte von Penicillium, Leipz., 1874, 98 pp., 8 pls.

BREFELD, O., 'Methoden zur Untersuchung der Pilze', *Verhandl. d. phys.-med. Gesellsch. in Würzburg*, 1875, N.F., Bd. viii. 43–62.
—— *Botanische Untersuchungen über Schimmelpilze*, Heft III, Basidiomyceten I, Leipz., 1877, 226 pp., 11 pls.
—— *Botanische Untersuchungen über Schimmelpilze*, Heft IV, Culturmethoden zur Untersuchungen der Pilze, etc., Leipz., 1881, 190 pp., 10 pls.

BRUCE, D., 'Note on the discovery of a micro-organism in Malta Fever', *Practitioner*, 1887, xxxix. 161–70.

BUCHNER, H., 'Beiträge zur Kenntniss des Neapeler Cholerabacillus und einiger demselben nahestehender Spaltpilze', *Arch. f. Hyg.*, 1885, iii. 361.
—— 'Eine neue Methode zur Kultur anaerober Mikroorganismen', *Centralbl. f. Bakteriol.*, etc., 1888, iv. 149.

BUMM, E., *Der Mikro-Organismus der gonorrhoischen Schleimhaut-Erkrankungen*. '*Gonococcus Neisser*', Wiesb., 1885, 164 pp., 4 pls.

BURRI, R., *Das Tuschverfahren*, Jena, 1909.

CHAMBERLAND, CH., 'Résistance des germes de certains organismes à la température de 100 degrés; conditions de leur développement', *Compt. rend. Acad. d. sc.*, 1879, lxxxviii. 659–61.
—— 'Sur un filtre donnant de l'eau physiologiquement pure', *Compt. rend. Acad. d. sc.*, Paris, 1884, xcix. 247.
—— 'Sur la filtration parfaite des liquides', *Compt. rend. Soc. de biol.*, 1885, 8e sér., xii. 117–20.

COCHIN, 'De la fermentation alcoolique et de la vie de la levure de bière privée de l'air', *Ann. chimie et phys.*, 1880, 5e sér., xxi. 551.

COHN, F., 'Untersuchungen über Bacterien', *Beitr. z. Biol. d. Pflanzen*, 1872, i, Heft 2, 126–224.
—— 'Untersuchungen über Bacterien', *Beitr. z. Biol. d. Pflanzen*, 1876, ii, Heft 2, 249–76.

COZE, L., and FELTZ, V., 'Recherches expérimentales sur la présence des infusoires et l'état du sang dans les maladies infectieuses', *Gaz. méd. de Strasbourg*, 1866, xxvi. 61, 115, 208.

CROOKSHANK, E. M., *Photography of Bacteria*, Lónd., 1887, 22 pls.

DAVAINE, C., 'Études sur la contagion du charbon chez les animaux domestiques', *Bull. Acad. de méd.*, 1870[1], xxxv. 215–35.
—— 'Études sur la genèse et la propagation du charbon', *Bull. Acad. de méd.*, 1870[2], xxxv. 471–98.
—— 'Recherches sur quelques questions relatives à la septicémie', *Bull. Acad. de méd.*, Paris, 1872, 2e sér., i. 907, 976.

EBERTH, C. J., *Zur Kenntniss der bacteritischen Mykosen*, Leipz., 1872, 28 pp.

EHRENBERG, C. G., 'Das seit alter Zeit berühmte Prodigium des Blutes im Brode und auf Speisen ... bedingt durch ein bisher unbekanntes, monadenartiges Thierchen (Monas? prodigiosa),' *Bericht d. Berliner Akad.*, 1848, 349–63.

EHRLICH, P., 'Beiträge zur Kenntnis der Anilinfärbung und ihrer

Verwendung in der mikroskopischen Technik', *Arch. f. mikr. Anat.*, 1877, xiii. 263–77, 1 pl.

EHRLICH, P., 'Beiträge zur Kenntniss der granulirten Bindegewebszellen und der eosinophilen Leucocythen', *Arch.f. Anat. u. Physiol.*, Physiol. Abt., 1879, 166–9.

—— 'Ueber die specifischen Granulationen des Blutes', *Arch. f. Anat. u. Physiol.*, Physiol. Abt., 1879, 571–9.

—— 'Ueber das Methylenblau und seine klinisch-bakterioskopische Verwerthung', *Ztschr. f. klin. Med.*, 1881, ii. 710–13.

—— [Ueber die Färbung der Tuberkelbazillen], *Deutsche med. Wchnschr.*, 1882, viii. 269–70.

—— 'Ueber eine neue Methode der Färbung von Tuberkelbacillen', *Berl. klin. Wchnschr.*, 1883, xx. 13.

—— 'Beiträge zur Theorie der Bacillenfärbung', *Charité-Annalen*, Berl., 1886, xi. 123–38.

—— 'Ueber Neutralroth', *Ver. f. inn. Med.*, Berl., 18 Dec. 1893; abstract in *Ztschr. f. wissensch. Mikroskop.*, 1894, xi. 250.

—— *Farbanalytische Untersuchungen zur Histologie und Klinik des Blutes*; gesammelte Mittheilungen hrsg. v. P. Ehrlich, Berl., 1891, Theil 1, 137 pp.

VAN ERMENGEM, E., 'Nouvelle méthode de coloration des cils des bactéries', *Ann. Soc. de méd. de Gand*, 1893, lxxii. 231–6.

—— 'Ueber einen neuen anaëroben Bacillus und seine Beziehungen zum Botulismus', *Ztschr. f. Hyg. u. Infektionskrankh.*, 1897, xxvi. 1–56, 3 pls.

ESCHERICH, TH., 'Die Darmbacterien des Neugeborenen und Säuglings', *Fortschr. der Med.*, 1885, iii. 515, 547, 1 pl.

ESMARCH, E., 'Ueber eine Modification des Koch'schen Plattenverfahrens zur Isolirung und zum quantitativen Nachweis von Mikroorganismen', *Ztschr. f. Hyg.*, 1886, i. 293–301.

FITZ, A., 'Ueber Spaltpilzgährungen', *Ber. d. deutsch. chem. Gesellsch.*, 1882, xv. 867.

FRAENKEL, A., 'Bakteriologische Mittheilungen', *Ztschr. f. klin. Med.*, 1886, x. 401–61, 1 pl.

FRAENKEL, C., 'Ueber die Kultur anaerober Mikroorganismen', *Centralbl. f. Bakteriol.*, etc., 1888, iii. 735, 763.

FRAENKEL, C., and PFEIFFER, R., *Mikrophotographischer Atlas der Bakterienkunde*, Berl., 1889, 10 pls., 2. Aufl., 1893, 12 plates.

GAFFKY, 'Zur Aetiologie des Abdominaltyphus', *Mitth. a. d. kaiserl. Gesundheitsamte*, 1884, ii. 372–420.

GAUTIER, A., 'Sterilisation à froid des liquides fermentescibles', *Bull. de la Soc. chim. de Paris*, 1884, xlii. 146–50.

GEPPERT, J., 'Zur Lehre von den Antisepticis. Eine Experimentaluntersuchung', *Berl. klin. Wchnschr.*, 1889, xxvi. 789, 819.

—— 'Ueber desinficierende Mittel und Methoden', *Berl. klin. Wchnschr.*, 1890, xxvii. 246, 272, 297.

GERLACH, J., 'Ueber die Einwirkung von Farbstoff auf lebende

Gewebe', *Wissenschaftl. Mitth. d. phys.-med. Soc. zu Erlangen*, Erlang., 1858, Bd. I, Heft 1, 5–12.

GIERKE, H., 'Färberei zu mikroskopischen Zwecken', *Ztschr. f. wissenschaftl. Mikr.*, 1884, i. 62, 372, 497; 1885, ii. 13, 164.

GOEPPERT, H. R., and COHN, F., 'Ueber die Rotation des Zellinhalts in Nitella flexilis', *Bot. Ztg.*, 1849, vi. 688.

GRAM, C., 'Ueber die isolirte Färbung der Schizomyceten in Schnitt- und Trockenpräparaten', *Fortschr. der Med.*, 1884, ii. 185.

—— 'Ueber die Färbung der Schizomyceten in Schnittpräparaten', *Cong. périod. internat. d. sc. med. Compt. rend.*, 1884, Copenh., 1886, i, sect. de path. gén., p. 116.

GRUBER, M., 'Eine Methode der Cultur anaërobischer Bacterien nebst Bemerkungen zur Morphologie der Buttersäuregährung', *Centralbl. f. Bakteriol.*, etc., 1887, i. 367.

GUNNING, J. W., 'Ueber sauerstoffgasfreie Medien', *J. f. prakt. Chemie*, 1877, N.F., xvi. 314.

—— 'Experimental-Untersuchung über Anaërobiose bei den Fäulnissbacterien', *J. f. prakt. Chemie*, 1878, N.F., xvii. 266.

—— 'Ueber die Lebensfähigkeit der Spaltpilze bei fehlendem Sauerstoff', *J. f. prakt. Chemie*, 1879, N.F., xx. 434.

HANSEN, E. C., 'Recherches sur la physiologie et la morphologie des ferments alcooliques', *Résumé du compte-rendu des travaux du Lab. de Carlsberg*, Kjøbenhavn, 1883, ii, liv. 2, 13–47.

—— 'Ueber das Zählen mikroskopischer Gegenstände in der Botanik', *Ztschr. f. wissensch. Mikrosk.*, 1884, i. 191–210.

—— *Practical studies in fermentation, being contributions to the life-history of micro-organisms*, transl. by A. K. Miller, Lond., 1896, 277 pp.

HARTIG, TH., 'Ueber das Verfahren bei Behandlung des Zellenkerns mit Farbstoffen', *Bot. Ztg.*, 1854, xii. 877–81.

HAUSER, G., *Ueber Fäulnissbacterien und deren Beziehungen zu Septicaemie, ein Beitrag zur Morphologie der Spaltpilze*, Leipz., 1885, 94 pp.

HESSE, W. and R., 'Ueber Züchtung der Bacillen des malignen Oedems', *Deutsche med. Wchnschr.*, 1885, xi. 214.

HEYDENREICH, L., 'Sur la stérilisation des liquides au moyen de la marmite de Papin', *Compt. rend. Acad. d. sc.*, 1884, xcviii. 998.

HOFFMANN, H., 'Recherches sur la nature végétale de la levure', *Compt. rend. Acad. d. sc.*, 1865, lx. 633; also with figure in *Bot. Ztg.*, 1865, xxiii. 348.

—— 'Ueber Bacterien', *Bot. Ztg.*, 1869, xxvii. 252.

HOPPE-SEYLER, 'Ueber die Einwirkung von Sauerstoff auf die Lebensthätigkeit niederer Organismen', *Ztschr. f. physiol. Chemie*, 1884, viii. 214.

HUEPPE, F., *Die Methoden der Bakterien-Forschung*, Wiesb., 1886, 3. Aufl., 244 pp., 2 pls.

HUFNER, G., 'Ueber eine neue einfache Versuchsform zur Entscheidung der Frage ob sich die niedere Organismen bei Abwesenheit von gasförmigem Sauerstoffe entwickeln können', *J. f. prakt. Chemie*, 1876, N.F., xiii. 475.

ITZEROTT, G., and NIEMANN, F., *Mikrophotographischer Atlas der Bakterienkunde*, Leipz., 1895.

JOHNE, A., 'Ein zweifelloser Fall von congenitaler Tuberculose', *Fortschr. d. Med.*, 1885, iii. 198–202 (footnote, p. 200).

KITASATO, S., 'Ueber den Tetanusbacillus', *Ztschr. f. Hyg.*, 1889, vii. 225.

—— 'The bacillus of bubonic plague', *Lancet*, 1894, ii. 428–30.

KLEBS, 'Beiträge zur Kenntniss der Micrococcen', *Arch. f. exp. Path. u. Pharmakol.*, 1873, i. 31–64.

—— 'Note sur la cause du charbon', *Compt. rend. Acad. d. sc.*, Paris, 1877, lxxxv. 760–1.

KOCH, R., 'Die Aetiologie der Milzbrand-Krankheit, begründet auf die Entwicklungsgeschichte des Bacillus anthracis', *Beitr. z. Biol. d. Pflanzen*, 1876, Bd. II, Heft 2, 277–310, 1 pl.

—— 'Verfahren zur Untersuchung, zum Conserviren und Photographiren der Bacterien', *Beitr. z. Biol. d. Pflanzen*, 1877, Bd. II, Heft 3, 399–434, 24 photos.

—— *Untersuchungen über die Aetiologie der Wundinfectionskrankheiten*, Leipz., 1878, 80 pp., 5 pls.

—— 'Zur Untersuchung von pathogenen Organismen', *Mitth. a. d. kaiserl. Gesundheitsamte*, 1881[1], i. 1–48, 14 pls.

—— 'Ueber Desinfection', *Mitth. a. d. kaiserl. Gesundheitsamte*, 1881[2], i. 234–82.

—— 'Die Aetiologie der Tuberculose', *Berl. klin. Wchnschr.*, 1882, xix. 221–30.

—— 'Ueber die neuen Untersuchungsmethoden zum Nachweis der Mikrokosmen in Boden, Luft und Wasser', *Aerztliches Vereinsblatt f. Deutschland*, 1883, no. 237; also *Gesammelte Werke von R. Koch*, Leipz., 1912, i. 274–84.

—— [Bericht des Leiters der deutschen wissenschaftlichen Commission zur Erforschung der Cholera], *Deutsche med. Wchnschr.*, 1883, ix. 615, 743; 1884, x. 63, 111, 191, 221.

—— 'Die Aetiologie der Tuberkulose', *Mitth. a. d. kaiserl. Gesundheitsamte*, 1884, ii. p. 57.

—— *Gesammelte Werke von Robert Koch*, Leipz., 1912, Bd. I, 706 pp., Bd. II, 1,216 pp., 43 pls.

KOCH, GAFFKY, and LOEFFLER, 'Versuche über die Verwerthbarkeit heisser Wasserdämpfe zu Desinfectionszwecken', *Mitth. a. d. kaiserl. Gesundheitsamte*, 1881, i. 322–41.

KOCH and WOLFFHÜGEL, 'Untersuchungen über die Desinfection mit heisser Luft', *Mitth. a. d. kaiserl. Gesundheitsamte*, 1881, i. 301–21.

KRÖNIG, B., and PAUL, T., 'Die chemischen Grundlagen der Lehre von der Giftwirkung und Desinfection', *Ztschr. f. Hyg. u. Infectionskrankh.*, 1897, xxv. 1–112.

LACHOWICZ and NENCKI, 'Die Anaërobiosefrage', *Arch. f. d. ges. Physiol.*, 1884, xxxiii. 1.

LEEUWENHOEK, 32nd letter, 14 June 1680. Manuscript in Royal Soc. London. Printed in Leeuwenhoek's *Werken*, vol. i (Dutch), pp. 1–8 (2nd pagination), and in *Opera omnia*, vol. ii (Latin), pp. 1–5 (1st pagination). Engl. trans. in Dobell's *Antony van Leeuwenhoek and his 'little animals'*, 1932, 197–8.

LEMAIRE, J., *Du coal tar saponiné, désinfectant énergique, arrêtant les fermentations*, Paris, 1860, 92 pp.

LIBORIUS, P., 'Beiträge zur Kenntniss des Sauerstoffbedürfnisses der Bacterien', *Ztschr. f. Hyg.*, 1886, i. 115.

LISTER, J., 'An address on the antiseptic system of treatment in surgery', *Brit. M. J.*, 1868, ii. 53, 101, 461, 515; 1869, i. 301.

—— 'Observations on ligatures of arteries on the antiseptic system', *Lancet*, 1869, i. 451.

—— 'On lactic fermentation', *Trans. Path. Soc.*, Lond., 1878, xxix. 425–67.

LOEFFLER, F., 'Zur Immunitätsfrage', *Mitth. a. d. kaiserl. Gesundheitsamte*, 1881, i. 134–87.

—— 'Untersuchung über die Bedeutung der Mikroorganismen für die Entstehung der Diphtherie beim Menschen, bei der Taube und beim Kalbe', *Mitth. a. d. kaiserl. Gesundheitsamte*, 1884, ii. 421–99.

—— 'Weitere Untersuchungen über die Beizung und Färbung der Geisseln bei den Bakterien', *Centralbl. f. Bakteriol.*, etc., 1890, vii. 625–39.

MALASSEZ, 'Coloration des bactéries par le violet de méthyle', *Bull. Soc. anat. de Par.*, 1881, lvi. 670–4.

MAYER, ADOLF, *Untersuchungen über die alkoholische Gährung, den Stoffbedarf und den Stoffwechsel der Hefepflanzen . . .*, Heidelberg, 1869, 81 pp., 7 pls.

MIQUEL, P., *Les organismes vivants de l'atmosphère*, Paris, 1883, 310 pp.

MIQUEL, P., and BENOIST, L., 'De la stérilisation à froid des liquides animaux et végétaux réputés les plus altérables', *Bull. Soc. chim. de Paris*, 1881, xxxv. 552–7.

VON NÄGELI, C., 'Ernährung der niederen Pilze durch Kohlenstoff und Stickstoffverbindungen', in von Nägeli's *Untersuchungen über niedere Pilze*, München u. Leipz., 1882, 1–75.

VON NÄGELI, C., and SCHWENDENER, S., *Das Mikroskop. Theorie und Anwendung desselben*, Leipz., 1877, 2. Aufl., 644 pp.

NENCKI, 'Ueber die Lebensfähigkeit der Spaltpilze bei fehlendem Sauerstoff', *J. f. prakt. Chemie*, 1879, N.F., xix. 337–58.

NORDTMEYER, H., 'Ueber Wasserfiltration durch Filter aus gebrannter Infusorienerde', *Ztschr. f. Hyg.*, 1891, x. 145–54.

OBERMEIER, O., 'Vorkommen feinster, eine Eigenbewegung zeigender Fäden im Blute von Recurrenskrankheiten', *Centralbl. f. d. med. Wissensch.*, 1873, xi. 145–7.

PAPIN, DENYS, *A new digester or engine for softning bones, containing the description of its make and use in these particulars, viz. cookery voyages at sea, confectionary, making of drinks, chymistry and dying with an account of the price a good big engine will cost and of the profit it will afford,* Lond., 1681, 54 pp.

PASCHUTIN, 'Einige Versuche über Fäulniss und Fäulnissorganismen', *Arch. f. path. Anat.,* etc., 1874, lix. 490.

PASTEUR, L., 'Mémoire sur la fermentation appelée lactique', *Compt. rend. Acad. d. sc.,* 1857, xlv. 913–16.

—— 'Animalcules infusoires vivant sans gas oxygène libre et déterminant des fermentations', *Compt. rend. Acad. d. sc.,* 1861[1], lii. 344–7.

—— 'Expériences et vues nouvelles sur la nature des fermentations', *Compt. rend. Acad. d. sc.,* 1861[2], lii. 1260–4.

—— 'Nouvel exemple de fermentation déterminée par des animalcules infusoires pouvant vivre sans gas oxygène libre, et en dehors de tout contact avec l'air de l'atmosphère', *Compt. rend. Acad. d. sc.,* 1863[1], lvi. 416–21.

—— 'Recherches sur la putréfaction', *Compt. rend. Acad. d. sc.,* 1863[2], lvi. 1189–94.

—— *Études sur la bière,* Paris, 1876, 387 pp.

PASTEUR and JOUBERT, 'Étude sur la maladie charbonneuse', *Compt. rend. Acad. d. sc.,* 1877[1], lxxxiv. 900–6.

—— —— 'Charbon et septicémie', *Compt. rend. Acad. d. sc.,* 1877[2], lxxxv. 101–15; also, *Bull. Acad. de méd.,* 1877, 2e sér., vi. 781–98.

PASTEUR, JOUBERT, and CHAMBERLAND, 'La théorie des germes et ses applications à la médecine et à la chirurgie', *Compt. rend. Acad. d. sc.,* 1878, lxxxvi. 1037–43.

PEDERSEN, R., 'Undersøgelser over de Factorer der havde Inflydelse paa Formeringen af Undergjærsformen af Saccharomyces cerevisiae', *Medd. f. Carlsberg Laboratoriet,* Kjøbenhavn, 1878, i, Hefte 1, 40–71.

PETRI, R. J., 'Eine kleine Modification des Koch'schen Plattenverfahrens', *Centralbl. f. Bakteriol.,* etc., 1887, i. 279–80.

ROSENBACH, J., *Mikroorganismen bei den Wund-Infections-Krankheiten des Menschen,* Wiesb., 1884, 122 pp., 5 pls.

ROUX, E., 'Sur la culture des microbes anaérobies', *Ann. de l'Inst. Pasteur,* 1887, i. 49–62.

SALOMONSEN, C. J., 'Zur Isolation differenter Bacterienformen', *Bot. Ztg.,* 1876, xxxiv. 609–22.

—— *Studier over Blodets Forraadnelse,* Kjøbenhaven, 1877, 172 pp., 3 pls.

—— 'Lebenserinnerungen aus dem Breslauer Sommersemester 1877', *Berl. klin. Wchnschr.,* 1914, li. 485–90.

—— *Smaa-Arbejder,* Kjøbenhavn, 1917, 272 pp.

SALOMONSEN, C. J., and LEVISON, F., 'Versuche mit verschiedenen Desinfections-Apparaten', *Ztschr. f. Hyg.,* 1888, iv. 94–142.

SCHEURLEN, 'Geschichtliche und experimentelle Studien über den Prodigiosus', *Arch. f. Hyg.*, 1896, xxvi. 1–31.

SCHROETER, J., 'Ueber einige durch Bacterien gebildete Pigmente', *Beitr. z. Biol. d. Pflanzen*, 1872, i, Heft 2, 109–26.

SETTE, V., 'Memoria storico-naturale sull'arrosimento straordinario di alcune sostanse alimentose, osservato nella provincia di Padova, l'anno 1819', Venezia, 1824, abstract in *Mem. scient. et lit. dell'ateneo di Treviso*, Treviso, 1824, iii. 56; also *J. der Chemie u. Physik*, 1827, l. 396–419.

SHIGA, K., 'Ueber den Erreger der Dysenterie in Japan', *Centralbl. f. Bakteriol.*, etc., 1. Abt., 1898, xxiii. 599–600.

SMITH, THEOBALD, 'Das Gährungskolbchen in der Bacteriologie', *Centralb. f. Bakteriol.*, etc., 1890, vii. 502–6.

SPALLANZANI, L., *Opuscoli di fisica animale e vegetabile*, Modena, 1776, vol. i, chap. vii, p. 117.

STRAUS, I., 'De la stérilisation et de la désinfection par la chaleur', *Arch. de méd. expér. et d'anat. path.*, 1890, ii. 307–35.

STRUCK, 'Vorläufige Mittheilung über die Arbeiten des kaiserl. Gesundheitsamtes welche zur Entdeckung des Bacillus der Rotzkrankheit geführt haben', *Deutsche med. Wchnschr.*, 1882, viii. 707.

TIEGEL [and KLEBS], 'Die Ursache des Milzbrandes', *Correspondenz-Bl. f. schweizer. Aerzte*, 1871, i. 275–80.

UNNA, P. G., 'Die Entwickelung der Bacterienfärbung. Eine historisch-kritische Uebersicht', *Centralbl. f. Bakteriol.*, etc., 1888, iii. 22, 61, 93, 120, 153, 189, 218, 254, 285, 312, 345.

VIGNAL, 'Sur un moyen d'isolation et de culture des microbes anaérobies', *Ann. de l'Inst. Pasteur*, 1887, i. 358.

VITTADINI, C., 'Della natura del calcino o mal del segno', *Mem. dell'I.R. Istituto Lombardo di sc., lett. ed arti*, 1852, iii. 447–512, 2 pls.

WALDEYER, W., 'Untersuchungen über den Ursprung und den Verlauf des Axencylinders bei Wirbellosen und Wirbelthieren sowie über dessen Endverhalten in der quer gestreiften Muskelfaser', *Ztschr. f. rationelle Med.*, 1863, 3. R. xx. 193–256.

WEICHSELBAUM, A., 'Ueber die Aetiologie der akuten Meningitis cerebro-spinalis', *Fortschr. d. Med.*, 1887, v. 573, 620, 1 pl.

WEIGERT, C., 'Ueber Bacterien in der Pockenhaut', *Centralbl. f. d. med. Wissensch.*, 1871, ix. 609–11.

—— 'Uber eine Mykose bei einem neugeborenen Kinde', *Jahresber. d. schlesischen Gesellsch. f. vaterländ. Cultur* (1875), Breslau, 1876, liii. 229–30.

—— 'Bismarckbraun als Färbemittel', *Arch. f. mikr. Anat.*, 1878, xv. 258–60.

—— 'Zur Technik der mikroskopischen Bakterienuntersuchungen', *Arch. f. path. Anat.*, etc., 1881, lxxxiv. 275–315.

WHITTAKER, H. A., 'The source, manufacture and composition of commercial agar-agar', *J. Amer. Pub. Health Assocn.*, 1911, i. 632–9.

WOLFF, M., and ISRAEL, J., 'Ueber Reincultur des Actinomyces und seine Uebertragbarkeit auf Thiere', *Arch. f. path. Anat.*, 1891, cxxvi. 11–59, 8 pls.

YERSIN, 'Sur la peste de Hong-Kong', *Compt. rend. Acad. d. sc.*, 1894, cxix. 356.

ZETTNOW, 'Ueber Geisselfärbung bei Bakterien', *Ztschr. f. Hyg. u. Infectionskrankh.*, 1899, xxx. 95–106.

CHAPTER X

BERNSTEIN, R., 'Ueber die Ergebnisse des Pasteur'schen Immunisierungsverfahren gegen Tollwuth', *Fortschr. d. Med.*, 1905, xxiii. 157–61.

CHAMBERLAND, C., *Le Charbon et la vaccination charbonneuse d'après les travaux récents de M. Pasteur*, Paris, 1883, 316 pp.

—— 'Résultats pratiques des vaccinations contre le charbon et le rouget en France', *Ann. de l'Inst. Pasteur*, 1894, viii. 160–5.

CHAUVEAU, A., 'De la préparation en grandes masses des cultures atténuées par le chauffage rapide pour l'inoculation préventive du sang de rate', *Compt. rend. Acad. d. sc.*, 1884, xcviii. 73–7.

—— 'De l'atténuation des cultures virulentes par l'oxygène comprimé', *Compt. rend. Acad. d. sc.*, 1884, xcviii. 1232–5.

GALTIER, V., 'Les injections de virus rabique dans le torrent circulatoire ne provoquent pas l'éclosion de la rage et semblent conférer l'immunité; la rage peut être transmise par l'ingestion de la matière rabique', *Compt. rend. Acad. d. sc.*, 1881, xciii. 284.

—— [Transmission du virus rabique], *Bull. Acad. de méd.*, Paris, 1881, 2ᵉ sér., x. 90–4.

HÖGYES, A., *Die experimentelle Basis des antirabischen Schutzimpfungen Pasteurs, nebst einigen Beiträgen zur Statistik der Wuthbehandlung*, Stuttgart, 1889, 108 pp.

Inauguration de l'Institut Pasteur, *Ann. de l'Inst. Pasteur*, 1888, ii. 5–30.

KITT, T., 'Beiträge zur Kenntniss der Geflügelcholera und deren Schutzimpfung', *Deutsche Ztschr. f. Thiermed.*, 1886–7, xiii. 1–30.

KOCH, R., *Ueber die Milzbrandimpfung. Eine Entgegnung auf den von Pasteur in Genf gehaltenen Vortrag*, Kassel u. Berl., 1882.

KOCH, R., GAFFKY, and LOEFFLER, 'Experimentelle Studien über die künstliche Abschwächung der Milzbrandbacillen und Milzbrandinfection durch Fütterung', *Mitth. a. d. kaiserl. Gesundheitsamte*, 1884, ii. 147–81.

LOEFFLER, F., 'Zur Immunitätsfrage', *Mitth. a. d. kaiserl. Gesundheitsamte*, 1881, i. 134–87.

MAGENDIE, 'Expérience sur la rage', *J. de physiol. expér.*, Paris, 1821, i. 40–6.

PASTEUR, L., 'Sur l'étiologie de l'affection charbonneuse', *Bull. Acad. de méd.*, 1879, 2ᵉ sér., viii. 1063–4.

PASTEUR, L., 'Sur les maladies virulentes et en particulier sur la maladie appelée vulgairement choléra des poules', *Compt. rend. Acad. d. sc.*, 1880[1], xc. 239–48; also *Bull. Acad. de méd.*, 1880, 2e sér., ix. 121–34.

—— 'Sur le choléra des poules. Études des conditions de la non-récidive de la maladie et de quelques autres de ces caractères', *Compt. rend. Acad. d. sc.*, 1880[2], xc. 952, 1030; also *Bull. Acad. de méd.*, 1880, 2e sér., ix. 390–401.

—— 'De l'extension de la théorie des germes à l'étiologie de quelques maladies communes', *Compt. rend. Acad. d. sc.*, 1880[3], xc. 1033–44; also *Bull. Acad. de méd.*, 1880, 2e sér., ix. 435–47.

—— 'Expériences tendant à démontrer que les poules vaccinées pour le choléra sont réfractaires au charbon', *Compt. rend. Acad. d. sc.*, 1880[4], xci. 315.

—— 'Sur l'étiologie des affections charbonneuses', *Compt. rend. Acad. d. sc.*, 1880[5], xci. 455–7.

—— 'De l'atténuation du virus du choléra des poules', *Compt. rend. Acad. d. sc.*, 1880[6], xci. 673–80; also *Bull. Acad. de méd.*, 1880, 2e sér., ix. 1119–27.

—— 'Nouvelles observations sur l'étiologie et la prophylaxie du charbon', *Compt. rend. Acad. d. sc.*, 1880[7], xci. 697–701; also *Bull. Acad. de méd.*, 1880, 2e sér., ix. 1138–43.

—— 'Vaccination in relation to chicken-cholera and splenic fever', *Trans. Internat. Med. Congr.*, 7 Sess., Lond., 1881, i. 85–90; also *Brit. M. J.*, 1881, ii. 283; *Lancet*, 1881, ii. 271.

—— 'Résultats des vaccinations charbonneuses pratiquées pendant les mois de juillet, août et septembre 1881', *Arch. vét. Paris*, 1882[1], vii. 177.

—— 'Une statistique au sujet de la vaccination préventive contre le charbon portant sur quatre-vingt-cinq mille animaux', *Compt. rend. Acad. d. sc.*, 1882[2], xcv. 1250–2.

—— 'De l'atténuation des virus', *Cong. internat. d'hyg. et de démog.*, *Compt. rend.*, 1882[3], Genève, 1883, i. 127–49.

—— [La vaccination charbonneuse] Réponse au docteur Koch, *Rev. scient.*, Paris, 1883[1], 2e sér., xii. 509–14.

—— 'Les doctrines dites microbiennes et la vaccination charbon-neuse', *Bull. Acad. de méd.*, 1883[2], N.S., xii. 509–14.

—— 'Sur la vaccination charbonneuse', *Compt. rend. Acad. d. sc.*, 1883[3], xcvi. 979–82.

—— 'Microbes pathogènes et vaccins', *Congr. périod. internat. d. sc. méd.*, *Compt. rend.*, 1884, Copenhagen, 1886, I, séance gén. 19–28.

—— 'Méthode pour prévenir la rage après morsure', *Compt. rend. Acad. d. sc.*, 1885, ci. 765–74.

—— 'Résultats de l'application de la méthode pour prévenir la rage après morsure', *Compt. rend. Acad. d. sc.*, 1886[1], cii. 459–69; also *Bull. Acad. de méd.*, 1886, 2e sér., xv. 294–303.

—— 'Note complémentaire sur les résultats d'application de la méthode de prophylaxie de la rage après morsure', *Compt. rend. Acad. d. sc.*, 1886², cii. 835–8.

PASTEUR, L., 'Nouvelle communication sur la rage', *Compt. rend. Acad. d. sc.*, 1886³, ciii. 777–85; also *Bull. Acad. de méd.*, 1886, 2ᵉ sér., xvi. 370–9.

—— 'Sur les résultats de l'application de la méthode de prophylaxie de la rage', *Bull. Acad. de méd.*, 1886⁴, 2ᵉ sér., xv. 664–5.

—— 'Sur la destruction des lapins en Australie et dans la Nouvelle-Zélande', *Ann. de l'Inst. Pasteur*, 1888, ii. 1–8.

—— 'Sur la méthode de prophylaxie de la rage après morsure', *Compt. rend. Acad. d. sc.*, 1889, cviii. 1228.

PASTEUR, L., and CHAMBERLAND, 'Sur la non-récidive de l'affection charbonneuse', *Compt. rend. Acad. d. sc.*, 1880, xci. 531–8; also *Bull. Acad. de méd.*, 1880, 2ᵉ sér., ix. 983–91.

PASTEUR, L., CHAMBERLAND, and ROUX, 'Sur l'étiologie du charbon', *Compt. rend. Acad. d. sc.*, 1880, xci. 86–94; also *Bull. Acad. de méd.*, 1880, 2ᵉ sér., ix. 682–92.

—— —— —— 'Sur une maladie nouvelle provoquée par la salive d'un enfant mort de la rage', *Compt. rend. Acad. d. sc.*, 1881¹, xcii. 159–65.

—— —— —— 'Sur la longue durée de la vie de germes charbonneux et leur conservation dans les terres cultivées', *Compt. rend. Acad. d. sc.*, 1881², xcii. 209–11.

—— —— —— 'De l'atténuation des virus et de leur retour à la virulence', *Compt. rend. Acad. d. sc.*, 1881³, xcii. 429–35.

—— —— —— 'Le vaccin du charbon', *Compt. rend. Acad. d. sc.*, 1881⁴, xcii. 666–8.

—— —— —— 'Compte rendu sommaire des expériences faites à Pouilly-le-Fort, près Melun, sur la vaccination charbonneuse', *Compt. rend. Acad. d. sc.*, 1881⁵, xcii. 1378–83; also *Bull. Acad. de méd.*, 1881, 2ᵉ sér., x. 782–95.

—— —— —— 'Nouvelle communication sur la rage', *Bull. Acad. de méd.*, Paris, 1884¹, 2ᵉ sér., xiii. 337–44; also *Compt. rend. Acad. d. sc.*, 1884, xcviii. 457–63.

—— —— —— 'Sur la rage', *Bull. Acad. de méd.*, 1884², 2ᵉ sér., xiii. 661–4; also *Compt. rend. Acad. d. sc.*, 1884, xcviii. 1229–31.

PASTEUR, L., CHAMBERLAND, ROUX, and THUILLIER, 'Sur la rage', *Compt. rend. Acad. d. sc.*, 1881, xcii. 1259–60; also *Bull. Acad. de méd.*, Paris, 1881, 2ᵉ sér., x. 717–19.

—— —— —— —— 'Nouveaux faits pour servir à la connaissance de la rage', *Compt. rend. Acad. d. sc.*, 1882, xcv. 1187–92; also *Bull. Acad. de méd.*, Paris, 1882, 2ᵉ sér., xi. 1440–5.

PASTEUR, L., and THUILLIER, 'Sur le rouget ou mal rouge des porcs', *Compt. rend. Acad. d. sc.*, 1882¹, xcv. 1120–1.

—— —— 'Le rouget du porc', *Bull. Acad. de méd.*, 1882², 2ᵉ sér., xi. 1438–40.

—— —— 'La vaccination du rouget des porcs à l'aide du virus mortel atténué de cette maladie', *Compt. rend. Acad. d. sc.*, 1883, xcvii. 1163–9; also *Bull. Acad. de méd.*, 1883, 2ᵉ sér., xii. 1359–66.

PERRONCITO, E., 'Ueber das epizootische Typhoide der Hühner', *Arch. f. wissenschaftl. Tierheilk.*, 1879, v. 22–51.

SUZOR, J.-R., *La rage, contenant la collection complète des communications de M. Pasteur sur ce sujet avec un court résumé historique de la question, la technique*, etc., 4°, Paris, 1887; Engl. trans., Lond., 1887.

TOUSSAINT, H., 'De l'immunité pour le charbon acquise à la suite d'inoculations préventives', *Compt. rend. Acad. d. sc.*, 1880, xci. 135–7.

CHAPTER XI

ARLOING, S., 'Aperçu sur les théories actuelles de l'immunité', *Lyon méd.*, 1901, xcvi. 501–13.

ARONSON, H., 'Experimentelle Untersuchungen über Diphtherie und die immunisirende Substanz des Blutserums', *Berl. klin. Wchnschr.*, 1893, xxx. 592, 625.

ARRHENIUS, S., and MADSEN, TH., 'Physical chemistry applied to toxins and antitoxins', *Contributions . . . to celebrate the inauguration of the State Serum Institute*, Copenhagen, 1902, Article III.

BABES, V., 'Sur la première constatation de la transmission des propriétés immunisantes et curatives par le sang des animaux immunisés', *Bull. Acad. de méd.*, Paris, 1895, 3ᵉ sér., xxxiii. 10, 95, 155.

BABES, V., and LEPP, 'Recherches sur la vaccination antirabique', *Ann. de l'Inst. Pasteur*, 1889, iii. 384–90.

BANG, I., and FORSSMAN, J., 'Ist die Ehrlichsche Seitenkettentheorie mit den tatsächlichen Verhältnissen vereinbar?' *München. med. Wchnschr.*, 1909, lvi. 1769–72; 1910, lvii. 851–3.

BAUMGARTEN, P., 'Zur Kritik der Metchnikoff'schen Phagocytentheorie', *Ztschr. f. klin. Med.*, 1889[1], xv. 1–41.

—— 'Ueber das "Experimentum crucis" der Phagocytenlehre', *Beitr. z. path. Anat. u. z. allg. Path.*, 1889[2], vii. 3–10.

BEHRING, E., 'Ueber die Ursache der Immunität von Ratten gegen Milzbrand', *Centralbl. f. klin. Med.*, 1888, ix. 681–90.

—— 'Thatsächliches, Historisches und Theoretisches aus der Lehre von der Giftimmunität', *Deutsche med. Wchnschr.*, 1898, xxiv. 661–6.

BEHRING, E., BOER, and KOSSEL, 'Zur Behandlung diphtheriekranker Menschen mit Diphtherieheilserum', *Deutsche med. Wchnschr.*, 1893, xix. 389, 415.

BEHRING, E., and EHRLICH, 'Zur Diphtherieimmunisirungs- und Heilungsfrage', *Deutsche med. Wchnschr.*, 1894, xx. 437.

BEHRING, E., and KITASATO, 'Ueber das Zustandekommen der Diphtherie-Immunität und der Tetanus-Immunität bei Thieren', *Deutsche med. Wchnschr.*, 1890, xvi. 1113, 1145.

BEHRING, E., and NISSEN, F., 'Ueber bacterienfeindliche Eigenschaften verschiedener Blutserumarten; ein Beitrag zur Immunitäts-frage', *Ztschr. f. Hyg.*, 1890, viii. 412–33.

BEHRING, E., and WERNICKE, 'Ueber Immunisirung und Heilung von Versuchstieren bei der Diphtherie', *Ztschr. f. Hyg. u. Infektions-krankh.*, 1892, xii. 10–44.

BELFANTI, S., and CARBONE, T., 'Produzione di sostanze tossiche nei siero di animali inoculati con sangue eterogeneo', *Gior. d. r. Accad. di med. di Torino*, 1898, 4ᵉ ser., xlvi. 321–4.

BIRCH-HIRSCHFELD, 'Die neuern pathologisch-anatomischen Unter-suchungen über krankmachende Schmarotzerpilze', Schmidt's *Jahrb.*, 1872, clv. 97–109.

BITTER, H., 'Kommt durch die Entwickelung von Bacterien im lebenden Körper eine Erschöpfung desselben an Bacterien-nährstoffen zu Stande?' *Ztschr. f. Hyg.*, 1888¹, iv. 291–8.

—— 'Kritische Bemerkungen zu E. Metschnikoff's Phagocytenlehre', *Ztschr. f. Hyg.*, 1888², iv. 318–52.

BORDET, J., 'Les leucocytes et les propriétés actives du sérum chez les vaccinés', *Ann. de l'Inst. Pasteur*, 1895, ix. 462–506.

—— 'Recherches sur la phagocytose', *Ann. de l'Inst. Pasteur*, 1896, x. 104–18, 1 pl.

—— 'Sur l'agglutination et la dissolution des globules rouges par le sérum d'animaux injectés de sang défibriné', *Ann. de l'Inst. Pasteur*, 1898, xii. 688–95.

—— 'Le mécanisme de l'agglutination', *Ann. de l'Inst. Pasteur*, 1899, xiii. 225–50.

—— 'Sur le mode d'action des antitoxines sur les toxines', *Ann. de l'Inst. Pasteur*, 1903, xvii. 162–86.

BORDET, J., and GENGOU, O., 'Sur l'existence de substances sensibili-satrices dans la plupart des sérums antimicrobiens', *Ann. de l'Inst. Pasteur*, 1901, xv. 289–303.

BRIEGER, L., and COHN, G., 'Untersuchungen über Tetanusgift', *Ztschr. f. Hyg. u. Infektionskrankh.*, 1893, xv. 1–10.

BRIEGER, L., and EHRLICH, 'Ueber die Uebertragung von Immunität durch Milch', *Deutsche med. Wchnschr.*, 1892, xviii. 393–4.

—— —— 'Beiträge zur Kenntniss der Milch immunisirter Thiere', *Ztschr. f. Hyg. u. Infectionskrankh.*, 1893, xiii. 336–46.

BRIEGER, L., KITASATO, S., and WASSERMANN, A., 'Ueber Immunität und Giftfestigung', *Ztschr. f. Hyg. u. Infectionskrankh.*, 1892, xii. 137–82.

BUCHNER, H., 'Ueber die bakterientödtende Wirkung des zellen-freien Blutserums', *Centralbl. f. Bakteriol.*, 1889¹, v. 817–23; vi. 1–11.

—— 'Ueber die nähere Natur der bakterientödtenden Substanz im Blutserum', *Centralbl. f. Bakteriol.*, 1889², vi. 561–5.

—— 'Ueber die bakterientödtenden Wirkungen des Blutserums', *Centralbl. f. Chir.*, 1889³, xvi. 851–2.

BUCHNER, H., 'Untersuchungen über die bacterienfeindlichen Wirkungen des Blutes und Blutserums', *Arch. f. Hyg.*, 1890, x. 84–101.
—— 'Die Forschungsmethoden in der Immunitätsfrage', *Centralbl. f. Bakteriol.*, 1891, x. 727–36.
—— 'Die neuen Gesichtspuncte der Immunitätsfrage', *Fortschr. d. Med.*, 1892¹, x. 319, 363.
—— 'Ueber die bakterientödtende Wirkung des Blutserums', *Centralbl. f. Bakteriol.*, 1892², xii. 855–8.
—— 'Weitere Untersuchungen über die bacterienfeindlichen und globuliciden Wirkungen des Blutserums', *Arch. f. Hyg.*, 1893¹, xvii. 112–37.
—— 'Ueber Bacteriengifte und Gegengifte', *München. med. Wchnschr.*, 1893², xl. 449, 480.
—— 'Ueber die Phagocytentheorie', *München. med. Wchnschr.*, 1897, xliv, 1320–3.
—— 'Zur Kenntniss der Alexine sowie der specifisch-bactericiden und specifisch-haemolytischen Wirkungen', *München. med. Wchnschr.*, 1900, xlvii. 277–83.
BUCHNER, H., and ORTHENBERGER, M., 'Versuche über die Natur der bakterientödtenden Substanz im Serum', *Arch. f. Hyg.*, 1890, x. 149–73.
BUCHNER, H., and SITTMANN, G., 'Welchen Bestandtheilen des Blutes ist die bakterientödtende Wirkung zuzuschreiben?' *Arch. f. Hyg.*, 1890, x. 121–49.
BUCHNER, H., and VOIT, FR., 'Ueber den bacterientödtenden Einfluss des Blutes', *Arch. f. Hyg.*, 1890, x. 101–20.
BULLOCH, W., 'Ueber die Beziehung zwischen Hämolysis und Bacteriolysis', *Centralbl. f. Bakteriol.*, Abt. I, 1901, xxix. 724–32.
CALMETTE, A., 'Contribution à l'étude des venins, des toxines et des sérums antitoxiques', *Ann. de l'Inst. Pasteur*, 1895, ix. 225–51.
CAMUS, L., and GLEY, E., 'Nouvelles recherches sur l'immunité contre le sérum d'anguille', *Ann. de l'Inst. Pasteur*, 1899, xiii. 779–87.
CASTELLANI, A., 'Ueber das Verhältniss der Agglutinine zu den Schützkorpern', *Ztschr. f. Hyg. u. Infektionskrankh.*, 1901, xxxvii. 381–92.
—— 'Die Agglutination bei gemischter Infection und die Diagnose der letzteren', *Ztschr. f. Hyg. u. Infektionskrankh.*, 1902, xl. 1–19.
CHARRIN, A., 'Sur des procédés capables d'augmenter la résistance de l'organisme à l'action des microbes', *Compt. rend. Acad. d. sc.*, 1887, cv. 756–9.
CHARRIN, A., and GLEY, E., 'Recherches sur la transmission héréditaire de l'immunité', *Arch. de physiol. norm. et path.*, 1893, 5ᵉ sér., v. 75–82.
—— —— 'Nouvelles recherches expérimentales sur la transmission héréditaire de l'immunité', *Arch. de physiol. norm. et path.*, 1894, 5ᵉ sér., vi. 1–6.

CHARRIN, A., and ROGER, 'Note sur le développement des microbes pathogènes dans le sérum des animaux vaccinés', *Compt. rend. Soc. de biol.*, 1889, 9ᵉ sér., i. 667–9.

—— —— 'Le rôle du sérum dans le mécanisme de l'immunité', *Compt. rend. Soc. de biol.*, 1892, 9ᵉ ser., iv. 924–8.

CHAUVEAU, A., 'Du renforcement de l'immunité des moutons algériens à l'égard du sang de rate, par les inoculations préventives', *Compt. rend. Acad. d. sc.*, 1880, xci. 148–51.

—— 'Sur le mécanisme de l'immunité', *Compt. rend. Acad. d. sc.*, 1888, cvi. 392–8; also *Ann. de l'Inst. Pasteur*, 1888, ii. 66–74.

WATSON CHEYNE, W., 'On the relation of organisms to antiseptic dressings', *Trans. Path. Soc.*, Lond., 1879, xxx. 557–82.

DE CHRISTMAS, J., 'Étude sur les substances microbicides du sérum et des organes d'animaux à sang chaud', *Ann. de l'Inst. Pasteur*, 1891, v. 487–505.

CREITE, A., 'Versuche über die Wirkung des Serumeiweisses nach Injection in das Blut', *Ztschr. f. rationelle Med.*, 1869, xxxvi. 90–108.

DANYSZ, J., 'Contribution à l'étude de l'immunité; propriétés des mélanges des toxines avec leurs antitoxines; constitution des toxines', *Ann. de l'Inst. Pasteur*, 1899, xiii. 581–95.

—— 'Contribution à l'étude des propriétés et de la nature des mélanges des toxines avec leurs antitoxines', *Ann. de l'Inst. Pasteur*, 1902, xvi. 331–45.

DAREMBERG, G., 'De l'action destructive du sérum du sang sur les globules rouges', *Arch. de méd. expér. et d'anat. path.*, 1891, iii. 720–33.

DEAN, GEORGE, 'An experimental enquiry into the nature of the substance in serum which influences phagocytosis', *Proc. Roy. Soc.*, Lond., 1907, lxxix. 399–412.

DENYS, J., 'Observations sur l'immunité', *Cong. internat. d'hyg. et de démog.*, *Compt. rend.*, 1894, Budapest, 1896, viii, pt. 2, 32–4.

DENYS, J., and HAVET, J., 'Sur la part des leucocytes dans le pouvoir bactéricide du sang de chien', *La Cellule*, 1894, x. 1–35.

DENYS, J., and LECLEF, J., 'Sur le mécanisme de l'immunité chez le lapin vacciné contre le streptocoque pyogène', *La Cellule*, 1895, xi. 175–221.

DENYS, J., and MARCHAND, L., 'Du mécanisme de l'immunité conférée au lapin par l'injection de sérum antistreptococcique de cheval et d'un nouveau mode d'application de ce sérum', *Bull. Acad. roy. de méd. de Belgique*, 1896, 4ᵉ sér., x. 249–70.

DEUTSCH, L., 'Contribution à l'étude de l'origine des anticorps typhiques', *Ann. de l'Inst. Pasteur*, 1899, xiii. 689–727.

VON DUNGERN, E., 'Globulicide Wirkungen des thierischen Organismus', *München. med. Wchnschr.*, 1899, xlvi. 405, 449.

—— *Die Antikörper. Resultate früherer Forschungen und neue Versuche*, Jena, 1903, 114 pp. ·

DURHAM, H. E., 'On a special action of the serum of highly immunised animals and its use for diagnostic and other purposes', *Proc. Roy. Soc.*, Lond., 1896, lix. 224–6.

—— 'Some theoretical considerations upon the nature of agglutinins together with further observations upon Bacillus typhi abdominalis, Bacillus enteritidis, Bacillus coli communis, Bacillus lactis aerogenes and some other bacilli of allied character', *J. Exper. Med.*, 1900–1, v. 353–88.

EHRLICH, P., *Das Sauerstoff-Bedürfniss des Organismus, eine farbanalytische Studie*, Berl., 1885, 167 pp.

—— 'Experimentelle Untersuchungen über Immunität, I. Ueber Ricin, II. Ueber Abrin', *Deutsche med. Wchnschr.*, 1891, xvii. 976, 1218.

—— 'Ueber Immunität durch Vererbung und Säugung', *Ztschr. f. Hyg. u. Infektionskrankh.*, 1892, xii. 183–203.

—— 'Zur Kenntniss der Antitoxinwirkung', *Fortschr. d. Med.*, 1897[1], xv. 41.

——'Die Wertbemessung des Diphtherieheilserums und deren theoretische Grundlagen', *Klin. Jahrb.*, 1897[2], vi. 299–326.

—— 'Ueber die Constitution des Diphtheriegiftes', *Deutsche med. Wchnschr.*, 1898, xxiv. 597.

—— 'On immunity with special reference to cell life', *Proc. Roy. Soc.*, Lond., 1900, lxvi. 424–48, 2 pls.

—— 'Die Schutzstoffe des Blutes', *Deutsche med. Wchnschr.*, 1901, xxvii. 865, 888, 913.

—— 'Toxin und antitoxin; Entgegnung auf den neuesten Angriff Grubers', *München. med. Wchnschr.*, 1903, l. 1428, 1465, 2295.

—— : Eine Darstellung seines wissenschaftlichen Wirkens, von H. Apolant, H. Aronson, H. Bechhold [*et al.*] Festschrift zum 60. Geburtstage des Forschers (14 März 1914), Jena, 1914, 668 pp.

EHRLICH, P., KOSSEL, H., and WASSERMANN, A., 'Ueber Gewinnung und Verwendung des Diphtherieheilserums', *Deutsche med. Wchnschr.*, 1894, xx, 353–5.

EHRLICH, P., and MORGENROTH, J., 'Zur Theorie der Lysinwirkung', *Berl. klin. Wchnschr.*, 1899, xxxvi, 6–9.

—— —— 'Ueber Haemolysine', *Berl. klin. Wchnschr.*, 1899, xxxvi. 481; 1900, xxxvii. 453, 681.

EISENBERG, P., and VOLK, R., 'Untersuchungen über die Agglutination', *Ztschr. f. Hyg. u. Infektionskrankh.*, 1902, xl. 155–95.

EMMERICH, R., and LÖW, O., 'Die Ursache der künstlichen Immunität und die Heilung von Infectionskrankheiten', *München. med. Wchnschr.*, 1898, xlv. 1433.

—— —— 'Bacteriolytische Enzyme als Ursache der erworbenen Immunität und die Heilung von Infektionskrankheiten durch dieselben', *Ztschr. f. Hyg. u. Infectionskrankh.*, 1899, xxxi. 1–65.

VON FODOR, J., 'Bakterien im Blute lebender Thiere', *Arch. f. Hyg.*, 1886, iv. 129–48.

VON FODOR, J., 'Die Fähigkeit des Blutes Bakterien zu vernichten', *Deutsche med. Wchnschr.*, 1887, xiii. 745–7.

—— 'Neuere Untersuchungen über die bakterientödtende Wirkung des Blutes und über Immunisation', *Centralbl. f. Bakteriol.*, 1890, vii. 753–66.

FORSSMAN, J., 'Sind das Antigen und die amboceptorfixierende Substanz der Blutkörperchen identisch oder verschieden?' *Biochem. Ztschr.*, Berl., 1908, ix. 330–51.

FORSSMAN, J., and LUNDSTROM, E., 'Sur la marche de la courbe d'antitoxine dans l'immunisation active contre le botulisme', *Ann. de l'Inst. Pasteur*, 1902, xvi. 294–303.

FRAENKEL, C., 'Untersuchungen über Bacteriengifte. Immunisirungsversuche bei Diphtherie', *Berl. klin. Wchnschr.*, 1890, xxvii. 1133.

FRAENKEL, C., and SOBERNHEIM, 'Versuche über das Zustandekommen der künstlichen Immunität', *Hyg. Rundschau*, Berl., 1894, iv. 97, 145.

GAMALÉIA, N., 'Vibrio Metchnikovi; vaccination chimique', *Ann. de l'Inst. Pasteur*, 1889, iii. 542–55.

GENGOU, 'Sur les sensibilisatrices des sérums actifs contre les substances albuminoïdes', *Ann. de l'Inst. Pasteur*, 1902, xvi. 734–55.

GROHMANN, W., *Ueber die Einwirkung des zellenfreien Blutplasma auf einige pflanzliche Microorganismen, Schimmel-, Spross-, pathogene und nicht pathogene Spaltpilze*, Dorpat, 1884, 34 pp.

GRUBER, M., 'Ueber active und passive Immunität gegen Cholera und Typhus', *Verhandl. d. Cong. f. inn. Med.*, Wiesb., 1896, xiv. 207–27.

—— 'Theorie der activen und passiven Immunität gegen Cholera, Typhus und verwandte Krankheitsprocesse', *München. med. Wchnschr.*, 1896, xliii. 206; trans., *Lancet*, 1897, ii. 910.

GRUBER, M. (and DURHAM, H. E.), 'Ueber active und passive Immunität gegen Cholera und Typhus', *Wiener klin. Wchnschr.*, 1896, ix. 183, 204.

GRUBER, M., and DURHAM, H. E., 'Eine neue Methode zur raschen Erkennung des Choleravibrio und des Typhusbacillus', *München. med. Wchnschr.*, 1896, xliii. 285–6.

GRUBER, M., and VON PIRQUET, CL., 'Toxin und Antitoxin', *München. med. Wchnschr.*, 1903, l. 1193, 1259.

GRÜNBAUM, A. S., 'Un mot sur l'histoire du sérodiagnostic', *Ann. de l'Inst. Pasteur*, 1897, xi. 670.

GÜNTHER, C., 'Die Blutserumtherapie; ihre geschichtliche Entwickelung und ihr gegenwärtiger Stand', *Deutsche med. Wchnschr.*, 1893, xix. 1162–5.

HANKIN, E. H., 'Report on the conflict between the organism and the microbe', *Brit. M. J.*, 1890, ii. 65–8.

—— 'On immunity', *Tr. VII. Internat. Congr. Hyg. and Demog.*, 1891, Lond., 1892, ii. 145–52.

HAVET, J., 'Du rapport entre le pouvoir bactéricide du sang de chien et sa richesse en leucocytes', *La Cellule*, 1894, x. 219–47.

HÉRICOURT, J., and RICHET, CH., 'De la transfusion péritonéale et de l'immunité qu'elle confère', *Compt. rend. Acad. d. sc.*, 1888, cvii. 748–50.

—— —— 'Immunité conférée à des lapins par la transfusion péritonéale de sang de chien', in Richet's *Physiol. Trav. du lab.*, Paris, 1895, iii. 289–316.

HEYMANN, B., 'Zur Geschichte der Seitenkettentheorie Paul Erhlichs', *Klin. Wchnschr.*, 1928, vii. 1257, 1305.

ISSAËFF, B., 'Contribution à l'étude de l'immunité acquise contre le pneumocoque', *Ann. de l'Inst. Pasteur*, 1893, vii. 260–85.

JETTER, P., 'Untersuchungen über die "bactericide" Eigenschaft des Blutserums', *Arb. a. d. Geb. d. path. Anat. . . . Inst. zu Tübingen*, 1891–2, i. 421–49.

—— 'Ueber Buchner's "Alexine" und ihre Bedeutung für die Erklärung der Immunität', *Centralbl. f. Bacteriol.*, 1893, xvi. 724–8.

JOOS, A., 'Untersuchungen über den Mechanismus der Agglutination', *Ztschr. f. Hyg. u. Infektionskrankh.*, 1901, xxxvi. 422–39; 1902, xl. 203–30.

KOCH, R., 'Die Aetiologie der Milzbrand-Krankheit begründet auf die Entwicklungsgeschichte des Bacillus anthracis', *Beitr. z. Biol. d. Pflanzen*, hrsg. v. F. Cohn, 1876, ii. Heft 2, 277–310.

KRAUS, R., 'Ueber spezifische Reaktionen in keimfreien Filtraten aus Cholera-Typhus-Pestbouillonculturen erzeugt durch homologes Serum', *Wiener klin. Wchnschr.*, 1897, x. 736–9.

KRAUS, R., and SCHIFFMANN, 'Sur l'origine des anticorps précipitines et agglutinines', *Ann. de l'Inst. Pasteur*, 1906, xx. 225–40.

LANDOIS, L., *Die Transfusion des Blutes . . .*, 1875, 358 pp., 4 pls.

LANDSTEINER, K., and JAGIĆ, N., 'Ueber die Verbindungen und die Entstehung von Immunkörper', *München. med. Wchnschr.*, 1903, l. 764–8.

LANGHANS, T., 'Beobachtungen über Resorption der Extravasate und Pigmentbildung in denselben', *Arch. f. path. Anat.*, 1870, xlix. 66–116.

LEISHMAN, W. B., 'Note on a method of quantitatively estimating the phagocytic power of the leucocytes of the blood', *Brit. M. J.*, 1902, i. 73–5.

LEWIS, T. R., and CUNNINGHAM, D. D., 'Microscopical and physiological researches into the nature of the agent or agents producing cholera', *8th Ann. Rep. San. Comm. Gov. of India* (1871), Calcutta, 1872, Appendix C, pt. 2, p. 159.

LUBARSCH, O., 'Ueber die bacterienvernichtenden Eigenschaften des Blutes und' ihre Beziehungen zur Immunität', *Centralbl. f. Bacteriol.*, 1889, vi. 481, 529.

—— 'Ueber Abschwächung der Milzbrandbacillen im Froschkörper', *Fortschr. d. Med.*, 1888, vi. 121–9.

—— 'Untersuchungen über die Ursachen der angeborenen und erworbenen Immunität', *Ztschr. f. klin. Med.*, 1890–1, xviii. 421–68; 1891, xix. 80, 215, 360.

MADSEN, T., and JØRGENSEN, A., 'The fate of typhoid and cholera agglutinins during active and passive immunisation', *Festskr. ved Indvielsen af Statens Serum Inst.*, Copenhagen, 1902, Art. VI.

MASSART, J., 'Le chimiotaxisme des leucocytes et l'immunité', *Ann. de l'Inst. Pasteur*, 1892, vi. 321–7.

MENNES, FR., 'Das Antipneumokokken-Serum und der Mechanismus der Immunität des Kaninchens gegen den Pneumococcus', *Ztschr. f. Hyg. u. Infektionskrankh.*, 1897, xxv. 413–38.

METCHNIKOFF, E., 'Untersuchungen über die intracelluläre Verdauung bei wirbellosen Thieren', *Arb. a. d. zool. Inst. der Univ. Wien*, 1883, v. 141–68, 2 pls.

—— 'Ueber eine Sprosspilzkrankheit der Daphnien. Beitrag zur Lehre über den Kampf der Phagocyten gegen Krankheitserreger', *Arch. f. path. Anat.*, 1884[1], xcvi. 177–95, 2 pls.

—— 'Ueber die pathologische Bedeutung der intracellulären Verdauung', *Fortschr. d. Med.*, 1884[2], ii. 558–69.

—— 'Ueber die Beziehung der Phagocyten zur Milzbrandbacillen', *Arch. f. path. Anat.*, 1884[3], xcvii. 502–26, 2 pls.

—— 'Sur la lutte des cellules de l'organisme contre l'invasion des microbes', *Ann. de l'Inst. Pasteur*, 1887[1], i. 321–36.

—— 'Ueber den Kampf der Zellen gegen Erysipelkokken', *Arch. f. path. Anat.*, 1887[2], cvii. 209–49.

—— 'Ueber das Verhalten der Milzbrandbakterien im Organismus; Beitrag zur Phagocytenlehre', *Arch. f. path. Anat.*, 1888, cxiv. 465–92.

—— 'Recherches sur la digestion intracellulaire', *Ann. de l'Inst. Pasteur*, 1889, iii. 25–9.

—— 'Études sur l'immunité', *Ann. de l'Inst. Pasteur*, 1889, iii. 289–301; 1890, iv. 65, 193; 1891, v. 465–78; 1892, vi. 289–320.

—— 'Atrophie des muscles pendant la transformation des batraciens', *Ann. de l'Inst. Pasteur*, 1892, vi. 1–12, 2 pls.

—— *Leçons sur la pathologie comparée de l'inflammation*, Paris, 1892; Engl. trans., Lond., 1893.

—— 'L'état actuel de la question de l'immunité', *Ann. de l'Inst. Pasteur*, 1894, viii. 706–21.

—— 'Études sur l'immunité, 6e mém. Sur la destruction extracellulaire des bactéries dans l'organisme', *Ann. de l'Inst. Pasteur*, 1895, ix. 369–461.

—— 'Ueber die Immunität bei Infectionskrankheiten mit besonderer Berücksichtigung der Cellulartheorie', *Ergeb. d. all. Path. u. path. Anat.*, 1896, Abt. I, 298–343.

—— *L'immunité dans les maladies infectieuses*, Paris, 1901, 600 pp.

—— 'Zur Geschichte der Phagocytenlehre', *Handb. d. Immunitätsforschung u. exper. Therapie*, 1914, 2. Aufl., Lief. 1, p. 124–9.

MORESCHI, C., 'Zur Lehre von den Antikomplementen', *Berl. klin. Wchnschr.*, 1905, xlii. 1181–5.

MORGENROTH, J., 'Ueber den Antikörper des Labenzyms', *Centralbl. f. Bacteriol.*, Abt. I, 1899, xxvi. 349–59.

NEISSER, M,. and SACHS, H., 'Ein Verfahren zum forensichen Nachweis der Herkunft des Blutes', *Berl. klin. Wchnschr.*, 1905, xlii. 1388.

NEUFELD and HÜNE, 'Untersuchungen über baktericide Immunität und Phagocytose nebst Beiträgen zur Frage der Komplementablenkung', *Arb. a. d. kaiserl. Gesundheitsamte*, 1907, xxv. 164.

NEUBURGER, M., *Die Vorgeschichte der antitoxischen Therapie der acuten Infectionskrankheiten*, Stuttg., 1901, 67 pp.

NISSEN, F., 'Zur Kenntniss der bacterienvernichtenden Eigenschaft des Blutes', *Ztschr. f. Hyg.*, 1889, vi. 487–520.

NUTTALL, G., 'Experimente über die bacterienfeindlichen Einflüsse des thierischen Körpers', *Ztschr. f. Hyg.*, 1888, iv. 353–94, 1 pl.

PANUM, P. L., 'Das putride Gift, die Bakterien, die putride Infection oder Intoxication und die Septicämie', *Arch. f. path. Anat.*, 1874, lx. 301–52.

PASTEUR, L., 'Sur les maladies virulentes et en particulier sur la maladie appelée vulgairement choléra des poules', *Compt. rend. Acad. d. sc.*, 1880, xc. 239–48.

PETRUSCHKY, J., 'Untersuchungen über die Immunität des Frosches gegen Milzbrand', *Beitr. z. path. Anat. u. z. allg. Path.*, 1888, iii. 357–86.

PFAUNDLER, M., 'Ueber Gruppenagglutination und über das Verhalten des Bacterium coli bei Typhus', *München. med. Wchnschr.*, 1899, xlvi. 472–5.

PFEIFFER, R., 'Weitere Untersuchungen über das Wesen der Choleraimmunität und über specifisch bactericide Processe', *Ztschr. f. Hyg. u. Infektionskrankh.*, 1894, xviii. 1–16.

—— 'Ein neues Grundgesetz der Immunität', *Deutsche med. Wchnschr.*, 1896, xxii. 97, 119.

PFEIFFER, R., and FRIEDBERGER, E., 'Ueber Antikörper gegen die bacteriolytischen Immunkörper der Cholera', *Berl. klin. Wchnschr.*, 1902, xxxix. 4–7.

PFEIFFER, R., and ISSAËFF. 'Ueber die specifische Bedeutung der Choleraimmunität', *Ztschr. f. Hyg. u. Infektionskrankh.*, 1894, xvii. 354–400.

PFEIFFER, R., and MARX, 'Die Bildungsstätte der Choleraschutzstoffe', *Ztschr. f. Hyg. u. Infektionskrankh.*, 1898, xxvii. 272–97.

PORTIER and RICHET, CH., 'De l'action anaphylactique de certains venins', *Compt. rend. Soc. de biol.*, 1902, 2ᵉ sér., iv. 170–2.

RICHET, C., 'De l'anaphylaxie en général et de l'anaphylaxie par la mytilo-congestine en particulier', *Ann. de l'Inst. Pasteur*, 1907, xxi. 497–524.

RÖMER, P., 'Untersuchungen über die intrauterine und extrauterine

Antitoxinübertragung von der Mutter auf ihre Descendenten', *Berl. klin. Wchnschr.*, 1901, xxxviii. 1150–7.

RÖMER, P., 'Experimentelle Untersuchungen über Abrin-(Jequiritol)-Immunität als Grundlagen einer rationellen Jequirity-Therapie', *Arch. f. Ophthalmol.*, 1901, lii. 72–142.

ROSER, K., *Beiträge zur Biologie niederster Organismen*, Marburg, 1881, 23 pp., 1 pl.

ROUX, E., 'De l'immunité; immunité acquise et immunité naturelle', *Ann. de l'Inst. Pasteur*, 1891, v. 517–33.

—— 'Sur les sérums antitoxiques', *Ann. de l'Inst. Pasteur*, 1894, viii. 722–7.

ROUX, E., and CHAMBERLAND, 'Sur l'immunité contre le charbon conferée par des substances chimiques', *Ann. de l'Inst. Pasteur*, 1888, ii. 405–25.

ROUX, E., and VAILLARD, L., 'Contribution à l'étude du tétanos: prévention et traitement par le sérum antitoxique', *Ann. de l'Inst. Pasteur*, 1893, vii. 65–140.

SALMON, D. E., and SMITH, THEOBALD, 'On a new method of producing immunity from contagious diseases', *Proc. Biol. Soc. Wash.*, 1884–6, iii. 29–33.

—— —— 'Experiments on the production of immunity by the hypodermic injection of sterilized cultures', [Abstr.] *Tr. Internat. M. Congr.*, Wash., 1887, iii. 403–7.

SALOMONSEN, C. J., 'Undersøgelserne over Kemotaksis og deres Betydning for Patologien', *Biblioth. f. Laeger*, 1894, 7 R. v. 21–53.

SALOMONSEN, C. J., and MADSEN, T., 'Sur la reproduction de la substance antitoxique après de fortes saignées', *Ann. de l'Inst. Pasteur*, 1898, xii. 762–73.

—— —— 'Undersøgelser over Immunitet og Predisposition', *Overs. o. d. k. Danske Vidensk. Selsk. Forh.*, 1898, 227–35.

SAUERBECK, E., *Die Krise in den Immunitätsforschungen*, Leipz., 1909, 91 pp.

SCHOTTMÜLLER, 'Ueber eine das Bild des Typhus bietende Erkrankung hervorgerufen durch typhusähnliche Bacillen', *Deutsche med. Wchnschr.*, 1900, xxvi. 511.

SIROTININ, 'Ueber die entwickelungshemmenden Stoffwechselprodukte der Bacterien und die sog. Retentionshypothese', *Ztschr. f. Hyg.*, 1888, iv. 262–90.

STERN, R., 'Ueber die Wirkung des menschlichen Blutes und anderer Körperflüssigkeiten auf pathogene Mikroorganismen', *Ztschr. f. klin. Med.*, 1891, xviii. 46–71.

VON SZÉKELY, A., and SZANA, A., 'Experimentelle Untersuchungen über die Veränderungen der sog. mikrobiciden Kraft des Blutes während und nach der Infection des Organismus', *Centralbl. f. Bacteriol.*, 1892, xii. 61, 139.

TCHISTOVITCH, T., 'Études sur l'immunisation contre le sérum d'anguilles', *Ann. de l'Inst. Pasteur*, 1899, xiii. 406–25.

TRAUBE, M., and GSCHEIDLEN, 'Ueber Fäulniss und den Widerstand der lebenden Organismen gegen dieselbe', *Jahresber. d. schlesischen Gesellsch. f. vaterländ. Cultur* (1874), Breslau, 1875, lii. 179.

WASSERMANN, A., 'Experimentelle Beiträge zur Kenntniss der natürlichen und künstlichen Immunität', *Ztschr. f. Hyg. u. Infectionskrankh.*, 1901, xxxvii. 173-204.

WASSERMANN, A., and CITRON, J., 'Ueber die Bildungsstätten der Typhusimmunkörper. Ein Beitrag zur Frage der localen Immunität der Gewebe', *Ztschr. f. Hyg. u. Infektionskrankh.*, 1905, l. 331-48.

WASSERMANN, A., NEISSER, A., and BRUCK, C., 'Eine serodiagnostische Reaktion bei Syphilis', *Deutsche med. Wchnschr.*, 1906, xxxii. 745.

WASSERMANN, A., NEISSER, A., BRUCK, C., and SCHUCHT, A., 'Weitere Mittheilung über den Nachweis spezifisch luetischer Substanzen durch Komplementverankerung', *Ztschr. f. Hyg. u. Infektionskrankh.*, 1906, lv. 451-77.

WASSERMANN, A., and PLAUT, F., 'Ueber das Vorhandensein syphilitischer Antistoffe in der Cerebrospinalflüssigkeit von Paralytikern', *Deutsche med. Wchnschr.*, 1906, xxxii. 1769-72.

WEIGERT, C., 'Ueber Chemotaxis', *Hyg. Rundschau*, 1891, i. 589-97.

—— 'Neue Fragestellungen in der pathologischen Anatomie', *Verh. d. Gesellsch. deutscher Naturf. u. Ärzte. Frankf. a. M.*, Leipz., 1896, Theil I, 121-39.

WERNICKE, E., 'Die Immunität bei Diphtherie', *Handb. d. path. Mikroorg. v. Kolle-Wassermann*, 1913, 2. Aufl., v. 1020 (footnote).

WIDAL, F., 'Sérodiagnostic de la fièvre typhoïde', *Bull. et mém. Soc. méd. d'hôp. de Par.*, 1896, 3ᵉ sér., xiii. 561-6.

WIDAL, F., and SICARD, A., 'Étude sur le sérodiagnostic et sur la réaction agglutinante chez les typhiques', *Ann. de l'Inst. Pasteur*, 1897, xi. 353-432.

WRIGHT, A. E., *Studies on immunisation and their application to the diagnosis and treatment of bacterial infections*, Lond., 1909, 490 pp.

WRIGHT, A. E., and DOUGLAS, S. R., 'An experimental investigation of the rôle of the blood fluids in connection with phagocytosis', *Proc. Roy. Soc.*, Lond., 1903, lxxii. 357-70.

—— —— 'Further observations on the rôle of the blood fluids in connection with phagocytosis', *Proc. Roy. Soc.*, Lond., 1904, lxxiii. 128-42.

WYSSOKOWITSCH, W., 'Ueber die Schicksale der in's Blut injicirten Mikroorganismen im Körper der Warmblüter', *Ztschr. f. Hyg.*, 1886, i. 3-46, 1 pl.

BIOGRAPHICAL NOTICES

OF SOME OF THE EARLY WORKERS IN BACTERIOLOGY

ABBE, ERNST (born 1840, died 1905). Mathematician and optician. Studied in Jena and Göttingen. Privat Doc. in mathematics, physics, and astronomy in Jena 1863. In 1866 became associated with Carl Zeiss, manufacturer of optical instruments in Jena. On the death of Zeiss in 1888 Abbe became proprietor of the optical works. In 1891 founded Carl Zeiss Stiftung, to which he ceded all his proprietary rights. Homogeneous immersion lenses worked out in 1878.

ABBOTT, ALEXANDER CREVER (born 1860). American bacteriologist and hygienist. Born in Baltimore and became Prof. of Bacteriology and Director of Lab. of Hygiene in Philadelphia. Wrote successful bacteriological text-book.

ABEL, RUDOLF (born 1868). German bacteriologist and hygienist. Born at Frankfurt a. O. Was assistant to Loeffler in Greifswald and later professor in Jena. Published several important bacteriological discoveries such as the existence of antitoxin in normal persons.

ANDREWES, FREDERICK WILLIAM (born 1859, died 1932). English pathologist and bacteriologist. Educated at Oxford and St. Bartholomew's Hospital, London. Prof. of Pathology, Univ. of London, 1912. Knighted 1920. Wrote extensively on bacteriology of streptococci and immunological subjects, and was a lucid teacher and writer.

AOYAMA, TANEMICHI (born 1859, died 1917). Prof. of pathology in Tokyo. Was associated with Kitasato on plague work in Hong-Kong 1894. (Obituary, *J. Am. M. Ass.* 1918, lxx. 110.)

APPERT. Until recently nothing known about him, except that he was 'ancien confiseur et distillateur, élève de la bouche de maison ducale de Christian IV' as stated in his book, *L'art de conserver pendant plusieurs années toutes les substances animales et végétales*, Paris, 1810. He was said by some to have been the brother of Benj. Appert, the French philanthropist. It is now known, however, that he was Nicolas Appert and was born at Châlons-sur-Marne about 1750. He had a large business in Paris but died in poverty at Massy (Seine-et-Oise) 1 June 1841, aged 91.

ARKWRIGHT, JOSEPH ARTHUR (born 1864). English bacteriologist. Educated at Trinity College, Cambridge, and St. Bartholomew's Hospital, London. Practised medicine for some years and then became bacteriologist in the Lister Institute, London. Has published important work on many bacteriological subjects. Knighted 1937.

ARLOING, SATURNIN (born 1846, died 1911). Was Prof. of Physiology in Lyons and, later, succeeded Chauveau as Prof. of Experimental and Comparative Pathology and Director of the Lyons Veterinary School. Wrote extensively on the bacteriology of symptomatic anthrax, pleuro-pneumonia, tuberculosis, and other diseases. (Biog.: *J. de méd. vét. et zootech.*, Lyon, 1911, 5th s., xv. 129–39; Portrait and list of works: *Arch. internat. de pharmacod.*, 1912, xxii. 1–25.)

AUDOUIN, JEAN VICTOR (born 1797, died 1839). Eminent French naturalist and entomologist. Born in Paris. Became an authority on Arthropoda. His researches gave impetus to the study of entomology. He published important observations on muscardine of silk-worms. Gruby's *Microsporon Audouini* is called after him.

BABES, VICTOR (born 1854, died 1926). Bacteriologist and pathologist. Born in Vienna. Graduated in Budapest. Studied with Virchow and Koch in Berlin and with Cornil in Paris. In 1887 was appointed Prof. of Pathology and Bacteriology in Bucharest. He was a voluminous writer on many pathological and bacteriological subjects. With Cornil he wrote a large book, *Les bactéries* (1885), which had a wide circulation. (Pagel, *Biog. Lex. Berlin-Wien*, 1901 [portrait].)

BAIL, OSCAR (born 1869, died 1927). German bacteriologist and hygienist, born in Tillisch (Böhmen) and trained as zoologist. Studied medicine in Vienna and became Prof. of Hygiene in Prague. Worked chiefly at immunology and developed the theory of aggressins. (*Ztschr. f. Immunitätsforsch. u. exper. Therap.*, 1928, lv, pp. i–iv; *Klin. Wchnschr.*, 1928, vii. 719; *Deutsche med. Wchnschr.*, 1928, liv. 286.)

BAKER, HENRY (born 1698, died 1774). English naturalist. Taught the deaf and dumb. Son-in-law of Defoe. By the publication of *The microscope made easy* (1742) and *Employment for the microscope* (1753) he did much to concentrate interest in the work of van Leeuwenhoek and the other early microscopists and their instruments. (*Dict. Nat. Biog.* iii. 9.)

BANG, BERNHARD LAURITS FREDERICK (born 1848, died 1932). Eminent Danish veterinary pathologist and bacteriologist. Born in Sorö (Sjaelland). Studied in Copenhagen. Became Director of the Veterinary High School in Copenhagen. He carried out important researches on tuberculosis and discovered *Bacillus abortus*.

DE BARY, ANTON (born 1831, died 1888). Mycologist and botanist. Born in Frankfort-am-Main. Studied medicine there and at Heidelberg and Marburg. Was a pupil of von Mohl and became Prof. at Freiburg, Halle, and ultimately (1872) in Strassburg. Made numerous contributions to the science of mycology, of which he was one of the founders. Devoted much attention to the study of bacteria. De Bary died of sarcoma of the jaw in his 57th year. (*Deutsche med.*

Wchnschr., 1888, xcviii. 118 [F. Cohn]; Jost. L., *Zum hundertsten Geburtstag Anton de Barys*; [portrait] *Ztschr. f. Botanik*, 1930, xxiv.)

BASSI, AGOSTINO (born 1773, died 1856). By his demonstration (1835) of the parasitic nature of the muscardine disease of silkworms Bassi may be regarded as the founder of the doctrine of pathogenic microbes. He was born in Mairago, near Lodi, and was educated for the law but studied natural science in Pavia. He held various civil posts under the French and Austrian Governments but had to give them up on account of persistent ill health and failure of vision. Reduced to great financial straits he tried farming but was unsuccessful. By means of a legacy inherited from a relative he was able to discharge his debts and to help the needy and infirm. He wrote on the cultivation of potatoes (1812, 1817), on cheese (1820), vinification (1823), contagion (1844), pellagra (1846), cholera (1849). His great work, *Del mal del segno calcinaccio o moscardino*, was published in Lodi in 1835 and 1836. (Riquier, *Agostino Bassi e la sua opera*, Pavia, 1924, 63 pp. [portrait]; *Opere di Agostino Bassi*, Pavia, 1925, 673 pp. [portrait].)

BASTIAN, HENRY CHARLTON (born 1837, died 1915). Born in Truro and educated at University College, London, where he became Prof. of Pathology and Physician to University College Hospital. He was a distinguished neurologist but is principally known as the great champion of the doctrine of abiogenesis and as a doughty opponent of Pasteur. He wrote a number of books in support of his views. He died poor. (Biog.: *Brit. M. J.*, 1915, ii. 795; *Lancet*, 1915, ii, 1220–4, with portrait; *Nature*, 1915, xcvi. 347; *Proc. Roy. Soc.*, 1916, B, lxxxix, pp. xxi–xxiv.)

VON BAUMGARTEN, PAUL (born 1848, died 1928). German pathologist and bacteriologist. Born in Dresden. In 1874 became Prof. of Pathological Anatomy in Königsberg. In 1889 Prof. in Tübingen, where he remained until he retired. Baumgarten was a pioneer in bacteriology and did much to further the science by the publication of papers and books. His *Lehrbuch der pathologischen Mykologie* (*1886–90*) remains one of the best books on the subject. His accurate and critical *Jahresbericht über die Fortschritte in der Lehre von den pathogenen Mikroorganismen* (1885–1911) has been indispensable to bacteriologists. (*Deutsche med. Wchnschr.*, 1918, xliv. 1111; *München. med. Wchnschr.*, 1928, lxxv. 1507 [portrait].)

BEALE, LIONEL SMITH (born 1828, died 1906). Was educated at King's College, London, where he became Prof. of Physiology in 1852. He was a voluminous writer on microscopy and held untenable views on the nature of living matter. Opposed the germ theory of disease. He was F.R.S. (1857). (Biog.: *Brit. M. J.*, 1906, i. 836, and *Lancet*, Lond., 1906, i. 1004–7 [portrait]; *Dict. Nat. Biog.*, 2nd Suppl., vol. i, p. 120.)

BÉCHAMP, [Pierre Jacques] Antoine (born 1816, died 1908). French chemist and implacable opponent of Pasteur and the germ theory. Born at Bassing (Lorraine). Spent some years in Rumania. Returned to France and studied in Strassburg, where he met Pasteur. Béchamp became Prof. of Medical Chemistry and Pharmacy in Montpellier, and in 1874 Dean of the Free Faculty of Medicine in Lille. Resigned after eleven years and settled first in Havre and then in Paris, where he died in a room in the Quartier Latin aged 92. For some years B. attained great notoriety with his theory of micro-zymas but this is now forgotten. (Biog. and portrait in Hume, E. D., *Béchamp or Pasteur? A lost chapter in the history of biology*, Chicago, 1923, 295 pp.)

von BEHRING, Emil [Adolf] (born 1854, died 1917). Discoverer of antitoxin and principle of serotherapy. Born in Deutsch-Eylau. Studied in Berlin. Entered Army Medical Corps and was lecturer in Army Medical College, Berlin, 1888. Assistant in Koch's Institute 1889. Prof. of Hygiene in Halle 1894, and in Marburg from 1895. Behring was a most diligent worker and made his great discovery in 1890. Received many distinctions and several monetary prizes. He established works in Marburg for the manufacture of antitoxins and a remedy for tuberculosis of cattle. (*Med. Klinik*, 1917, xiii. 407; Biog.: *Berl. klin. Wchnschr.*, 1917, liv. 471; *Lancet*, London, 1917, i. 890; *Parasitology*, 1924, xvi. 214 [portrait].)

BEIJERINCK, Martinus Willem (born 1851, died 1931). Great Dutch microbiologist. Born in Amsterdam. Educated in Haarlem. Lecturer on botany, physiology, and physics at Agricultural School at Warffum (Groningen). Lecturer in Utrecht and in Wageningen for ten years. In 1897 bacteriologist to Yeast works in Delft, and from 1898 to 1921 as Prof. of General Bacteriology in the High School at Delft. B. published a large number of highly important researches dealing with fundamental problems in the physiology of bacteria, the bacteria of soil and plants, and plant infective diseases. Foreign member of the Royal Society 1926. His *Verzamelde Geschriften van M. W. Beijerinck ter Gelegenheid van zijn 70sten Verjaardag*, publ. at Delft, 1921–2, 5 parts. Biog.: *Proc. Roy. Soc.* B, cix, pp. i–iii.

BENNETT, John Hughes (born 1812, died 1875). Physiologist and physician. Prof. of the Institutes of Medicine in Edinburgh, 1848–74. Born in London, educated in Exeter and in France. Studied medicine in Edinburgh and graduated there 1839. He was one of the first teachers of histology and wrote extensively on pathology and medicine. In 1868 he published experiments in support of spontaneous generation of bacteria. Died in Norwich. (*Edinb. M.J.*, 1875, xxi. 466; *Brit. M.J.*,. 1875, ii. 473; Comrie, J., *History of Scottish Medicine*, Lond., 1927, 276 [portrait].)

BERNARD, CLAUDE (born 1813, died 1878). Great French physiologist. Born at St. Julien, near Villefranche (Rhône). Studied in Lyons and Paris. Became Preparateur to Magendie in the Collège de France and ultimately succeeded him as Prof. From a bacteriological point of view it is interesting that Bernard left notes of experiments which indicated that he considered that alcoholic fermentation could take place apart from living ferments, a view which greatly perturbed Pasteur. Bernard received the Copley medal of the Royal Society in 1876. (Foster, M., *Claude Bernard*, Lond., 1899, 245 pp.)

BERT, PAUL (born 1833, died 1886). French physiologist and politician. Educated as an engineer, lawyer, and doctor. Succeeded Claude Bernard in Paris. Took to politics and was minister of public instruction under Gambetta 1881. Returned to scientific pursuits but was made Resident-General in Tongking, where he died of dysentery. Wrote classical works on physiology and was an early but unsuccessful worker on protective inoculation against anthrax. (Biog.: *Compt. rend. Soc. de biol.*, Paris, 1887, 8th s. iv, pt. 2, pp. 17–24.)

BILLROTH, [CHRISTIAN ALBERT] THEODOR (born 1829, died 1894). Surgeon and pathologist. Born at Bergen in Rügen. Studied in Greifswald, Göttingen, and Berlin. Assistant to Langenbeck 1853–60. Prof. of Surgery in Zürich. From 1867 he was Director of Surgical Clinic and Prof. of Surgery in Vienna and attained great fame. Wrote important works on surgical pathology and in bacteriology, but was unfortunate in his interpretation in regard to surgical infections of bacterial origin. (Portrait, *Arch. f. klin. Chir.*, 1894, xlvii, Heft 1 [Frontispiece].)

BIRCH-HIRSCHFELD, F[ELIX] V[ICTOR] (born 1842, died 1899). German pathological anatomist and pioneer in bacteriology. Born near Rendsburg (Holstein). Pupil of Wunderlich and E. Wagner in Leipzig. Prosector in Stadt-Krankenhaus in Dresden. Succeeded Cohnheim as Prof. of Pathological Anatomy in Leipzig. Voluminous writer on pathological and bacteriological subjects, especially Pyaemia. Died in Leipzig. (*Beitr. z. path. Anat. u. z. allg. Path.*, 1900, xxvii, pp. iii–xii; *München. med. Wchnschr.*, 1900, xlvii. 53–5 [portrait].)

VON BOLLINGER, OTTO (born 1843, died 1909). German morbid anatomist and bacteriologist. Born in Altenkirchen (Rheinpfalz). Studied in Munich, Vienna, and Berlin. From 1880 he was Prof. of Pathological Anatomy in Munich. A voluminous writer and active worker, especially on animal diseases such as anthrax, symptomatic anthrax, actinomycosis, and tubercle. (*München. med. Wchnschr.*, 1909, lvi. 2058.)

BORDET, JULES [JEAN BAPTISTE VINCENT] (born 1870). Belgian bacteriologist born at Soignies, Belgium. Studied at Univ. of

Brussels. In 1894 went to Institut Pasteur, Paris, and was preparateur in the laboratory of Metchnikoff until 1901, when he founded
the Institut Pasteur in Brussels. In 1907 Prof. of Bacteriology in
Brussels. Foreign member of the Royal Society 1916. Nobel
Prizeman 1919. Bordet has enriched many branches of bacteriology and by common consent is regarded as the greatest living
exponent and worker on immunology. An original researcher of the
first rank.

BOSTROEM, EUGEN (born 1850, died 1928). German pathologist
and Prof. of Pathological Anatomy from 1883 to 1926 in Giessen.
He worked at the bacteriology of actinomycosis. (*München. med.
Wchnschr.*, 1928, lxxv. 1419 [portrait]; *Deutsche med. Wchnschr.*,
1928, liv. 1389.)

BOYCE, RUBERT WILLIAM (born 1863, died 1911). An Irishman,
educated in England, Ireland, France, and Germany. Was Assistant
Prof. of Pathology in University College, London. In 1894 became
Prof. of Pathology in Univ. of Liverpool and did a great deal to
build up the reputation of the medical faculty there. He also
founded the Liverpool School of Tropical Medicine. In 1906 he
had a hemiplegic attack which disabled him partially and he ceased
to do scientific work. Prior to this he published much on bacteriological work in reference to hygiene and tropical diseases. A man of
small stature but of great energy. (*J. Path. and Bacteriol.*, 1911–12,
xvi. 276–82 [portrait]; *Proc. Roy. Soc.* 1912, B, lxxxiv, pp. iii–ix.)

BRAUELL, FRIEDRICH AUGUST (born 1807, died 1882). Was the
first to discover *Bacillus anthracis* in man (1857) but did not believe
it was peculiar to the disease. Born 11 Dec. 1807 in Weimar.
Studied in Jena, Berlin, and Copenhagen. Dr. Phil. Erlangen 1834.
In 1838 was in Wilna, 1841 in Kasan. 1846 Extraordinary Prof. in
Kasan. 1848 Prof. in Dorpat. Retired 1868. Removed to Leipzig
and in 1871 was ordinary honorary professor there and lectured on
pathological anatomy of domestic animals. He died 10 Dec. 1882.
(*Centralbl. f. Bacteriol.*, I. Abt. Orig., 1929, cxv. 1–17.)

BREFELD, OSCAR (born 1839, died 1925). Born at Telgte (Westphalia) and early took an interest in botany. He began his famous
mycological studies in 1868 and at the outset realized the necessity
of sterilizing culture media and of studying microbes and spores as
individuals. He devised original methods for pure cultivation. He
was present at the siege of Paris. In 1872 he published the first
volume of his great mycological work which reached 18 volumes.
He was Prof. at the Forestry Academy in Eberswalde and later in
Münster and Berlin. In his later years this great investigator became
blind. He died in 1925, aged 86 years. (*Nature*, 1925, cxvi. 369.)

BRETONNEAU, PIERRE FIDÈLE (born 1778, died 1862). Great
French clinician of Tours. Born at St.-Georges-sur-Cher. Studied

at the École de Santé, Paris, 1795. Returned home on account of ill health, and on returning in 1799 failed to pass his examination for doctorate. As an 'Officier de Santé', he set up in practice at Chenonceaux and was subsequently appointed physician to the hospital of Tours. Obtained doctor's degree in 1814 and became médecin-en-chef at Tours, where he carried out his chief researches. He differentiated and named Diphtheria, and gave the first accurate account of Enteric fever—its pathology and specificity. Apart from his important work *Des inflammations spéciales du tissu muqueux* . . . (1826) he wrote comparatively little during his life. His views were disseminated mainly by his pupils, especially Trousseau. Bretonneau's masterly treatises on enteric fever (Dothinentérie) and specificity were published posthumously in 1922. He died in 1862 aged 84. (Biog.: *Proc. Roy. Soc. Med.*, 1924, xviii, Sect. Hist. of Med., pp. 1-12; *Progrès méd.*, 1937, Suppl. 1-8.)

BRIEGER, Ludwig (born 1849, died 1919). German physician and chemist. Born in Silesia. Worked with Nencki in Berne and with Frerichs and Koch in Berlin. Ultimately was Director of the University Institute for Hydrotherapy in Berlin. Wrote extensively on many subjects but obtained distinction chiefly for work on ptomaines and toxins. (Biog.: *Med. Klin.*, 1919, xv. 757; *Ztschr. f. exper. Path. u. Therap.*, 1909, vi. 913–16.)

BRUCE, David (born 1855, died 1931). English bacteriologist and protozoologist. Born in Melbourne, educated in Scotland. Studied medicine in Edinburgh. Entered Royal Army Medical Corps 1883 and rose to the rank of major-general. Served in Malta and there discovered the cause of Malta fever 1887. Worked out the aetiology of tsetse-fly disease in Zululand 1894–5, and investigated sleeping-sickness in Uganda 1903. He received many recognitions for his discoveries. Knighted in 1908. He died 27 Nov. 1931, four days after the death of his wife, who had actively collaborated with him in all his researches.

BUCHANAN, Robert MacNeil (born 1861, died 1931). City bacteriologist, Glasgow. Born at Drymen, Stirlingshire. Educated at Glasgow Univ. M.B. Glasgow, 1888. He also studied with Weigert and Koch, and was assistant to Dr. Joseph Coats of Glasgow Western Infirmary. In 1900 Buchanan investigated plague in Glasgow.

BUCHNER, Edouard (born 1860, died 1907). Brother of Hans Buchner. Born in Munich and became Director of the Analytical Department of the Univ. Chemical Institute in Munich. In 1896 Prof. in Tübingen and later in Berlin. Discovered Zymase and opened up new field in fermentation. Received Nobel Prize (1907) for his work. (Biog.: *München. med. Wchnschr.*, 1908, lv. 342.)

BUCHNER, Hans (born 1850, died 1902). German bacteriologist and immunologist. Educated in Munich and Leipzig. Military

Surgeon 1875, Staff Surgeon 1885, and finally Surgeon-general. In 1894 he succeeded Pettenkofer as Prof. of Hygiene at Munich. H. Buchner was a prolific worker, and at first supported the doctrines of Nägeli in opposition to those of Koch. Studied the action of light upon bacteria, but chiefly devoted himself to the study of the bactericidal action of serum, and the humoral theory of immunity. Brother of Edouard Buchner. (Biog.: *München. med. Wchnschr.* 1902, xlix. 844–7.)

BUDD, WILLIAM (born 1811, died 1880). Born in North Tawton. Studied in Paris, London, and Edinburgh. Settled in Bristol and was on the staff of the hospital there. A man of great ability and a strong and early upholder of the specificity of infection and the germ theory of disease. (*Brit. M.J.*, 1880, i. 163; *Lancet*, 1880, i. 148; *Med. Times and Gaz.*, 1880, i. 79; Goodall, E. W., *William Budd, the Bristol physician and epidemiologist*, Bristol, 1936, 159 pp.)

BUFFON, GEORGES LOUIS LECLERC, COMTE DE (born 1707, died 1788). French writer on Natural History. Born at Montbard. Travelled in France, Italy, and England. Director of the Jardin du Roi in Paris 1739–88. Began to publish his famous *Histoire naturelle* in 1749. Along with Needham he developed the doctrine of organic molecules in reference to the origin of life.

VON BUMM, ERNST (born 1858, died 1925). Gynaecologist. Born in Würzburg. Was Prof. in Basel, Halle, and finally Director of the Frauenklinik in the Charité Hospital in Berlin. He wrote a classical monograph on the micro-organism of gonorrhoea, which he was the first successfully to cultivate (1885). (*Arch. f. Gynäk.*, 1924–5, cxxiii, pp. i–iv [portrait].)

BUXTON, BERTRAM HENRY (born 1852, died 1934). Elder brother of Viscount Buxton. In his earlier years Bertram Buxton took up bacteriology and pathology and was for a time instructor in bacteriology at Cornell University, and later was Prof. of Bacteriology there until he retired in 1912, when he settled in England and worked at the Royal Horticultural Society's Laboratory at Wisley, where he investigated the colours of plants. He became stone deaf.

CAGNIARD-LATOUR, CHARLES (born 1777, died 1859). French physicist and engineer. Born in Paris. Studied at the École polytechnique and became one of the 'ingénieurs géographiques'. Was ennobled with rank of baron in 1818. He made many inventions, including the blowing machine, called 'cagniardelle' after him. In 1836 he announced his conviction that yeast globules are organized bodies and probably of a vegetable nature, and he described the reproduction by budding.

CALMETTE, ALBERT (born 1863, died 1933). French bacteriologist and hygienist. Born in Nice. Educated at Clermont-Ferrand and in Paris. Entered naval and colonial service as surgeon. Founded Pasteur Institute in Saigon. Prof. in Lille. Sub-director Pasteur Institute, Paris. Published important work on snake venom, plague serum, tubercle and many other subjects. He died 29 Oct. 1933. (*Ann. de l'Inst. Pasteur*, 1933, li. 559–64.)

DA CAMARA-PESTANA, LUIZ (born 1863, died 1899). Portuguese bacteriologist. Educated in Lisbon. Director of bacteriological institute there. While investigating plague in the Oporto epidemic (1899) he contracted the disease from a post-mortem wound and died. The Istituto bacteriologico da Camara-Pestana was named after him. (*Brit. M.J.*, 1899, ii. 1713.)

CARBONE, TITO (born 1863, died 1904). Was assistant to Lombroso, and later Prof. in Cagliari, Modena, and Pisa, where he died of Malta fever contracted in the course of his investigations on this disease.

CARLE, ANTONIO (born 1854, died 1927). Born in Chiusa di Pesio (Cuneo) and studied in Turin. He became Prof. of Surgery in Turin. With G. Rattone, Carle first showed the infectivity of tetanic pus in 1884.

CARTER, HENRY VANDYKE (born 1831, died 1897). Eminent tropical pathologist. Educated at University College and at St. George's Hospital, where he drew the illustrations for Gray's *Anatomy*. Joined the Bombay Medical Service 1858, and was Prof. of Anatomy and Physiology in Grant Med. Coll. Returned to Europe on leave and visited Norway, Italy, Greece, Algiers, and Crete to study leprosy, which he afterwards investigated in India. He also studied mycetoma, surra, and malaria, and worked at relapsing fever in the great Indian famine 1877–8. Died of phthisis aged 66. (*Brit. M.J.*, 1897, i. 1256; *Lancet*, 1897, i. 1381.)

CATTANI, GIUSEPPINA (born ?, died ?). Italian bacteriologist. Libera docente in Bologna. Collaborated with Tizzoni in many important researches on tetanus and its antitoxin.

CHABERT, PHILIBERT (born 1737, died 1814). French veterinary surgeon. Born at Lyons. He was Prof. at Alfort, and among other works wrote *Traité du charbon ou anthrax dans les animaux*, Paris, 1783—an important work.

CHAMBERLAND, CHARLES (born 1851, died 1908). French bacteriologist and one of Pasteur's chief collaborators on anthrax, rabies, and other diseases. He perfected the autoclave and the bacterial filter, to both of which his name has been applied. He was Chef de service and Sub-director in the Pasteur Institute in Paris. He was deputy for Jura and was an ardent politician. (*Ann. de l'Inst. Pasteur*, 1908, xxii. 369–80 [portrait].)

CHANTEMESSE, ANDRÉ (born 1851, died 1919). French pathologist, bacteriologist, and hygienist. Born at Puy. Was a pupil of Cornil in Paris, and was an associate of Pasteur. He finally became Prof. of Hygiene in the Faculty of Medicine in Paris. Chantemesse was a voluminous writer on pathological subjects and made special study of enteric fever. (*Ann. de l'Inst. Pasteur*, 1919, xxxiii. 137; *Brit. M.J.*, 1919, i. 431.)

CHARRIN, ALBERT (born 1856, died 1907). French physiologist and pathologist. Born at Condrieu. Studied in Lyons and Paris. Chef de laboratoire de pathologie générale. Physician. Finally Prof. of General and Comparative Pathology in the Collège de France. Charrin wrote on glanders, 'maladie pyocyanique', bacterial poisons, immunity, bactericidal action of serum, and many other subjects. Died of a lingering disease aged 52. (*Presse méd.*, 1907, xv, annexes, 329 [portrait]; *Compt. rend. Soc. de biol.*, 1907, lxii. 926–8; *Bull. et mém. Soc. méd. d. hôp. de Par.*, 1907, 3rd s. xxiv. 525.)

CHAUVEAU, [JEAN BAPTISTE] A[UGUSTE] (born 1827, died 1917). One of the chief French physiologists and pathologists of the nineteenth century. Born at Villeneuve-le-Guyard (Yonne). Studied veterinary medicine at Alfort and at Lyons, where he became Director of the Vet. School (1875). Inspector-General of Veterinary services in France and Prof. of Comparative Pathology at the Natural History Museum in Paris. In addition to physiological researches Chauveau was a pioneer worker on infective diseases and wrote on virus and its attenuation, cow-pox, contagion, tuberculosis, and many other subjects connected with bacteriology and pathology. (Biog.: *J. de physiol. et de path. gén.*, 1917, xvii, pp. i–ii [portrait]; *Bull. Acad. de méd.*, Paris, 1917, 3rd s., lxxvii. 56.)

CHEYNE, WILLIAM WATSON (born 1852, died 1932). English surgeon and bacteriologist. Born in Fetlar (Shetland). Educated in Aberdeen and in Edinburgh, where he became associated with Lister and went with him to King's College, London. Here he became Prof. of Surgery, President of the Royal College of Surgeons, and a baronet. Retired to Shetland. In his earlier days Cheyne was one of the most prominent bacteriologists in England and published important work.

COHN, FERDINAND (born 1828, died 1898). Great German botanist and one of the founders of bacteriology. Born in Breslau, where he was for many years Prof. of Botany. He early took to the study of microscopic algae and fungi and made many important discoveries. From 1860 onwards devoted himself particularly to the study of bacteria and became the leading authority on the subject. He was one of the first to hold that bacteria can be arranged in genera and species which exhibit a high degree of constancy. Much of our knowledge is based on his work. He supported Pasteur's ideas on spontaneous generation in opposition to Pouchet and Bastian, and first clearly described bacterial spores. He wrote a great deal and most of it was accurate. He discovered Robert Koch and befriended him. In Cohn's *Beiträge zur Biologie der Pflanzen* appeared many of the classical papers on bacteriology by Cohn, Schroeter, Koch, and others. Cohn was a man of great diligence and talent and personally a fine character. (Cohn, Pauline, *Ferdinand Cohn, Blätter der Erinnerung*, Breslau, 1901, 258 pp.; *München. med. Wchnschr.*, 1898, xlv. 1005 [portrait].)

COHNHEIM, JULIUS (born 1839, died 1884). One of the greatest pathologists. Born in Demmin in Pomerania. Worked in Berlin and was Prof. in Kiel (1868), Breslau (1872), and Leipzig (1878–84), where he died of chronic renal disease, aged 45. Famous among his works were those on inflammation, embolism, and his *Lectures on General Pathology*. Cohnheim did much to further bacteriology in the early days and was a staunch supporter of Robert Koch. (Cohnheim's *Ges. Abhandl.*, Berlin, 1885, pp. vii–li [portrait]; *Berlin klin. Wchnschr.*, 1884, xxi. 564–6 [by Weigert].)

CONN, HERBERT WILLIAM (born 1859, died 1917). American dairy bacteriologist and writer of bacteriological text-books. (*J. Bact.*, 1917, ii. 501 [portrait]; *Med. Rec. N.Y.*, 1917, xci. 779.)

CORNET, GEORGE (born 1858, died 1915). German physician and bacteriologist. Born in Eichstatt (Bavaria). Studied in Munich. Developed tuberculosis and took up this subject scientifically. Was assistant to Brehmer in Görbersdorf, and then became a pupil of Koch in Berlin, where he carried out extensive researches on the bacteriology of tuberculosis. The modern ideas of prophylaxis against this disease were largely the outcome of his work. He died of typhus contracted in a prisoners' camp during the Great War. (*München. med. Wchnschr.*, 1915, lxii. 711.)

CORNIL, ANDRÉ-[VICTOR] (born 1837, died 1908). Eminent French pathological anatomist and bacteriologist. Born in Cusset (Allier). Educated in Paris, where he became (1882) Prof. of Pathology. Wrote extensively on morbid anatomy and was an early worker in the bacteriological field. With Ranvier he wrote an important work on pathological histology, and with V. Babes, *Les bactéries* (1885). (*Bull. Acad. de méd.*, Paris, 1908, 3rd s., lix. 480; *Brit. M.J.*, 1908, i. 1150.)

COUNCILMAN, WILLIAM THOMAS (born 1854, died 1933). American pathologist. Born in Pikesville, Maryland, 1 Jan. 1854. M.D. 1878. Associate Prof. of Pathology at Johns Hopkins Univ. 1886–91. Shattuck Prof. of Pathological Anatomy at Harvard University Medical School, 1892–1922. He published important researches on dysentery, cerebro-spinal meningitis, diphtheria, and small-pox. He died at York village, Maine, 26 May 1933, aged 79.

COURMONT, JULES (born 1865, died 1917). French bacteriologist and hygienist. Born and studied in Lyons, where he worked for years in the laboratory of S. Arloing. In 1900 he became Prof. of Hygiene in Lyons. In his earlier years he published much bacteriological and experimental work and later on hygiene. Served in the Great War and died of apoplexy in the Hôtel-Dieu, Paris, while visiting his patients there (1917). (Biog.: *Compt. rend. Soc. de biol.*, 1917, lxxx. 245–8; *J. de physiol. et de path. gén.*, 1917, xvii, no. 2, pp. i–ii [portrait].)

COZE, LÉON (born 1817, died 1896). Born in Strassburg, where he became Prof. of Pharmacy. Later he was Prof. in Nancy. Along with V. Feltz he was one of the pioneers of the germ theory of disease, 1866–70. (Biog.: *Rev. méd. de l'est*, Nancy, 1896, xxviii. 639–44.)

CROOKSHANK, EDGAR M[ARCH] (born 1858, died 1928). Studied in London at King's College, under Lister. Was on special service at battle of Tel-el-Kebir. In 1885 studied with Koch in Berlin. 1886 published *Manual of Bacteriology* and became Prof. of Bacteriology in King's College, where he founded the first bacteriological laboratory in England. Published important work on *Photography of Bacteria* (1887) and *The History and Pathology of Vaccination* (1889). Became interested in politics and big-game hunting and retired from his professorial chair and bacteriological work.

CUNNINGHAM, DAVID DOUGLAS (born 1843, died 1914). English pathologist, naturalist, and bacteriologist. Born in Prestonpans. Graduated in Edinburgh. Entered Indian Medical Service. Studied with Hallier and De Bary. With Timothy Lewis, Cunningham made exhaustive study of cholera in India. Became Prof. of Physiology in Calcutta but retired in 1898 owing to ill health. Lived in Torquay, where he died. Cunningham was a high-class worker and did much in the study of tropical disease causes. F.R.S. 1889. (Biog.: *Brit. M.J.*, 1915, i. 89, 141; *Proc. Roy. Soc. Lond.*, 1916, B., lxxxix, pp. xv–xx.)

DALLINGER, WILLIAM HENRY (born 1842, died 1909). Wesleyan minister and biologist. President of Wesleyan College, Sheffield. He made classical observations on the life-history of flagellates and monads and worked on spontaneous generation. He did much to advance microscopy. (*Proc. Roy. Soc., Lond.*, 1910, B., lxxxii, pp. iv–vi.)

DANYSZ, JEAN (born 1860, died 1928). Born in Poland but early settled in France, where he worked in the Pasteur Institute in Paris. Discovered virus for destroying rodents, and visited Portugal, S. Africa, and Australia in connexion with this work. He discovered the 'Danysz phenomenon' in immunity reactions and later worked on chemotherapy. He held unusual views on the bacterial treatment of disease. (*Bull. de l'Institut Pasteur*, 1928, xxvi. 97, *Presse méd.*, 1928, xxxvi. 158.)

DAVAINE, CASIMIR-JOSEPH (born 1812, died 1882). French pathologist, parasitologist, and experimenter. With Rayer he first saw (1850) the bacillus of anthrax. Born at St. Amand-les-Eaux (Nord). Studied in Paris, where he entered the service of Rayer at the Charité Hospital. Assisted Rayer in the foundation of Soc. de biologie 1848. Wrote important treatise on animal parasites of man and domestic animals 1860. He was one of the early writers on infective diseases and published classical work on anthrax and septicaemia. He was in general practice and never held any public appointment nor had he a laboratory. He kept his experimental animals in the garden of a friend. He died at Garches (Seine-et-Oise), 14 Oct. 1882. (Jeanson, G., *Le docteur Casimir Davaine; sa vie; son œuvre, 1812–82*, Paris, 1912; *Gaz. méd. de Paris*, 1882, 6th s., iv. 521; 533; *Progrès méd.*, 1928, xliii. 2157 [portrait in Salomonsen's *Smaa-Arbejder*, Kjøbenhavn, 1917, p. 20].)

DEAN, GEORGE (born 1863, died 1914). Born in Aberdeenshire. Educated in Aberdeen, Berlin, and Vienna. Was assistant to the Prof. of Pathology, Aberdeen Univ., and in 1897 Superintendent of antitoxin department, Lister Institute of Preventive Medicine in Elstree and in London. In 1908 he became Prof. of Pathology in the Univ. of Aberdeen, but in four years had to give up work on account of ill health. Died in Aberdeen, 1914. Dean was an active research worker and published accurate work on diphtheria immunization, phagocytosis, antitrypsin, rat leprosy, and tuberculosis. (*J. Path. and Bacteriol.*, 1914–15, xix. 114 [portrait]; *Brit. M.J.*, 1914, i. 1389, 1435.)

DOLD, HERMANN (born 1882). German bacteriologist. Born in Stuttgart. Demonstrator of Bacteriology in Royal Institute of Public Health in London 1908, and in Reichsgesundheitsamt in Berlin 1910. Privat Docent in Strassburg, 1912. German Medical School in Shanghai, 1914. Prof. in Strassburg (1917), Halle (1919), Marburg (1921). Director of Behring Institute in Marburg.

DÖNITZ, [FRIEDRICH KARL] WILHELM (born 1838, died 1912). German bacteriologist and naturalist. Studied in Berlin, and worked at anatomy and histology. Was appointed Lecturer on Anatomy in Tokyo and organized anatomical teaching in Japan. He then took up Hygiene and on returning to Europe (1886) became associated with the Koch School. In 1896 was on the staff of Ehrlich's Institute for Serum-forschung in Steglitz, and did much work on standardization of diphtheria antitoxin and tuberculin. He rejoined the Institute of Infectious Diseases in Berlin, and was acting Director in Koch's absence in Africa. In old age he worked at Ticks in relation to disease. He died in Berlin, 1912. (*Deutsche med. Wchnschr.*, 1912, xxxviii. 718 [Gaffky]; *Parasitology*, Cambridge, 1912–13, v. 253–8, 1 pl. [Nuttall].)

DONNÉ, ALFRED (born 1801, died 1878). Born in Noyon. Studied in Paris. Was chef de clinique in the Charité Hospital in Paris and gave courses on microscopy. Later he was sub-librarian in the Faculty of Medicine in Paris, and Inspector-General of the University. He was an admirable microscopist and made several important discoveries.

DOUGLAS, STEWART RANKEN (born 1871, died 1936). Studied medicine at St. Bartholomew's Hospital, London, and entered Indian Medical Service. Lieutenant 1898; Captain 1901. Served in China Expedition 1900–1. He contracted dysentery and retired from the I.M.S. with rank of Captain. Took up bacteriological work with Sir A. E. Wright at St. Mary's Hospital, London, and afterwards became Director of Bacteriological Department of the National Institute of Medical Research. Douglas did a great deal of research work and was especially a highly skilled technician. He was F.R.S.

DUCLAUX, ÉMILE [PIERRE] (born 1840, died 1904). French chemist and bacteriologist. Born at Aurillac (Cantal). Studied at École normale, Paris, where he was preparateur to Pasteur and helped him in silk-worm work. Prof. at Tours, Clermont, and Lyons. Finally Prof. of Biological Chemistry at the Sorbonne. Succeeded Pasteur as Director of the Pasteur Institute. Duclaux was a man of the highest intellectual standing and published much on chemistry and bacteriology. (*Ann. de l'Inst. Pasteur*, 1904, xviii. 273 [portrait], 337–62; *Nature*, 1904, lxx. 34 [C. J. Martin].)

DUCREY, AUGUSTO (born 1860). Prof. of Dermatology in Pisa and Rome. In 1889 announced discovery of specific bacillus of soft chancre, now called *Bacillus ducreyii*.

DUJARDIN, FÉLIX (born 1801, died 1860). Eminent French zoologist. Born at Tours. Taught science there and at Toulouse. Later Prof. of Zoology in Rennes. In his famous *Histoire naturelle des Zoophytes* (1841) he attempted a classification of bacteria.

DUNBAR, WILLIAM PHILIPPS (born 1863, died 1922). Born in St. Paul, Minn., U.S.A. Educated in Germany. Pupil of Gaffky in Giessen. Became Director of Hygienic Institute in Hamburg. Published important works on hay fever and Asiatic cholera.

DUNHAM, EDWARD KELLOGG (born 1862, died 1922). American bacteriologist. Studied in Germany. Became Prof. of Pathology and Bacteriology in Univ. and Bellevue Hosp. Med. Coll., N.Y. City. During his stay in Germany he worked to improve the early bacteriological diagnosis of cholera, and introduced the peptone water method. (*J. Am. M. Ass.*, 1922, lxxviii. 1332.)

DURHAM, HERBERT EDWARD (born 1866). English bacteriologist and pioneer on immune serum reactions. Educated at Cambridge and Guy's Hospital, London. Worked with Max Gruber in Vienna

and there, conjointly with his teacher, discovered bacterial agglutination (1896) and its significance in bacteriological diagnosis. Later
he studied Trypanomiasis, Yellow Fever, and Beri-beri. Made
several scientific expeditions to the tropics, but developed visual
trouble and took to horticulture and the manufacture of cider in the
west of England.

VON DUSCH, THEODOR (born 1824, died 1890). Born in Karlsruhe.
Pupil of Henle. Was extraordinarius in Heidelberg 1854, and ordi-
narius in 1870. With Schröder he carried out important researches
on filtration of air through cotton-wool. Became a well-known clinician, and wrote medical works. (Portrait in Pagel, *Biog. Lex.*, 1901.)

EASTWOOD, ARTHUR (born 1867, died 1936). He studied at Oxford
and St. Bartholomew's Hospital. M.B. Lond. 1901. He served
under the Tuberculosis Commission from 1903 to 1909, and afterwards was in charge of the Laboratory of the Local Government
Board in London. He wrote extensively on immunological problems. He died 5 May 1936, aged 69.

EBERTH, CARL JOSEPH (born 1835, died 1926). A pathologist,
anatomist, and pioneer in bacteriology. Born in Würzburg and
educated there under Virchow and Kolliker. Eberth became Prof.
in Zürich 1865, and Halle 1881. Made important contributions to
bacteriology in connexion with typhoid fever, septicaemia of rabbits,
pseudo-tuberculosis of guinea-pigs. Died in Berlin, aged 91. (*Berl.
klin. Wchnschr.*, 1905, lii. 1010; *Deutsche med. Wchnschr.*, 1905,
xxxi. 1511; *Centralbl. f. allg. Path. u. path. Anat.*, 1927, xxxix. 226.)

EHRENBERG, CHRISTIAN GOTTFRIED (born 1795, died 1876).
Famous scientist of the nineteenth century. Born at Delitzsch.
Naturalist on African Expedition, and in Russia and Siberia.
Became Prof. in Berlin and spent the rest of his life there. Ehrenberg
was a man of amazing industry and he largely helped to found
Bacteriology and Protozoology, on which he left his mark. He published many works illustrated by himself, one of the most famous
being *Die Infusionsthierchen als vollkommene Organismen* (1838)—
a monumental folio. (*Parasitology*, 1923, xv. 320 [portrait].)

EHRLICH, PAUL (born 1854, died 1915). The greatest scientific
worker in medicine in the last fifty years. Born in Strehlen (Silesia).
Educated at Breslau and Strassburg. Assistant in Frerich's clinic
in Berlin 1878–85, and in Koch's laboratory 1890. Director of
Institut für Serum-Prüfung in Steglitz, 1896. Director of Kgl. Inst.
f. exp. Therapie in Frankfurt-a.-M. 1899-1915. Nobel Prizeman
1908. Raised to dignity of Wirklicher Geheimrath with title of
'Excellenz' 1911. Paul Ehrlich was a most original worker and
during his life added a great mass to our knowledge of medical
science. He founded, and was the greatest exponent of, the science

of haematology, and was largely responsible for the direction taken by immunology. He created the new branch of Chemotherapy. He introduced a large number of technical methods which are in daily use in bacteriology and medical chemistry. He wrote himself more than 150 papers in each of which there are new evidences of his genius, and he inspired many other workers with whom he was associated. His work may be classified in three groups: (1) the application of stains to the differentiation of cells and tissues for the purpose of revealing their function (1877–90); (2) immunity studies (1890–1900); (3) chemotherapeutic discoveries (1907–15). (*J. Path. and Bacteriol.*, 1915–16, xx. 350–60 [portrait] [R. Muir]; *Arch. Derm. u. Syph.*, 1915–16, cxxi, Orig., 559–78 [A. Neisser]; *Deutsche med. Wchnschr.*, 1915, xli. 1103, 1135 [A. v. Wassermann]; *Hosp.-Tid., Københ.*, 1915, 5. R. viii. 877–93 [portrait] [O. Thomsen]; *München. med. Wchnschr.*, 1915, lxii. 1357–61 [H. Sachs]; *Paul Ehrlich, eine Darstellung seines wissenschaftlichen Wirkens*, Jena, 1914, 668 pp., Martha Marquardt, *Paul Ehrlich als Mensch und Arbeiter*, Berl. u. Leipz. [1924], 112 pp.)

EMMERICH, RUDOLF (born 1852, died 1914). Hygienist and bacteriologist in Munich. Born in the Rheinpfalz, and studied in Munich with Pettenkofer. Published many papers on bacteriology but much of his work has suffered with time. (*München. med. Wchnschr.*, 1914, lxi. 2342.)

EPPINGER, HANS (born 1846, died 1916). Austrian morbid anatomist. Prof. of pathology in the Univ. of Graz. Wrote on anthrax and described the streptothrix called after him (1890).

VAN ERMENGEM, ÉMILE-PIERRE-MARIE (born 1851, died 1932). Belgian bacteriologist. Born in Louvain. M.D. Louvain 1875. He studied in Paris (with Claude Bernard and Ranvier), London, Edinburgh, and Vienna 1876–8. In 1883 he worked with Koch in Berlin. On his return to Belgium he settled in Brussels as a medical practitioner and also conducted a private bacteriological laboratory. In 1885 he was sent to Spain to study the cholera inoculation method of Ferran. Van Ermengem was elected Prof. of Bacteriology in the Univ. of Ghent 1888, and held this post till 1919, when he was made perpetual secretary of the Académie royale de Médecine de Belgique. Van Ermengem published a long series of researches on bacteriology and hygiene and discovered, 1895, *Bacillus botulinus*. Van Ermengem was a highly cultured man and spoke six languages. He died at Ixelles-lez-Bruxelles 29 Sept. 1932, aged 81. (*Revue d'hygiène*, Paris, 1900, xxii. 837–40; *Le Scalpel*, 1932, no. 41, 637.)

ERNST, HAROLD CLARENCE (born 1856, died 1922). American bacteriologist. Born in Cincinnati. Studied in Harvard and became Prof. of Bacteriology there. Was editor of the *Journal of Medical Research*. (*Boston M. and S. J.*, 1922, clxxxvii. 424.)

ESCHERICH, THEODOR (born 1857, died 1911). Well-known paediater and bacteriologist. Born in Munich. Was Docent for children's diseases in Munich. Prof. in Graz and finally in Vienna. An excellent worker; published classical work on intestinal bacteria and diphtheria. Discovered *Bacterium coli commune*. (*Brit. M.J.*, 1911, i. 472; *München. med. Wchnschr.*, 1911, lviii. 521 [portrait]; *Wien. med. Wchnschr.*, 1911, lxi. 497; *Mitth. d. Gesellsch. f. inn. Med. u. Kinderheilk. in Wien*, 1911, ix. 82–93 [v. Pirquet].)

VON ESMARCH, ERWIN (born 1855, died 1915). Son of the famous surgeon J. F. von Esmarch. Born in Kiel. Pupil of Koch. Prof. in Göttingen. Wrote many bacteriological papers. (*Deutsche med. Wchnschr.*, 1915, xli. 504 [Reichenbach]; *Hyg. Rundschau*, 1915, xxv. 117 [C. Fraenken].)

FEHLEISEN, FRIEDRICH (born 1854, died 1924). German surgeon. Gave the first detailed account of *Streptococcus erysipelatis* (1883). Born in Reutlingen. Was assistant in surgical clinic in Würzburg. Assistant to von Bergmann in Berlin but gave up an academic career. He died in San Francisco, 1924.

FELTZ, VICTOR TIMOTHÉE (born 1835, died 1893). Studied in Strassburg, but after the Franco-German war settled in Nancy, where he was Prof. of Anatomy and Physiological Pathology. Along with L. Coze, Feltz was one of the first to study the germs of diseases (1866–70). (*Rev. méd. de l'est*, Nancy, 1893, xxv. 257–60.)

FERRÁN Y CLUA, JAIME (born 1849, died 1929). Spanish bacteriologist. Born in Tortosa, Catalonia. Attained notoriety by carrying out active immunization against cholera in man. In Valencia (1881–5) he actively immunized 50,000 persons against cholera, and later advocated similar treatment for other infections. He met with much criticism outside Spain, but was held in high esteem in his own country. He founded a bacteriological institute in Barcelona. (*Deutsche med. Wchnschr.*, 1930, lvi. 114.)

FISCHER, BERNHARD (born 1852, died 1915). Born in Coburg. Educated as army and naval surgeon. Assistant to Koch in Berlin, and went with him on the Cholera Commission to Egypt and India. In 1899 became Prof. of Hygiene in Kiel. Studied particularly marine bacteria and other subjects. Served in Great War and fell dead in Flanders 1915. (*Deutsche med. Wchnschr.*, 1915, xli. 1165 [Gaffky].)

FLEXNER, SIMON (born 1863). American pathologist and bacteriologist. Born in Louisville. Educated at Harvard, Yale, Baltimore. Prof. of Pathological Anatomy Johns Hopkins Med. School, Baltimore, and in Univ. of Pennsylvania. Director of Rockefeller Institute for Medical Research since 1903. Has made many contributions to pathology and bacteriology.

FLÜGGE, CARL (born 1847, died 1923). Noted hygienist. Born in Hanover. Studied in Göttingen. Took up hygiene in Berlin. Went to Göttingen and established himself in Meissner's Institute. In 1883 founded the first hygienic institute in Germany in Göttingen. In 1885 Prof. of Hygiene in Göttingen. In 1887 Prof. in Breslau. Finally Prof. in Berlin. Flügge was an admirable worker and teacher and did much to advance the science of bacteriology by his lucid writings. With Koch he founded and edited the *Zeitschrift für Hygiene*—a famous periodical. (Festschr. with portrait, *Ztschr. f. Hyg. u. Infektionskr.*, 1922, xcviii; *Deutsche med. Wchnschr.*, 1917, xliii. 1542–4 [W. Kruse]; *Ztschr. f. Tuberkulose*, 1923–4, xxxix. 356–62 [B. Heymann].)

FODOR, JOSEF (born 1843, died 1901). Hungarian hygienist and bacteriologist. Was Prof. in Klausenburg 1872, and from 1874 in Budapest. He became a hygienist and was one of the first to investigate the bactericidal action of the blood. There is a statue of him in Budapest. (*Alienist and Neurol.*, St. Louis, 1909, xxx. 600–7, 1 pl.)

FORD, WILLIAM WEBBER (born 1872). American bacteriologist. Born in Norwalk, O. Studied at Johns Hopkins Med. School, Baltimore. Prof. of Bacteriology there 1904–18. Prof. in the Johns Hopkins School of Hygiene and Public Health. Worked at poisonous fungi, and at bacteria especially in relation to public health.

FORDOS, MATHURIN JOSEPH (born 1816, died 1878). Pharmacien-en-chef at the Hôp. du Midi and later at the Charité in Paris. Isolated pyocyanine from cultures of *B. pyocyaneus*.

FRACASTORO, GIROLAMO (born *circa* 1478, died 1553). Physician, astronomer, geographer, poet, and humanist. Born at Verona and studied in Padua. Lived for a time in his native town but mostly at his country seat at Incaffi (Lake Garda). He was chief physician to the Council of Trent. He published his poem *Syphilis sive Morbus Gallicus* (1530) and *De sympathia et antipathia rerum* and *De contagione* in 1546. (*Ann. Medical History*, 1917, i. 1–34 [C. and D. Singer].)

FRAENKEL, ALBERT (born 1848, died 1916). Physician to the Urban Hospital, Berlin, and an authority on respiratory diseases. Born at Frankfurt a. O. Educated in Berlin. Discovered and accurately described the pneumococcus, since called after him. (*Deutsche med. Wchnschr.*, 1916, xlii. 923 [portrait]; *Lancet*, Lond., 1916, ii. 344.)

FRAENKEL, CARL (born 1861, died 1915). Born in Berlin and became assistant to R. Koch in 1885. He was a successful teacher and did much original work on various bacteriological and immunological problems. In association with R. Pfeiffer he published an admirable *Mikrophotographischer Atlas der Bakterienkunde* (1889). He was Prof. of Hygiene in the Univ. of Halle until shortly before his death. After 1912 he called himself Fraenken. (*Deutsche med. Wchnschr.*, 1916, xlii. 362, 392; *Hyg. Rundschau*, 1916, xxvi. 1.)

FRAENKEL, EUGEN (born 1853,died 1925). Pathologist and bacteriologist in Eppendorf Hospital in Hamburg. Pupil of Cohnheim and Virchow. He was a voluminous writer and contributed much that is important on enteric fever, cholera, pneumonia, influenza, typhus fever, and gas gangrene. (*Verh. d. deutsch. path. Gesellsch.*, 1926, xxi. 466.)

FRANKLAND, PERCY FARADAY (born 1858). English chemist and bacteriologist. Was lecturer on chemistry Royal School of Mines, London, 1880–8. Prof. of Chemistry in Dundee 1888–94, and Birmingham 1894–1900. Frankland was one of the early English workers on bacteriology and made many valuable contributions on the bacteria of water supplies, bacterial filtration, and sewage treatment. With his wife (*née* Grace C. Toynbee) he wrote *Microorganisms in Water* and other works.

FRIEDBERGER, ERNST (born 1875, died 1932). German bacteriologist, immunologist, and hygienist. Born in Giessen, Privat Docent in Königsberg. Prof. in Greifswald 1915. Wrote voluminously on several subjects but especially on anaphylaxis. Edited *Ztschr. f. Immunitätsforschung* . . . (*Ztschr. f. Immunitätsforschung*, 1932, Heft 5/6, lxvii. pp. i–iv.)

FRIEDLÄNDER, CARL (born 1847, died 1887). Born at Brieg (Silesia) and studied in Breslau, Würzburg, Zürich, and Berlin. Was assistant to Heidenhain and subsequently to von Recklinghausen. Later he became prosector in the Berlin-Friedreichshain hospital. Founder and editor of *Fortschritte der Medizin*. Was an early student of bacteriology and discovered *Bacillus pneumoniae*. He developed pulmonary tuberculosis and after five years illness died at Meran, aged 40. (*Fortschr. d. Med.*, 1887, v. 321–9 [Weigert].)

FROSCH, PAUL (born 1860, died 1928). He was assistant to Koch and accompanied him in several scientific expeditions. Prof. of Bacteriology in Veterinary School, Berlin. With Loeffler he discovered the filter-passing character of the virus of foot-and-mouth disease.

GABRITCHEWSKY, GEORGIY NORBERTOVICH (born 1860, died 1907). Russian bacteriologist. Educated in Moscow. Studied in Germany and France. Founded bacteriological institute in Moscow, of which he was director. He published important researches especially on relapsing fever, and on scarlet fever in relation to streptococci. Died of pneumonia, aged 47. (*Berl. klin. Wchnschr.*, 1907, xliv. 586.)

GAFFKY, GEORG (born 1850, died 1918). Pioneer in bacteriology. Pupil, assistant, and successor of Robert Koch in Berlin. Born in Hanover. Educated in Army Medical Department, Berlin.

Accompanied Koch on German Cholera Commission to Egypt and India. In 1888 became Prof. of Hygiene in Giessen. Studied cholera in Hamburg, 1892. Head of the German Plague Commission in India, 1897. Director of Institut für Infektionskrankheiten in Berlin. Edited Koch's Collected Works. Gaffky made many important contributions to bacteriology and was the first to cultivate the typhoid bacillus. Throughout life he was held in the highest esteem as a man and scientist. (*Berl. klin. Wchnschr.*, 1918, lv. 1062 [Neufeld]; *Deutsche med. Wchnschr.*, 1918, xliv. 1199 [R. Pfeiffer]; *München. med. Wchnschr.*, 1918, lxv. 1191 [H. Kossel].)

GAMALÏEYA, (GAMALEIA) N[IKOLAÏ FYODOROVICH] (born 1859). Russian bacteriologist. Pupil of Pasteur. Privat Docent for pathology in Odessa. Practised Pasteur's methods of inoculation for anthrax and other diseases in Russia. Discovered *Vibrio Metchnikowi* in Odessa, 1887. Wrote *Les Poisons bacteriens*, 1892.

GARRÉ, KARL (born 1857, died 1928). Swiss surgeon and bacteriologist. Born in St. Gallen (Switzerland). Educated at Zürich, Leipzig, and Bern. Taught surgery in Basle, Tübingen, Rostock, and finally in Bonn. In his earlier days Garré did successful bacteriological work, especially on wound infections and osteomyelitis. Inoculated himself with cultures of staphylococcus with severe results. (*Schweiz. med. Wchnschr.*, 1928, lviii. 415; *Deutsche med. Wchnschr.*, 1928, liv. 709.)

GÄRTNER, AUGUST [ANTON HIERONYMUS] (born 1848, died 1934). Eminent hygienist. Prof. in Jena from 1886. Born in Westphalia. Educated in Army Medical Department, Berlin. Later, naval surgeon. Studied under Koch in Berlin. Investigated many hygienic problems including meat poisoning. Gärtner's bacillus called after him. (*Centralbl. f. Bakteriol.*, 1. Abt., 1928, cvii, pp. ii–xvi; *Deutsche med. Wchnschr.*, 1935, lxi. 233.)

GASPARD, MARIE HUMBERT BERNARD (born 1788, died 1871). Was a doctor and justice of the peace in Montret (Saône-et-Loire). A voluminous writer on physiology, meteorology, statistics, agriculture, and medicine. He was one of the first to make experimental studies on pyaemia following injection of putrid fluids.

GAUTIER, ARMAND (born 1837, died 1920). Was Prof. of Mineral and Organic Chemistry in the Faculty of Medicine in Paris for 45 years. Wrote much on toxins, ptomaines, and other food poisons. (*Bull. Acad. de méd.*, Paris, 1920, lxxxiv. 74; *Bull. Soc. chim. biol.*, Paris, 1921, iii. 248; *Lancet*, 1920, ii. 675.)

GAY, FREDERICK PARKER (born 1874). Born in Boston, and studied at Johns Hopkins Univ. at Baltimore. For a time he also studied with Bordet in Brussels. From 1923 he was Prof. of Bacteriology in Columbia Univ., N.Y.

GEPPERT, August Julius (born 1856, died 1937). Prof. of Pharmacology in Giessen. Born in Berlin. Contributed important papers on the subject of testing disinfectants on bacteria, 1889–90. (Pagel, J., *Biograph. Lexicon hervorrhag. Aerzte*, 1901, p. 592 [portrait].)

GHON, Anton (born 1866, died 1936). Born in Villach (Carinthia) and studied in Vienna. For many years he was Prof. of Pathological Anatomy in the German Univ. of Prague. He published many valuable researches in bacteriology and pathological anatomy.

van GIESON, Ira (born 1866, died 1913). Born and lived in New York. Devoted himself to neurology. Described his picro-fuchsin stain in 1889. (Laboratory notes on technical methods for the nervous system. *New York Med. Journ.*, 1889, l. 57–60.) He died of chronic nephritis, aged 47.

GLEICHEN, Wilhelm Friedreich (called Russworm) (born 1717, died 1783). Born in Bayreuth. He killed a man in a duel and had to flee disguised in female clothing. Joined the army. An accident brought Ledermüller's book into his hands and he forthwith devoted himself to microscopy, and wrote *Abhandlung über die Saamen und Infusionsthierchen und über die Erzeugung*, 1778. He prepared the plates himself. He was admired for his upright character.

GOODSIR, John (born 1814, died 1867). Prof. of Anatomy in Edinburgh, 1846–67. Wrote many important works in anatomy and pathology. Virchow dedicated his *Cellular Pathology* to him. In 1842 he discovered and accurately described *Sarcina ventriculi*. (*Edinb. M.J.*, 1867, xii. 959; *Brit. M.J.*, 1867, i. 307.)

GRAM, [Hans] Christian [Joachim] (born 1853). Danish physician. Born in Copenhagen. Was Director of the Medical Department of Frederick's Hospital there. In the early eighties of last century he worked with Friedländer in Berlin and published (1884) his famous method of the isolated staining of bacteria.

GRUBER, Max (born 1853, died 1927). Distinguished Austrian bacteriologist and hygienist. Born and educated in Vienna. Prof. in Graz 1884, and Vienna 1887. In 1902 he migrated to Munich. A pioneer in serodiagnosis, by means of agglutinating reaction of which, with Durham, he was the real discoverer. Opponent of Ehrlich's immunity theory. In later life devoted himself largely to purely hygienic matters. (*Ztschr. f. Immunitätsforsch. u. exper. Therap.*, 1927, liv, pp. i–x; *Wien. klin. Wchnschr.*, 1927, xl. 1304.)

GRUBY, David (born 1810, died 1898). A pioneer worker on the vegetable parasitic diseases of man. Discovered *Microsporon Audouini* 1843, and *Trichophyton tonsurans* 1844. An eccentric character. Born of Jewish parents in South Hungary. Worked as a labourer in poverty. Received some education in Vienna and graduated as doctor in 1839. Settled in London and then in Paris.

Taught physiology and pathology. Claude Bernard was one of his pupils. Gruby entered practice in Paris and attended many notables including George Sand, Marshal MacMahon, Alex. Dumas, Chopin, Liszt, and Heine on his death-bed. Lived life of a recluse making watches and clocks. Founded an observatory in Montmartre where he died alone, aged 89, in a house that nobody had been allowed to enter for years. (*Brit. M.J.*, 1898, ii. 1705; *Lancet*, 1898, ii. 1511; *Deutsche med. Wchnschr.*, 1900, xxvi. 118–20; *Presse méd.*, 1909, xvii. annexe, 113–17; *Ann. Med. Hist.*, 1932, iv. 339 [portrait]; Le Leu, L., *Le David Gruby*, Paris, 1908.)

VON GRUITHUISEN, FRANZ PAULA (born 1774, died 1852). German surgeon and scientist. Born at Haltenberg am Lech (Bavaria). Obtained doctor's degree in 1807. In 1826 became Prof. of Astronomy in Munich and devoted himself to the formation of theories terrestrial and celestial. Invented instrument for crushing vesical calculi. He was a staunch believer in spontaneous generation of infusion animalcules.

GRÜNBAUM (afterwards LEYTON), ALBERT SIDNEY FRANKAU (born 1869, died 1921). English bacteriologist. Educated at Cambridge and St. Thomas's Hospital, London. Worked with Gruber in Vienna and was the first to study the agglutinating reaction of serum in enteric fever. Was anticipated in publication by F. Widal. Later G. worked in Liverpool as a physiologist. 1904 Prof. of Pathology in Leeds. Served in Great War. Changed his name to Leyton, 1915. He died of chronic disease in 1921. (*J. Path. and Bacteriol.*, 1922, xxv. 109–12 [portrait].)

GSCHEIDLEN, RICHARD (born 1842, died 1889). German physiologist. Born in Augsburg. Assistant to Heidenhain in Breslau, 1869. Later Director of Board of Health in Breslau. Wrote against spontaneous generation doctrine 1874, and in association with M. Traube on putrefaction.

GÜNTHER, CARL [OSCAR] (born 1854, died 1929). German bacteriologist and hygienist. Born at Naumburg a. d. Saale. Became Director of Prussian Institute for Water Hygiene in Berlin. Worked chiefly on bacteriology in reference to public health. Wrote an excellent introduction to the *Study of Bacteriology*, 1890 (six editions). He was also one of the founders of *Hygienische Rundschau*.

GUÉRIN, JULES-RENÉ (born 1801, died 1886). French surgeon and bitter opponent of Pasteur, whom he challenged to a duel. Born in Belgium, became orthopaedic surgeon in Paris.

HAFFKINE, W[ALDEMAR MORDECAI WOLFF] (born 1860, died 1930). Bacteriologist and immunizator. Born in Odessa and educated there. Went to Paris and was assistant in Pasteur Institute, 1888–93. Went to India (1893) and there introduced prophylactic inoculations against cholera and later against plague on a gigantic scale. Founded Government Research Laboratory (now Haffkine Institute), Bombay. Retired 1915.

HALLIER, ERNST (born 1831, died 1904). Studied in Jena and became Prof. of Botany there. Hallier attained great notoriety between 1866 and 1876 by his erroneous views on the pleomorphism of bacteria. His theories were completely discarded and he died forgotten in 1904. He was, however, one of the first to direct attention to organized elements of infective disease.

HANKIN, ERNEST HANBURY (born 1865). English bacteriologist and chemist. Born in Ware (Herts). Educated at St. Bartholomew's Hospital, London, Cambridge, Berlin, and Paris. Wrote early papers on nature of immunity and alexins. Went to India and became chemical examiner United Provinces of Agra and Oudh.

HANSEN, [GERHARD HENRIK] ARMAUER (born 1841, died 1912). Norwegian leprologist. Worked chiefly at Bergen, where he was in charge of leprosy hospital. Discovered the leprosy bacillus, and spent his life in studying leprosy on which he was a great authority. (*München. med. Wchnschr.*, 1912, lix. 540 [portrait]; *Lepra*, Leipz., 1901, ii. 121; *Berl. klin. Wchnschr.*, 1901, xxxviii. 871.)

HANSEN, EMIL CHRISTIAN (born 1842, died 1909). Danish bacteriologist and student of fermentation. Educated in Copenhagen. From 1887 was in Jacobsen's Ny Carlsberg brewery, then Director of the Physiology Department in the Ny Carlsberg Laboratories. In 1881 he obtained pure cultures of yeast and made an exhaustive and fundamental study of yeasts. (*Centralbl. f. Bakteriol.*, 2. Abt. 1909, xxv. 1–8; *Hygiea*, Stockholm, 1909, 2.f., ix. 1137–53 [portrait].)

HARTING, PIETER (born 1812, died 1885). Dutch naturalist and authority on the microscope. Born in Rotterdam. From 1843 to 1856 taught microscopic anatomy and vegetable physiology, and from 1856 to 1882 zoology in Utrecht. He was a prolific writer. His chief work was his excellent treatise on the microscope (1848–54).

HATA, SAHACHIRO (born 1873). Born in Tsumo, Japan. He studied with Kitasato in Japan and with Wassermann in Berlin. From 1908 to 1910 he was with Ehrlich in Frankfurt-a.-M. and with him worked out the therapeutic value of salvarsan. On his return to Japan he was again in Kitasato's Institute and later held a government appointment.

HAUSER, GUSTAV (born 1856). German morbid anatomist and bacteriologist. Born in Nordlingen. Pupil of Cohnheim and Weigert. Was Prof. of Pathology in Erlangen. In his work *Ueber Faulnissbacterien und deren Beziehungen zur Septicämie*, 1885, he gave the first account of the 'Proteus' group of bacteria.

HEIM, Ludwig (born 1857). German hygienist and bacteriologist. Born in Bavaria. Pupil of Koch and Gaffky in Berlin. Military surgeon. Became Prof. of Hygiene in Erlangen. Best known through his *Text-book of Bacteriology*.

HEKTOEN, Ludvig (born 1863). American pathologist and bacteriologist. Born in Westby, Wis. Studied in Chicago, Upsala, Prague, and Berlin. Prof. of Pathology in Univ. of Chicago, 1901. Director of John McCormick Institute for Infectious Diseases, Chicago. Has published many bacteriological researches, especially in connexion with immune reactions.

van HELMONT, Joannes Baptista (born 1577, died 1644). Alchemist, physician, and philosopher. Born in Brussels. Studied medicine and philosophy. Lectured on surgery at Louvain. He gave up medicine, divided his fortune among his relatives and set out on travel to Switzerland, Italy, England, and other parts. Everywhere he found 'sluggishness and ignorance'. In 1609 he married a rich heiress and settled in Vilvorde, where he applied himself to the study of chemistry and to the free treatment of the sick poor. Van Helmont wrote much and was a firm believer in spontaneous generation of animals.

HENLE, [Friedrich Gustav] Jacob (born 1809, died 1885). German pathologist, physiologist, and anatomist. Born in Fürth (Bavaria). Educated at Bonn and Heidelberg, 1827–32. Pupil of Johannes Müller, whom he followed to Berlin. Prosector of anatomy, Berlin. Prof. of Anatomy in Zürich, Heidelberg, and in Göttingen, where he spent the latter part of his life. In his earlier years he was chiefly known as a pathologist. His *Handbuch der rationellen Pathologie* (1844–53) was one of the first important works on pathology. In his *Pathologische Untersuchungen* (1840) he made a remarkable anticipation of the germ theory of disease. (*Arch. f. mikroskop. Anat.*, Bonn, 1885–6, xxvi, pp. i–xxxii [Waldeyer]; *Deutsche med. Wchnschr.*, 1885, xi. 463, 483; *Proc. Roy. Soc.*, Lond., 1885, xxxix, pp. iii–viii.)

d'HÉRELLE, Félix-Hubert (born 1873). Born in Montreal. Studied in Lille and in Leyden, where he graduated M.D. From 1902 to 1906 he was Director of a bacteriological laboratory in Guatemala, and 1907–8 in Mérida (Yucatan). In 1908 he went to Pasteur Institute, Paris, and 1914–21 was head of a department there. Prof. in Leyden 1922–3, and Director of International Sanitary Commission in Egypt 1923–7. In 1928 he became Prof. of Protobiology in Yale University. d'Hérelle is chiefly known for his work on bacteriophage.

HESSE, Walther (born 1846, died 1911). Born in Bischofswerda. An early associate of Koch in Berlin, where he published important investigations on the bacteriology of the air and many other papers, especially on bacteriology in relation to hygienic questions. Died as

Obermedizinalrat in Dresden, 1911. His wife first suggested (1882) the use of agar-agar for bacteriological purposes. Through Dutch friends she had obtained samples from Batavia, where it was extensively used for cooking and the making of fruit preserves.

von HIBLER, Emmanuel (born 1865, died 1911). Born in Lienz (Tyrol). Studied in Innsbruck, where he became professor extraordinarius. Wrote extensively on anaerobic bacteria including a large book (1908). Died of streptococcus sepsis from autopsy wound. (*Verh. d. deutsch. path. Gesellsch.*, Jena, 1913, xvi. 450 [Ghon].)

HIRSCH, August (born 1817, died 1894). Great German medical historian and epidemiologist. Born in Danzig. Graduated in Berlin, 1843. Practised in Elbing 1844, and Danzig, from whence he was called to Berlin as Prof. of Medicine in 1863 after he had made a name for himself as an author by the publication of his monumental *Handbuch der historisch-geographischen Pathologie* (1860–4), 2nd edit. 1881–6. (*Berl. klin. Wchnschr.*, 1894, xxxi. 129 [R. Virchow]; *Brit. M.J.*, 1894, i. 275.)

HISS, Philip Hanson, jr. (born 1868, died 1913). American bacteriologist and Prof. in the College of Physicians and Surgeons, Columbia Univ., New York. Born in Baltimore. Educated at Johns Hopkins Hospital. Made important investigations on bacteriology of Dysentery and introduced improvements in technique. (*J. Am. M. Ass.*, 1913, lx. 846.)

HÖGYES, Andreas (born 1847, died 1906). A Hungarian. Prof. of Pathology and Therapy in Budapest. Was an authority on rabies and founded a great institute for treatment of the disease in Budapest. Elaborated a new treatment for rabies on a different plan from Pasteur's. He was held in high repute in his country. (*Brit. M.J.*, 1906, ii. 1340.)

HOFFMANN, [Heinrich] Hermann (born 1819, died 1891). Botanist and mycologist. Prof. of Botany in Giessen. He was an early and diligent worker on bacteriology.

von HOFMANN-WELLENHOF, G. (born ?, died 1890?). Austrian bacteriologist. Hofmann's bacillus called after him. Worked in Graz. He died young in 1890(?) of a glanders infection contracted during experiments.

HOOKE, Robert (born 1635, died 1703). English experimental philosopher and mechanical genius. Born at Freshwater, Isle of Wight. Educated at Westminster and Oxford. Assisted Thomas Willis and Robert Boyle in their chemical and physical researches. Curator of experiments to the Royal Society, 1665. Secretary of Roy. Soc. 1677–82. Surveyor of Works after the fire of London, 1666. Hooke published his famous *Micrographia* in 1665 and did much to rouse interest in microscopy. He was crooked and low of stature and his temper 'melancholy, distrustful, and jealous'.

HOUSTON, ALEXANDER CRUICKSHANK (born 1865, died 1933). He was educated at Edinburgh Univ. M.B.C.M. Edin. 1889. He worked for the Local Government Board on lead poisoning 1893–1905, and was bacteriologist to the Royal Commission on Sewage disposal 1899–1905, when he became Director of Water Examination to the Metropolitan Water Board and held this appointment till his death. He was knighted K.B.E. in 1918. He published many valuable researches on the bacteriology of water supplies.

HUEPPE, FERDINAND (born 1852). Born in Heddesdorf. Studied as an army surgeon. In 1879 was in the Kaiserliche Gesundheitsamt; 1884, established a laboratory in Wiesbaden. Prof. in Prag 1889. Wrote extensively on bacteriological subjects and published valuable books, especially *Die Methoden der Bakterienforschung*, 1885, and *Die Formen der Bakterien*, 1886, which made him widely known.

ISRAEL, JAMES (born 1848, died 1926). Noted German surgeon. Director of surgical division of Jewish Hospital in Berlin. Was a specialist in renal surgery. Published early and important work on bacteriology of actinomycosis, 1878–83. (*Ztschr. f. urol. Chirurgie*, 1923, xii [portrait]; *Deutsche med. Wchnschr.*, 1926, lii. 541; *München. med. Wchnschr.*, 1926, lxiii. 448.)

JENSEN, CARL OLUF (born 1864, died 1934). Born in Copenhagen. Studied veterinary medicine and worked at bacteriology with Salomonsen, Koch, and Kitt. Later he became assistant to Bernhard Bang in the Veterinary High School in Copenhagen, and subsequently was Prof. and Director of Serological Institute in the Veterinary School. Jensen carried out many important researches in veterinary bacteriology and attained world-wide fame by his experiments on the transmission of mouse cancer. (*Brit. M.J.*, 1934, ii. 535.)

JOBLOT, LOUIS (born 1645, died 1723). Early French microscopist. Born at Bar-le-Duc (Meuse). In 1680 he was assistant Prof. of Mathematics, Geometry, and Perspective at the Royal Academy of Painting and Sculpture in Paris. He succeeded to the professorship in 1699 and held the post till 1721. He invented and constructed microscopes and examined microscopic animalcules and infusoria, which he described and figured. He was an opponent of spontaneous generation and showed by experiment that infusoria do not appear in infusions that have been boiled, and kept from the air. (*Mém. Soc. lett. sc. et arts, Bar-le-Duc*, 1895, 3rd s., iii. 205–333; *Parasitology*, 1923, xv. 314.)

JOEST, ERNST (born 1873, died 1926). German veterinary pathologist and bacteriologist. Was Prof. in the Veterinary School in

Dresden 1904, and Prof. and Director of the Veterinary Pathological Institute in Leipzig. Wrote extensively on veterinary pathology, especially on tuberculosis. (*Verhandl. d. deutsch. path. Gesellsch.*, 1927, xxii. 316.)

JOHNE, [HEINRICH] ALBERT (born 1839, died 1910). German veterinary pathologist and bacteriologist. Born in Dresden and became Prof. of Pathological Anatomy in the Kgl. tierärztl. Hochschule there. He was also a doctor of medicine and pupil of Cohnheim and Leuckhart in Leipzig. Johne made many important contributions to the bacteriology and pathology of actinomycosis, botryomycosis, Borna disease, Trichinosis, and was the first to give an extended account of the pseudotuberculosis disease of cattle now called after him 'Johne's disease'. He died near Pirna (Sachsen) in 1910. (*Ztschr. f. Tiermed.*, Jena, 1910, xiv, pp. i–viii.)

JORDAN, EDWIN OAKES (born 1866, died 1936). American bacteriologist. Born at Thomaston (Maine). Prof. of Bacteriology Chicago Univ. Published much bacteriological work especially in connexion with food poisoning.

JOUBERT, JULES (born 1834, died 1910). Prof. of Physics in the Collège Rollin in Paris. Collaborated with Pasteur in several researches. (*Ann. de l'Inst. Past.*, 1910, xxiv. 241.)

KANTHACK, A[LFREDO] A[NTUNES] (born 1863, died 1898). English pathologist and bacteriologist. Born in Bahia. Educated in Germany, where he studied with Koch and Virchow. Taught at St. Bartholomew's Hospital, London, and succeeded Roy as Prof. of Pathology in Cambridge. Wrote much on bacteriology and morbid anatomy and was an excellent teacher. Died aged 35. (*J. Path. and Bacteriol.*, 1899, vi. 89; *Lancet*, 1898, ii. 1817; *Brit. M.J.*, 1898, ii. 1941.)

KIRCHER, ATHANASIUS (born 1602, died 1680). German Jesuit, philologist, physicist. Born in Geisa (Eisenach). Lived in Münster, Cologne, Mainz, and was Prof. in Würzburg. Later he lived in Rome and Vienna. He wrote a vast amount of which little is valuable to-day. Said to have been one of the first to suggest that there was a world of microscopic creatures.

KITASATO, SHIBASABURO (born 1852, died 1931). Famous Japanese bacteriologist. Studied in the Imperial Univ. of Tokyo 1875. Graduated 1883. Went in 1885 to Berlin, where he worked with Koch and received title of Professor. Returned to Japan 1891 and founded a bacteriological institute. First cultivated *B. tetani* 1889. Discovered, with Behring, tetanus antitoxin 1890. Discovered plague bacillus 1894. Ennobled for his researches. Director of the Institute for Infectious Diseases, Tokyo. (*J. Path. and Bacteriol.*, 1931, xxxiv. 597–602 [portrait].)

KITT, THEODOR (born 1858). German veterinary pathologist and bacteriologist. Prof. in Munich. Author of many important papers on symptomatic anthrax, swine fever, and allied subjects. He also published excellent text-books.

KLEBS, [THEODOR ALBRECHT] EDWIN (born 1834, died 1913). German pathologist and pioneer in bacteriology. Born in Königsberg. Studied there under Rathke and Helmholtz, and in Würzburg under Kölliker and Virchow, following the latter to Berlin. Became extraordinary Prof. in Berne and adopted Swiss nationality. Prof. in Würzburg in 1872 and in Prague 1873. Migrated to Zürich in 1882 and remained there for 11 years. Later he resided in Karlsruhe and Strassburg. In 1895 emigrated to America and settled in Asheville, N.C. Was for a time Prof. of Pathological Anatomy in Rush Medical College, Chicago. In 1900 he returned to Europe and worked in a private laboratory in Hanover till 1905. At the age of 71 went to work in Orth's Institute in Berlin. From there he went to Lausanne and finally to Berne, where he died aged 79. Klebs was a most prolific writer and worker. Published important memoir on the pathology of gun-shot wounds 1872, and wrote on the bacteriology of enteric fever, rinderpest, vaccinia, diphtheria, syphilis, and tuberculosis. He also wrote but did not finish a large *Handbuch d. path. Anatomie*. Klebs was one of the first in every advance in bacteriology but had the misfortune to miss almost every discovery that has turned out to be correct. (*J. Path. and Bacteriol.*, 1913–14, xviii. 401 [portrait]; *München. med. Wchnschr.*, 1914, lxi. 193 [portrait]; *Verh. d. deutsch. path. Gesellsch.*, 1914, xvii. 588, 590–7.)

KLEIN, EMANUEL (born 1844, died 1925). Born in Osijek (Slavonia) and educated in Vienna, where he was assistant to S. Stricker. Klein went to England (1872) as a histologist and became attached to the Brown Institution, where he worked for many years, 1871–97. He also taught histology at St. Bartholomew's Hospital, London. He also worked on behalf of the Local Government Board. Soon after his arrival in England he took up bacteriology but failed to make any discovery of permanent value. He opposed Koch's views on the aetiology of cholera. (*J. Path. and Bacteriol.*, 1925, xxviii. 684 [portrait].)

KLENCKE, HERMANN (born 1813, died 1881). A doctor in Hanover who devoted much time to popularizing medicine. Wrote voluminous works (200) on many subjects, including novels. Klencke was one of the first to produce experimental tuberculosis (1843).

KOCH, ROBERT (born 1843, died 1910). By common consent the greatest pure bacteriologist. Born in Clausthal, Hanover, the son of a mining engineer. Studied in Univ. of Göttingen under Wöhler, G. Meissner, and J. Henle. Graduated as doctor of medicine 1866.

Practised in Niemegk and Rakwitz. Served as surgeon in Franco-Prussian war, and in 1872 became Kreisphysikus in Wollstein. Published his classical research on anthrax in 1876, on the technical methods of bacterial examination in 1877, and on the etiology of traumatic infective diseases 1878. Became associated with the Gesundheitsamt in Berlin and founded famous school of bacteriology there. In 1881 he solved the problem of pure bacterial cultures and his methods were universally employed. Discovered tubercle bacillus 1882, and cholera vibrio in 1883. In 1890 made known his discovery of tuberculin, and in 1900 his views on the non-identity of human and bovine tuberculosis. From 1891 to 1904 Koch was Director of the Institut f. Infektionskrankheiten in Berlin. In 1896 he investigated rinderpest in S. Africa and sleeping-sickness in Uganda. Travelled extensively studying protozoal diseases in the East. Received Nobel Prize 1905 and was ennobled with the title of Excellenz. Foreign member of Royal Society 1897. Died of cardiac failure 27 May 1910, aged 67. His ashes deposited in the Institut für Infektionskrankheiten, Berlin. Portrait in youth in Salomonsen's *Smaa-Arbejder*, Kjøbenh., 1917, p. 53. (Becher, W., *Robert Koch: Eine biographische Studie*, Berl. [1891], 3rd ed., 104 pp.; Wezel, K., *Robert Koch: Eine biographische Studie*, Berl., 1912, 148 pp. [portrait]; *Berl. klin. Wchnschr.*, 1910, xlvii. 1045 [R. Pfeiffer]; *Deutsche med. Wchnschr.*, 1910, xxxvi. 2321 [portrait] [Gaffky]; *J. Path. and Bacteriol.*, 1910–11, xv. 108 [portrait] [Woodhead]; *Parasitology*, 1924–5, xvi. 214–38 [4 portraits] [Nuttall]; *Proc. Roy. Soc.*, 1910–11, lxxxiii, Suppl. pp. xviii–xxiv [C. J. Martin]; Brieger, L., and Kraus, F., 'Krankheitsgeschichte Robert Kochs', *Deutsche med. Wchnschr.*, 1910, xxxvi. 1045; 'Enthüllung des Robert Koch-Denkmals', *Deutsche med. Wchnschr.*, 1916, xlii. 704; 'Monument to Koch in Japan', *J. Am. M. Ass.*, 1914, lxii. 213; Neufeld, F., 'Zur Enthüllung des Denkmals für Robert Koch', *Ztschr. f. aerztl. Fortbild.*, 1916, xiii. 349; Heymann, Bruno, *Robert Koch*, 1. Teil, Leipzig, 1932, 353 pp.)

KOLLE, Wilhelm (born 1868, died 1935). German bacteriologist and chemotherapeutist. Born in Lehrbach (Harz). Worked for several years in Koch's Institute in Berlin and was engaged on rinderpest work in S. Africa. Prof. of Hygiene and Bacteriology in Berne 1902. In 1906 succeeded Ehrlich as Director of Inst. f. exp. Therapie in Frankfurt a. M. Kolle was a voluminous writer and a successful worker in many branches of bacteriology, immunology, and chemotherapy. With Wassermann he edited the monumental *Handbuch der pathogenen Mikroorganismen*, 3 editions. (*Deutsche med. Wchnschr.*, 1935, lxi. 849.)

KOSSEL, [Alexander August Richard] Hermann (born 1864, died 1925). German bacteriologist and hygienist, born in Rostock, brother of the chemist A. Kossel. In 1891 Kossel became assistant in Koch's Institute in Berlin and worked with Behring, Brieger, and Ehrlich on immunity problems. Accompanied Koch and Pfeiffer on Malaria expedition to Italy and studied plague in Oporto. Published important work on bovine and human tubercle. Prof. of Hygiene in Giessen 1904, and in Heidelberg 1910. Died suddenly of heart failure, 1925. (*Deutsche med. Wchnschr.*, 1925, li. 999; *Med. Klin. Berl.*, 1925, xxi. 1105 [portrait].)

KRAUS, RUDOLF (born 1868, died 1932). Austrian bacteriologist and immunologist. Pupil and assistant of R. Paltauf in Vienna. Director of Bacteriological Institute in Buenos Ayres and of State Serum Institute at São Paulo (Brazil). Subsequently he was Director of the State Serotherapeutic Institute in Vienna. Discovered bacterial precipitins 1897, and was a voluminous writer on bacteriological and serological subjects. With Levaditi and Uhlenhuth respectively Kraus edited two large handbooks on immunity and bacteriological technique.

KRUSE, WALTHER (born 1864). Hygienist and bacteriologist. Born in Berlin. Assistant to Flügge in Breslau. Prof. in Bonn 1898, Königsberg 1909, Leipzig 1913. He wrote very extensively on bacteriology, and was associated with the discovery of *B. dysenteriae*.

KÜTZING, FRIEDRICH TRAUGOTT (born 1807, died 1893). Born in Ritterburg. Studied in Halle and died in Nordhausen. Originally a pharmacist, he became teacher of science and made elaborate investigations on Algae, on which he wrote extensively. In 1835 he studied the marine plants of the Mediterranean and Adriatic seas, and in 1837 published his classical paper on the yeast plant.

LAFAR, FRANZ (born 1865). Worker on fermentation and technical mycology. Prof. in technical school in Modling (Vienna). Edited a large *Technische Mykologie*.

LAMB, GEORGE (born 1870, died 1911). Educated in Glasgow. Entered Indian Medical Service and worked at military hospital in Netley with A. E. Wright. In 1894 went to India and on return studied in Paris (1897). Was, later, assistant to Haffkine in Bombay and made important studies of snake venom. Subsequently he became Director of Pasteur Institute in India and carried out extensive investigations on Malta fever in India, on transmission of plague by fleas, and on rabies. Returned to England and died in Edinburgh 1911. (*J. Hyg.* (Plague suppl.), 1912, xii. 2 [portrait]; *J. Path. and Bacteriol.*, 1911, xvi. 119 [portrait]; *Brit. M.J.*, 1911, i. 969, 1029.)

LANDSTEINER, KARL (born 1868). Born in Vienna and studied medicine there. He was associated especially with Max Gruber and Weichselbaum and became extraordinary Prof. of Pathological Anatomy in Vienna. After the Great War he was for a time in the Ziekenhuis in The Hague, and from 1922 held an important post in the Rockefeller Institute in New York. Landsteiner has published a large number of highly important discoveries in many branches of pathology, bacteriology, and immunology, and these have raised him to the foremost position in these sciences. Nobel Prizeman.

LANKESTER, EDWIN RAY (born 1847, died 1929). Eminent English zoologist. Was Director of Natural History Department of the British Museum, 1898–1907. In his earlier years he worked at bacteriology and was led to support the pleomorphic doctrine 1873, 1876. Knighted 1907. Copley medallist Roy. Soc. 1913.

LEDERMÜLLER, MARTIN FROBENIUS (born 1719, died 1769). German naturalist and early microscopist. Born in Nürnberg. Studied philosophy and law, but had a very varied career. In 1759 devoted himself to microscopic studies and published *Micro-scopische Gemüths- und Augen-Ergötzung* (1760–5) and other works.

VAN LEEUWENHOEK, ANTONY (born 1632, died 1723). Great Dutch microscopist and first discoverer of bacteria. Born in Delft. In youth served as a book-keeper in a draper's shop in Amsterdam, but returned to his native town when 22 years of age and died there aged 91. Remained in obscurity for 40 years, but discovered how to grind microscopic lenses and in 1673 was introduced to the Royal Society of London and became one of its most famous correspondents. He was elected F.R.S. in 1680 (N.S.), and wrote to the Society about 200 letters, containing accounts of hundreds of discoveries which he had made by his lenses. A man of amazing industry and pertinacity and with an unbounded love of the truth. His 'microscopes' were single biconvex glasses and of these he made with his own hands more than 400, of which some magnified about 300 diameters. He never divulged his method of manufacture. He first saw living protozoa in 1674 and bacteria in 1675. Dobell says 'he was the first bacteriologist and the first protozoologist, and he created bacteriology and protozoology out of nothing'. Van Leeuwenhoek was one of the world's greatest observers. (C. Dobell, *Antony van Leeuwenhoek and his 'little animals'*, London, 1932, 435 pp.)

LEDINGHAM, JOHN CHARLES GRANT (born 1875). Bacteriologist. Born in Boyndie (Banff). Educated in Aberdeen, where he studied medicine. In London he became attached to the Lister Institute and rose to be Director of that institution. Prof. of Bacteriology in the University of London. Wrote many papers on bacteriology, pathology, and immunity. Knighted 1937.

LEHMANN, KARL BERNHARD (born 1858). Hygienist, toxicologist, and bacteriologist. Born in Zürich. Prof. in Würzburg. Author (with Neumann) of important *Atlas und Grundriss der Bakteriologie* (1896; 7th ed. 1927).

LEISHMAN, WILLIAM BOOG (born 1865, died 1926). English bacteriologist and worker on tropical diseases. Educated in Glasgow. Entered the army in 1887 and rose to be Director-General of Army Medical Department. Distinguished for his work on the value of anti-typhoid inoculation, phagocytosis, and Leishmaniases. He also invented a modification of the Romanowsky stain which has been extensively used. (*Lancet*, London, 1926, i. 1171–3.)

LEVADITI, CONSTANTIN (born 1874). Rumanian by birth. Educated in Paris and became attached to the Pasteur Institute. Has published a very large number of researches on many bacteriological subjects.

LEWIS, TIMOTHY RICHARDS (born 1841, died 1886). Born in Llanboidy (Wales) and studied in London and Aberdeen. Entered Army Medical Service and was sent to India. Here he studied cholera and other Indian diseases. In cholera evacuations he found amoebae, 1870. In 1872 he gave the first account of 'Filaria sanguinis hominis'. Worked at bacteria and the spirochaetes of relapsing fever. He gave the first description of rat trypanosoma now called *T. lewisi*. He returned to England (1883) and became Assistant Prof. of Pathology at Netley, where he died of pneumonia, aged 45. Lewis was an original investigator and pioneer in the study of tropical diseases. (*Parasitology*, 1923, xiv. 413 [portrait].)

LICHTHEIM, LUDWIG (born 1845, died 1928). German physician. Born in Breslau. Was Prof. in Berne and for many years in Königsberg. In 1882 he investigated the pathological mould fungi, especially Aspergillus. (*Mitth. a. d. Grenzgeb. d. Med. u. Chir.*, 1928, xli, pp. i–iii.)

VON LIEBIG, JUSTUS (born 1803, died 1873). Great German chemist. Born in Darmstadt. Worked in Paris with Gay-Lussac. Became Prof. of Chemistry in Giessen 1824, and in Munich 1852, was created a baron. He was a bitter opponent of the vital theory of fermentation as developed by Schwann and Pasteur. He received the Copley medal of the Royal Society in 1840. (*Lancet*, 1873, i. 613; *München. med. Wchnschr.*, 1903, l. 826 [portrait]; Hofmann, A. W., *The Life Work of Liebig in Experimental and Philosophic Chemistry*, Lond. 1876.)

LISTER, JOSEPH (born 1827, died 1912). Famous English surgeon and scientist. Revolutionized surgery by his antiseptic treatment. Born at Upton Park (Essex). Educated at Univ. College, London. Taught in Edinburgh and in Glasgow, where he began to develop his antiseptic system. Prof. of Surgery in Edinburgh, 1869. In 1877 went as surgeon to King's College, London. In 1897 was raised to the peerage as Baron Lister and received the Order of Merit 1902. Pres. Roy. Soc., 1895–1900. Died at Walmer (Kent) in 1912. Lister was an early student of bacteriology and was the first to obtain a pure culture of a bacterium, 1878. (Godlee, R. J., *Lord Lister*, Lond. 1917; 3rd ed. 1924; Wrench, G. T., *Lord Lister, his life and work*, Lond., [1913]; 'Lister as a pathologist and bacteriologist', *Brit. M.J.*, 1927, i. 654 [W. Bulloch].)

LOEFFLER, FRIEDRICH [AUGUST JOHN] (born 1852, died 1915). German bacteriologist and associate of R. Koch. Born at Frankfurt a. O., the son of a military surgeon. Studied in the Kais. Wilhelm Akademie in Berlin. Served in Franco-German war. In 1879 became Koch's assistant in the Kais. Gesundheitsamt and remained till 1884. In 1888 was made Prof. of Hygiene in Greifswald and ultimately succeeded Gaffky as Director of the Institut für Infektionskrankheiten in Berlin. Served in Great War and obtained Iron Cross. Died in Berlin after an operation, 1915. Loeffler was a most painstaking and accurate bacteriologist and one of the principal names connected with bacteriology. He discovered the diphtheria and glanders bacilli, and worked successfully at the attenuation of pathogenic virus, disinfection, water bacteriology, typhoid of mice, and many other subjects. He was a great technician and introduced the method of staining cilia, the preparation of blood serum for diphtheria cultures, Loeffler's methylene blue. He recognized the filter-passing character of the virus of foot and mouth disease and spent years trying to effect a cure of it. Loeffler was an excellent teacher, linguist, and writer. He founded the 'Centralblatt für Bakteriologie'. (*Centralbl. f. Bakteriol.*, 1. Abt. 1915, lxxvi. 241 [Abel]; *Deutsche med. Wchnschr.*, 1915, xli. 593 [Gaffky]; *Parasitology*, Lond., 1924–5, xvi. 233 [portrait] [Nuttall].)

LUBARSCH, OTTO (born 1860, died 1933). Pathological anatomist. Underwent very wide training in medicine and pathology. In 1894 became Prof. of Pathology in Rostock, and finally succeeded Orth as Director of the Pathol. Anat. Institute in Berlin. Lubarsch was a voluminous writer and founded and edited with Ostertag, *Ergebnisse der allgemeinen Pathologie und pathologischen Anatomie*, 1896– ?. He published important work on immunity and phagocytosis.

LUSTGARTEN, SIGMUND (born 1858, died 1911). An Austrian dermatologist. For a short time (1884–5) gained notoriety with discovery of reputed 'Syphilis bacillus'. Went to America and was on the staff of Mount Sinai Hospital, N.Y. (*J. Am. M. Ass.*, 1911, lvi. 439; *J. Cut. Dis. incl. Syph.*, 1911, xxix. 254.)

LUSTIG, ALESSANDRO (born 1857, died 1937). Italian bacteriologist and pathologist. Born in Trieste. Prof. of General Pathology in Florence. Studied plague in India and introduced a method of preventive inoculation against this disease. Visited S. America for the study of leprosy, malaria, and other infective diseases. Served in the Great War.

MacCONKEY, ALFRED THEODORE (born 1861, died 1931). English bacteriologist. Born in Liverpool. Educated at Cambridge and Guy's Hospital, London. Worked at bacteriology in Liverpool and at Leeds. In 1901 he joined the staff of the Lister Institute and became head of the antitoxin department, ultimately retiring in 1926. He published a number of important bacteriological researches.

McFADYEAN, JOHN (born 1853). English veterinary pathologist and bacteriologist. Born in Wigtownshire and educated at Edinburgh University and Edinburgh Veterinary College. M.B. Edinburgh, Principal of Royal Veterinary College, London, 1894–1927. Carried out many important researches on diseases of domestic animals. Founded and edited the *Journal of Comparative Pathology and Therapeutics*, 1888–1938. Knighted 1905.

MACFADYEN, ALLAN (born 1860, died 1907). English bacteriologist. Born in Glasgow. Educated in Edinburgh and subsequently in Germany, where he was a pupil of Flügge, Nencki, and Pettenkofer. Became Director of the Lister Institute in London. Worked especially on the effect of low temperatures on bacteria and on the technique of disintegrating bacteria. (*J. Hyg.*, 1907, vii. 319 [portrait]; *Lancet*, 1907, i. 696; *Nature*, 1906–7, lxxv. 443.)

McWEENEY, EDMOND JOSEPH (born 1864, died 1925). Born in Dublin and educated at the Catholic University there. Studied medicine in Dublin, Vienna, and Berlin. Was Prof. of Pathology and Bacteriology in Catholic Univ. School of Medicine in Dublin. Wrote extensively on bacteriological subjects. Died of paralysis agitans, aged 61. (*J. Path. and Bacteriol.*, 1925, xxviii. 700 [portrait].)

MADSEN, THORVALD (born 1870). Born in Frederiksberg, Denmark. Studied in Copenhagen. Assistant in Univ. Laboratory for Medical Bacteriology in Copenhagen, and later worked with Ehrlich in Frankfort and at the Pasteur Institute, Paris. Director of Statens Serum Institute in Copenhagen since 1902, and President of the Hygienic Commission of League of Nations.

MAFFUCCI, ANGIOLO (born 1847, died 1903). Italian morbid anatomist and bacteriologist. Was Prof. of Path. Anatomy in Pisa. Best known through his work on the bacillus of avian tuberculosis, which he was the first to cultivate (1890). (*Sperimentale. Arch. di biol.*, Firenze, 1903, lvii. 817.)

MARTEN, BENJAMIN. Nothing is known of this writer except his remarkable book published in London in 1720 entitled *A new theory of Consumptions* . . . In this he anticipated almost every doctrine held on tuberculosis to-day. Marten was a doctor of medicine. (*Janus*, 1911, xvi. 81.)

MARTIN, CHARLES JAMES (born 1866). English physiologist and bacteriologist. Born in London. Educated at King's College and

St. Thomas's Hospital. Went to Australia and was Prof. of Physiology in Melbourne. Returned to London as Director of the Lister Institute. Studied plague in India and made many contributions to bacteriological subjects. Knighted 1928.

MARTIN, SYDNEY [HARRIS COX] (born 1860, died 1924). English physician and physiological chemist. Born in Jamaica. Educated at Univ. College, London, where he became Prof. of Pathology. Physician at Univ. College Hospital. He early investigated bacterial poisons and attempted to isolate toxins from the tissues. He served on the English tuberculosis commission. He was a man held in high repute. (*Brit. M. J.*, 1924, ii. 647; *J. Path. and Bacteriol.*, 1925, xxviii. 698.)

MEISSNER, GEORG (born 1829, died 1905). Born in Hanover and studied in Göttingen. In 1851 was in Trieste, but returned to Göttingen. In Berlin he was a pupil of Johannes Müller, and in 1855 became Prof. of Anatomy and Physiology in Basel. Becoming more and more a physiologist he succeeded Rudolf Wagner in Göttingen. Discovered the submucous nerve plexus in alimentary canal. Famous for his preservation of whole organs from putrefaction without the use of disinfectants. (*Deutsche med. Wchnschr.*, 1905, xxxi. 758 [portrait]; *Berl. klin. Wchnschr.*, 1905, xlii. 488; *Arch. f. d. ges. Physiol.*, 1905, cx. 351–99; *München. med. Wchnschr.*, 1905, lii. 1206.)

METCHNIKOFF, ÉLIE [Russian ILYA ILYICH] (born 1845, died 1916). Russian zoologist, embryologist, and pathologist. Founder of the phagocytic theory of immunity. Born in province of Charkow (Russia). Educated at Univ. of Charkow. Later, studied at Giessen and Naples. Became Prof. in Odessa. Worked in Messina. Became Director of Bacteriological Institute in Odessa 1886, but left in 1887 and went to Paris, where he resided till the end of his life. Subdirector of the Pasteur Institute. He received the Copley medal of the Royal Society in 1906. Metchnikoff was a prolific worker in many different branches and exercised a great influence on the development of the doctrines of immunity. He received the Nobel Prize 1908. (Metchnikoff, Olga, *Vie d'Élie Metchnikoff*, Paris, 1920, 272 pp.; *Ann. de l'Inst. Pasteur*, 1915, xxix. 357–63 [Roux]; *J. Path. and Bacteriol.*, 1917, xxi. 215 [portrait]; *Proc. Roy. Soc. Lond.*, 1917, B. lxxxix, pp. li–lix [Ray Lankester].)

MIGULA, WALTER (born 1863). German botanist and bacteriologist. Prof. of Botany in technical school at Karlsruhe. Wrote a large *System der Bakterien* (1897–1900) and several other works.

MILLER, W(ILLOUGHBY) D(AYTON) (born 1853, died 1907). American dentist. Born Ohio. Studied at Ann Arbor, Edinburgh, and Berlin, where he settled and became Professor. Established a large practice there and published many papers and a book *The Micro-organisms of the Human Mouth*. He was one of the pioneers on dental bacteriology. After twenty-three years' residence in Berlin he returned to the United States to superintend the dental department of the University of Michigan, but died (1907) after an operation for appendicitis at Newark, Ohio, before he took up his new duties. (*Cor.-Bl. f. Zahnärzte*, Berl., 1908, xxxvii. 1–23 [portrait]; *Dental Brief*, 1908, xiii. 131, 162.)

MIQUEL, PIERRE (born 1850, died 1922). French bacteriologist. Born at Montmiral (Tarn). Studied pharmacy at Toulouse 1874, and in Paris. Doctor of medicine, 1883. Made elaborate investigations on the bacteria of air, water, and soil, and became the authority on the subject. Wrote *Les Organismes vivants de l'atmosphère*, 1883. For many years he worked in Paris at the Montsouris Observatory and published important Annals from this institution. He edited the *Ann. de micrographie*. (*Paris méd.*, 1922, xliv [annexe], p. 259 [portrait].)

MORAX, VICTOR (born 1866, died 1935). Born in Switzerland. Studied at Freiburg i. B. and at Paris. Became ophthalmic surgeon at the Hôpital Lariboisière, Paris, and was also associated with the Pasteur Institute. The Morax diplobacillus of conjunctivitis is named after him.

MORGENROTH, JULIUS (born 1871, died 1924). German bacteriologist and immunologist. Born in Bamberg. Studied with Weigert in Frankfurt a. M. and came in touch with Ehrlich, with whom he was closely associated first at Steglitz and then in Frankfort. With Ehrlich he published classical researches on Haemolysis. Later he went to the Berlin Pathological Institute, and finally to the chemotherapeutic department of the Institute 'Robert Koch' in Berlin. Here he published important work on the chemotherapy of bacterial infections, and made several discoveries. Morgenroth was a man of high standing and enriched the science of immunology by his work. Died of pernicious anaemia, aged 53. (*Med. Klin.*, 1925, xxi. 38 [F. Kraus]; *Deutsche med. Wchnschr.*, 1925, li. 159 [Neufeld] [portrait]; *Naturwissenschaft*, 1925, xiii. 157 [H. Sachs].)

MÜLLER, OTTO FRIDERICH (born 1730, died 1784). Great Danish naturalist. Born in Copenhagen of German parents, his father being court trumpeter. Educated at Ribe (Jutland) and Copenhagen, where he studied theology and law. Began natural history studies in Denmark and travelled abroad. On his return he entered the Civil Service and in 1771 became Archive secretary. Retired in 1773 with title of Justitsrad (1773) and Konferensrad (1781). In 1773–4 he published his *Vermium terrestrium et fluviatilium ... succinta historia*. His principal work of bacteriological interest *Animalcula infusoria et marina* . . . was published in 1786 posthumously, and in it he described 379 species and figured them in 50 plates. This is an important work in systematology and some of the names he employed are in use to-day. (Christensen, C., 'Otto Friderich Müller specielt som Botaniker'. *Naturens Verden*, 1922, vi. 49.)

MYERS, WALTER (born 1872, died 1901). English writer on immunology. Born in Birmingham. Educated at Cambridge and St. Thomas's Hospital, London. Worked with Ehrlich and studied cobra venom and protein immunity. Went out to Brazil on Yellow Fever Expedition and died in Para of malignant yellow fever. (*J. Hyg.*, 1901, i. 285 [portrait]; *Brit. M. J.*, 1901, i. 310.)

VON NÄGELI, CARL [WILHELM] (born 1817, died 1891). Great and many-sided botanist. Born near Zürich. From 1841 to 1857 Prof. of Botany in Zürich, and from 1857 to 1880 in Munich. In bacteriology he opposed Cohn's doctrine of the constancy of species. (*Proc. Roy. Soc.*, Lond., 1892, li, pp. xxvii–xxxvi.)

NEEDHAM, JOHN TURBERVILLE (born 1713, died 1781). Catholic divine and man of science. Born in London, and educated at Douai. Lived for years in Paris, where he collaborated with Buffon and developed his ideas on spontaneous generation. He was an opponent of Spallanzani. In later life he was director of the Imperial Academy in Brussels and died there. F.R.S. 1746.

NEELSEN, FRIEDRICH [CARL ADOLF] (born 1854, died 1894). Born in Holstein. Studied in Leipzig, where he became assistant in Pathological Institute 1876–8. Later Prof. in Rostock 1878–85, and finally Prosector in Stadt-Krankenhaus in Dresden, where he died. Wrote a good deal on pathological anatomy and bacteriology. The so-called Ziehl-Neelsen method was in reality Neelsen's. (*Arch. f. path. Anat.*, 1895, cxxxix. 564.)

NEGRI, ADELCHI (born 1876, died 1912). Born in Perugia. Educated in Pavia. Assistant to Golgi. Negri wrote on haematology, cytology, malaria, dysentery, and protozoa. In 1903 he began his work on rabies which led him to his discovery of the bodies since known by his name. (*Parasitology*, 1912, v. 151 [portrait]; *München. med. Wchnschr.*, 1912, lix. 712.)

NEISSER, ALBERT (born 1855, died 1916). German dermatologist and bacteriologist. Born in Schweidnitz. Studied in Breslau, where he became Director of the Dermatological Institute. Travelled in Norway and Spain to study leprosy. Confirmed and extended Hansen's work on leprosy bacillus. Discovered the gonococcus. Later became a recognized authority on syphilis. Made experiments on syphilis in monkeys in Java, and was associated with Wassermann in the discovery of the 'Wassermann reaction'. (*Brit. M. J.*, 1916, ii. 410.)

NEISSER, MAX (born 1869, died 1938). Hygienist and bacteriologist. Educated in Breslau. Became Director of the municipal hygienic institute in Frankfurt-a.-M. Wrote much on bacteriology and introduced important technical processes in connexion with staining and bacterial cultivation.

NENCKI, MARCEL (born 1847, died 1901). Pioneer in bacteriological chemistry. Born near Kalisch (Poland) and studied in Cracow and in Germany. In 1872 was chemical assistant in pathological institute in Berne. Later Prof. in St. Petersburg, where he died. A voluminous and original worker in biochemistry. Made extended study of chemistry of putrefaction. (*Arch. internat. de pharmacod.*, 1902, x. 1–24 [portrait]; *München. med. Wchnschr.*, 1901, xlviii. 1971.)

NETTER, ARNOLD (born 1855, died 1936). French clinician and bacteriologist. Born in Strassburg. Prof. agrégé in the Faculty of Medicine, Paris. Physician to the Hôpital Trousseau, Paris. A voluminous writer and original worker on bacteriological subjects, such as pneumonia, otitis, meningitis, plague, cholera, typhus fever, &c. (*Compt. rend. Soc. de biol.*, 1936, cxxi. 923.)

NEUFELD, FRED (born 1869). Born in Neuteich. Studied in Tübingen, Königsberg, Berlin, Heidelberg. He worked in Koch's Institute and in the Reichsgesundheitsamt in Berlin. In 1917 he became Director of the Institut für Infektionskrankheiten, Berlin, but was deprived of his post although permitted to continue his work there. Neufeld published many important researches, especially on pneumococcus.

NICOLAIER, ARTHUR (born 1862). Generally regarded as discoverer of tetanus bacillus (1884). Born in Kosel (Silesia), was assistant and Prof. in the Medical Klinik in Göttingen. Titular Prof. in Berlin. N. introduced urotropin into therapeutics.

NICOLLE, CHARLES-[JULES-HENRI] (born 1866, died 1936). Born at Rouen, and studied medicine in Paris. After holding various medical appointments in Paris and Rouen, he was in 1903 appointed Director of the Pasteur Institute in Tunis and carried out much scientific work there. In 1909 he proved that typhus fever is transmitted by lice. He also studied rinderpest, Malta fever, measles, and scarlet fever. Received Nobel Prize 1928. (*Compt. rend. Soc. de biol.*, 1936, cxxi. 797.)

NOCARD, EDMOND-[ISIDORE-ÉTIENNE] (born 1850, died 1903). French bacteriologist and veterinary pathologist. Born at Provins (Seine-et-Marne). Studied at Veterinary School at Alfort, where he became professor and in 1887 Director. A prolific writer and worker in bacteriology in connexion with bovine peripneumonia, glanders, tuberculosis, rabies, foot-and-mouth disease, sheep-pox, and other diseases. (*J. Hygiene*, 1903, iii. 517 [portrait].)

NOGUCHI, HIDEYO (born 1876, died 1928). Eminent bacteriologist. Born near Fukushima (Japan). Graduated in Tokyo Imperial

University 1897. Studied in Pennsylvania and Copenhagen. Was an indefatigable and successful worker and added to our knowledge in connexion with syphilis, snake venom, rabies, infantile paralysis, and spirochaetoses. Died of yellow fever in W. Africa 1928, aged 52. (*J. Am. M. Ass.*, 1928, xc. 1727; *Noguchi*, by G. Eckstein, New York and London, 1931, 419 pp.)

OBERMEIER, OTTO HUGO FRANZ (born 1843, died 1873). Studied in Berlin and was assistant in the Charité Krankenhaus. Discovered the micro-organism of relapsing fever 1873. Went into practice in Berlin and died of cholera 20 August 1873. (*Arch. f. Geschichte der Med.*, 1923, xv. 161 [portrait].)

OGSTON, ALEXANDER (born 1844, died 1929). Pioneer worker on the bacteriology of suppuration 1880. Born in Aberdeen. Studied there and in Vienna, Prague, and Berlin. Assistant Surgeon, and Senior Surgeon to the Aberdeen Royal Infirmary. From 1882 Prof. of Surgery in the Univ. of Aberdeen. Ogston first identified and named Staphylococcus and differentiated it from Streptococcus. Knighted 1912. (*Lancet*, London, 1929, i. 309–10.)

OMELIANSKY, VASIL LEON (born 1867). Director of section of gen. microbiology in Inst. Exp. Med. in Leningrad. He worked chiefly on general microbiological subjects and made important contributions.

ORTH, JOHANNES (born 1847, died 1923). German pathologist and early worker in bacteriology. Born in Wallmerod (Nassau). Was assistant to Virchow and succeeded him in Berlin after having been many years Prof. in Göttingen. (*Centralbl. f. allg. Path. u. path. Anat.*, 1922–3, xxxiii. 425.)

PACINI, FILIPPO (born 1812, died 1883). Italian anatomist, histologist, and student of cholera. Born in Pistoja. Studied medicine in Florence, where, in 1835, at the age of 23, he made known the tactile corpuscles since called after him. In 1849 was Prof. of Histology in Florence. He carried out important microscopic examinations on cholera dejecta in the 1854–5 epidemic and found numerous bacteria which he alleged caused cholera. (*Arch. ital. de biol.*, 1883, iv. 123.)

PALTAUF, RICHARD (born 1858, died 1924). Austrian pathologist and bacteriologist. Born in Judenburg. Educated in Graz and Vienna, where he was assistant to Kundrat. Became Prosector in the Rudolf Hospital in Vienna. He took up bacteriology and founded a large institute for the study of immunity and production of therapeutic sera. A man of encyclopaedic learning, he wrote extensively and critically on agglutination and other subjects. (*Centralbl. f. allg. Path. u. path. Anat.*, 1923–4, xxxiv. 609; *Ztschr. f. Immunitätsforsch.*, 1924, xli. 1.)

PANUM, PETER LUDWIG (born 1820, died 1885). Danish pathologist, physiologist, and experimenter. Born in Rönne (Born-
holm). Studied in Kiel and Copenhagen. Investigated famous
measles epidemic in Faroë Isles 1846. Became pupil of Virchow.
Later he took up physiological chemistry and visited Würzburg,
Leipzig, and Paris, where he was for a time assistant to Claude
Bernard. He then became Prof. of Physiology in Kiel and in 1863
in Copenhagen. He published many important works on embolism,
transfusion of blood, and became widely known through his studies
of the nature of septicaemia which he attributed to a chemical poison.
He died of heart rupture from coronary embolism in 1885. (Portrait
in Salomonsen's *Smaa-Arbejder*, Kjøbenhavn, 1917, p. 345.)

PASTEUR, LOUIS (born 1822, died 1895). Great French chemist
and bacteriologist. Born, the son of a tanner, at Dôle (Jura) 27 Dec.
1822. Spent early life at Arbois, and was educated there and at
Besançon and the École normale, Paris. Was Prof. of physics at the
Lycée of Dijon 1848, and of chemistry in Strassburg 1852. Dean of
the faculty of science at Lille 1854. Director of Studies in the École
normale, Paris (rue d'Ulm). In 1848 he discovered the true nature
of tartaric acid and revealed the connexion between right and left
handedness of crystalline form (enantiomorphism) and optical
activity. Received Rumford medal of the Royal Society (1848) for
his discovery of the nature of racemic acid, and its relation to polar-
ized light. Pasteur carried out epoch-making researches demonstrating the connexion between various fermentations and the activity
of living micro-organisms. Lactic fermentation 1857, alcoholic
fermentation 1858-60, butyric fermentation 1861, acetic fermentation 1861-4. *Études sur le vin*, 1866; *Études sur la bière*, 1876. His
Les maladies des vers à soie dates from 1865 to 1870. By his researches
on spontaneous generation (1860-1) he destroyed this ancient
belief. In 1877 he turned his attention to the study of the causes
and prevention of infective diseases in man and animals. He discovered the protective properties of attenuated virus in fowl cholera
1880, anthrax 1881, swine erysipelas 1882, rabies 1884. He was
elected a foreign member of the Royal Society 1869 and received its
Copley medal 1874. In his honour the Pasteur Institute in Paris
was founded 1888. He died at Villeneuve l'Etang, near Garches, on
28 Sept. 1895, and received a State funeral with military honours.
He was buried in a magnificent crypt in the Institut Pasteur, Paris.
Statues of him have been erected in Dôle, Arbois, Besançon, Lille,
Alais, Melun, Chartres, Marnes, and three in Paris. To his memory
a great monument was inaugurated in Strassburg in 1923. [Vallery-
Radot, R.] M. *Pasteur: histoire d'un savant par un ignorant*, Paris,
1883, 5ᵉ ed. 1884, English transl. by Lady Claud Hamilton under
title *Louis Pasteur, his life and labours*, New York, 1885; Vallery-
Radot, R., *La Vie de Pasteur*, Paris, 1900, 692 pp., English transl.

by Mrs. R. L. Devonshire, Westminster, 1902, 2 vols.; another edit. 1919; Duclaux, E., *Pasteur, Histoire d'un esprit*, Paris, 1896, 400 pp., English transl. by Erwin F. Smith and Florence Hedges, Philad. and Lond., 1920, 383 pp.; Frankland, Percy, and Mrs. Frankland, *Pasteur*, Lond., Paris, and Melb., 1898, 224 pp.; Descour, L., *Pasteur et son œuvre*, Paris, 1921, 296 pp., English transl. by A. F. and B. W. Wedd, Lond., 1922, 256 pp.; Holmes, Samuel J., *Louis Pasteur*, Lond. [1925], 246 pp.; Bordet, J., *La vie et l'œuvre de Pasteur*, Bruxelles, 1902.)

PERTIK, OTTO (born 1852, died 1913). Prof. of Pathological Anatomy in Budapest. Born in Budapest. Pupil of Ranvier De Bary and Koch. Assistant to von Recklinghausen. Pertik published much of a high class character on pathology and bacteriology.

PERTY, [JOSEPH ANTON] MAXIMILIAN (born 1804, died 1884). German naturalist. Born in Ornbau (Bavaria). Studied science and medicine in Munich and Landshut. In 1833 became Prof. of Zoology, Psychology, and Anthropology in the Academy of Berne. Wrote a great deal on the most diverse subjects. In his *Zur Kenntniss kleinster Lebensformen* . . . 1852, he attempted a classification of bacteria and was one of the first to observe the presence of bacterial spores.

PETRI, RICHARD JULIUS (born 1852, died 1921). Born in Barmen. Was curator of the Hygiene Museum in Berlin 1886. Assistant in Koch's Institute, where he invented his famous 'plate'. Later he was a member of the Reichsgesundheitsamt.

PETRUSCHKY, JOHANNES [THEODOR WILHELM] (born 1863). German bacteriologist. Born in Königsberg. Studied there 1882–7. Later, he worked under Baumgarten and Koch. From 1891 to 1897 assistant in Koch's Institute in Berlin. Director of municipal institute of bacteriology in Danzig. Published papers on differential diagnosis of typhoid bacillus, on streptococcus, tuberculin, and immunological subjects.

PFEIFFER, AUGUST (born 1848, died 1919). German bacteriologist and hygienist. Pupil of Koch and Flügge. Was for many years medical officer of health in Wiesbaden. Published several important bacteriological observations on tubercle bacilli in lupus 1883, typhoid bacilli in stools 1885. Discovered *B. pseudo-tuberculosis rodentium* 1889.

PFEIFFER, RICHARD [FRIEDRICH JOHANNES] (born 1858). Born at Zduny (Posen) 27 March 1858. Educated at Schweidnitz and in Berlin. Military doctor, worked at bacteriology in Wiesbaden with A. Pfeiffer. At Hygienic Institute in Berlin with Koch 1887. Prof. 1894. Prof. of Hygiene in Königsberg 1899–1909. Prof. at Breslau 1909–26. Served as hygienist with rank of General in the Great

War 1914–17. Pfeiffer made many fundamental discoveries in bacteriology. Discovered influenza bacillus 1892. Traced the development of coccidium oviforme 1892. Studied immunity reactions of typhoid and cholera. Discovered specific lysis of typhoid and cholera. Discovered *M. catarrhalis* 1896. Immunized human beings against typhoid 1896. Served on the German Plague Commission in India 1896, and Malaria Commission in Italy. Published with C. Fraenkel photographic atlas of Bacteria. (*Centralbl. f. Bakteriol.*, 1. Abt., 1928, cvi. 1–6; *München. med. Wchnschr.*, 1928, lxxv. 524.)

PINEL, PHILIPPE (born 1755, died 1826). French physician. Born at Lavaur (Tarn). Studied theology and then medicine in Toulouse, Montpellier, and Paris. Became physician to the Bicêtre Hospital and the Salpetrière, and became famous for his humane treatment of lunatics. Later he was Prof. of Internal Pathology in Paris. His *Nosographie philosophique* (1789) was a landmark in the history of medical doctrines.

(VON) PIRQUET, CLEMENS (born 1874, died 1929). Austrian physician and experimental pathologist. Born in Vienna and educated there. Assistant under Escherich in clinic for children's diseases. Prof. in Johns Hopkins Hospital, Baltimore. Prof. in Breslau 1910 and Vienna 1911. An original investigator who greatly extended ideas on infectious diseases. Discovered the cutaneous tuberculin reaction known by his name. Introduced the idea of allergy. Invented a new system of alimentation 1917–19. He died of a misadventure.

PLAUT, HUGO CARL (born 1858, died 1928). Born in Leipzig. Studied agriculture and medicine in Leipzig, Kiel, Paris, and Vienna. Turned to exhaustive study of moulds and fungi of disease. In 1897 went to Dermatological Institute in Hamburg. In 1912 founded an institute for the study of moulds and fungi there. A recognized authority on the subject.

VON PLENCIZ, MARCUS ANTONIUS (born 1705, died 1786). Studied in Vienna and in Padua (under Morgagni). Practised in Vienna. Held advanced views on contagium animatum in a work published in 1762.

POLLENDER, FRANZ ALOYS ANTOINE (born 1800, died 1879). He gave the first clear account of the vegetable rods (*Bacillus anthracis*) in the blood and organs of animals dead from splenic fever. He was born in 1800, studied in Bonn and graduated there 1824. Practised first in Lindlar and then, till 1870, as Sanitätsrath in Wipperfürth (Elberfeld-Barmen). Retired to Schaerbeck (Brussels) and died in poverty in his native place in Barmen 1879. On 28 July 1929 a tablet was placed on his house, 22 Hochstrasse, Wipperfürth, with the inscription 'In diesem Hause entdeckte 1849 Dr. Aloys Pollender

(1800–1879) den Milzbrandbazillus'. (*Zentralbl. f. Bakteriol.*, &c., *I. Abt. Orig.*, 1929, cxv. 1–17 [portrait].)

POUCHET, Félix-Archimède (born 1800, died 1872). French physician and naturalist. Born, lived, and died in Rouen, where he was Director of the Natural History Museum. Wrote several works in natural history but is mainly known as the chief opponent of Pasteur on the question of spontaneous generation. It was by the publication of Pouchet's *Hétérogénie* (1859) that Pasteur took up the subject experimentally. Pouchet died in Rouen and a statue to him was erected there in 1877. (See Roger, Jules, *Les Médecins normands*, 1890, i. 221–9 [portrait of Pouchet].)

POWER, Henry (born 1623, died 1668). Physician and naturalist. Educated at Cambridge. F.R.S. 1663. Published in 1664 *Experimental Philosophy*, the first English work on microscopic observations.

PROSKAUER, Bernhard (born 1851, died 1915). German chemist, bacteriologist, and hygienist. Worked with Koch in the Kaiserl. Gesundheitsamt and Inst. f. Infektionskrankheiten in Berlin. In 1907 became Director of the Untersuchungsamt of Berlin. Published excellent work on bacteriology and hygiene. (*Deutsche med. Wchnschr.*, 1915, xli. 1224.)

RABINOWITSCH, Lydia (born 1871, died 1935). Born in Kovno. Prof. and Director of the Bacteriological Institute of the Moabit Hospital in Berlin-Lichterfeld. Worked extensively on the bacteriology of tuberculosis and other subjects. Married W. Kempner the bacteriologist in 1898. She died in Berlin.

RANSOM, Frederick Parlett Fisher (born 1850, died 1937). Studied at King's College, London. M.D. Edinburgh. Spent some years in Germany and was assistant to Behring and Hans Meyer in Marburg. He wrote on tetanus and diphtheria and was for a time Prof. of Pharmacology in the Univ. of London. Died, aged 87.

RATTONE, Georgio (born 1857, died 1930). Born at Moncalieri, Turin, 1857. In 1885 he became Prof. at Sassari and Prof. of General Pathology in Parma in 1886. He published work on tetanus and pneumococcus.

RAYER, Pierre-François-Olive (born 1793, died 1867). French physician and pathologist. Prof. of Comparative Medicine Univ. of Paris, and worked at pathological anatomy, epidemiology, and comparative medicine. Wrote classical work on diseases of the kidney. With Davaine, Rayer first made mention of anthrax bacillus 1850.

von RECKLINGHAUSEN, Friedrich Daniel (born 1833, died 1910). One of the chief morbid anatomists of the nineteenth century. Was assistant to Virchow and later Prof. in Königsberg,

Würzburg, and in Strassburg, where he spent the latter half of his life. He enriched almost every branch of morbid anatomy and in earlier years published important observations on the occurrence of bacteria in morbid processes.

REDI, FRANCESCO (born 1626, died 1697). Italian physician, naturalist, and poet. Born in Arezzo. Studied in Pisa. Travelled for five years and worked in the Vatican libraries as a bibliophile and linguist. Practised in Florence. Member of the Accademia del Cimento. Published his famous book *Esperienze intorno alla generazione degl'insetti* in 1688, and in it controverted the doctrine of the spontaneous generation of maggots in putrid flesh. Buried in Arezzo. (*Ann. med. History*, 1926, viii. 347.)

RIBBERT, HUGO (born 1855, died 1920). German morbid anatomist. Born in Westphalia. Educated in Bonn, Berlin, and Strassburg. Was Prof. of Pathology in Zürich, Marburg, Göttingen, and finally in Bonn, where he died of tuberculosis. In the early eighties of last century he published many papers on bacteriological subjects. (*München. med. Wchnschr.*, 1920, lxvii. 1476; *Centralbl. f. allg. Path. u. path. Anat.*, 1920, xxxi. 281.)

RICHET, CHARLES (born 1850, died 1935). Eminent French physiologist. Born in Paris. Became Prof. of Physiology there. A voluminous writer on physiological and general subjects. Nobel Prizeman 1913. In immunology he discovered and successfully worked at the problem of anaphylaxis 1902.

RITCHIE, JAMES (born 1864, died 1923). Educated at Edinburgh and taught pathology in Oxford for several years. In 1907 was Director of the Laboratory of the Royal College of Physicians in Edinburgh, and in 1913 Prof. of Bacteriology in the Univ. of Edinburgh. Wrote a good deal, especially on immunological subjects.

ROBERTS, WILLIAM (born 1830, died 1899). English physician. Educated at Univ. College, London. Was Prof. of Medicine at Owens College, Manchester, but ultimately settled in practice in London, where he was a recognized authority on renal diseases. Knighted in 1885. In 1874 he published important researches on biogenesis and sterilization by heat, and was an early and ardent supporter of the germ theory of disease. (*Brit. M. J.*, 1899, i. 1063; *Med. Chron.*, Manchester, 1899, 3rd s., i. 157–89; *Proc. Roy. Soc.*, Lond., 1905, lxxxv. 68.)

ROBIN, CH[ARLES-PHILIPPE] (born 1821, died 1885). French biologist and histologist. Was Prof. in Paris, a senator and politician. Robin was one of the early writers on histology and published a large number of papers and books on the microscope and microscopic structures. He founded the *J. de l'anat. et de physiol.*, and with Littré edited several editions of the monumental *Dict. de méd.* He was an opponent of the biological theory of fermentation and the germ theory of disease. *J. de l'anat. et de physiol.*, Paris, 1886, xxii, pp. i–clxxxiv [portrait].)

ROMANOWSKY, DIMITRI LEONIDOWITSCH (born 1861, died 1921). Born in Pskoff and studied medicine in St. Petersburg. M.D. 1891. He became Prof. of Internal Medicine and died 19 Feb. 1921 in Kislowdsk in the Caucasus. In 1891 he recommended eosin and methylene blue for staining the malaria parasite. (*Petersb. med. Zeitung*, 1891, N.F., Bd. VIII, 297, 307.)

RÖMER, PAUL HEINRICH (born 1876, died 1916). Born in Kirchhain near Marburg. He worked in Marburg with von Behring 1900–13. He took up hygiene and was Director of Hygiene Institute in Greifswald 1913, and in Halle 1915. He died of typhus fever on the Russian front in the Great War. He published much on tubercle bacilli and inoculation against cattle tuberculosis, and was author of a book on Epidemic infantile paralysis, 1913, and one entitled *Die Ehrlichsche Seitenketten Theorie*, 1904. (*Deutsche med. Wchnschr.*, 1916, xlii. 734 [portrait].)

ROSENBACH, [ANTON] J[ULIUS] [FRIEDRICH] (born 1842, died 1923.) German surgeon and bacteriologist. Born at Grohnde a. W. (Hanover). Studied in Heidelberg, Göttingen, Vienna, Paris. Prof. of Surgery in Göttingen for many years. Was an early and accurate worker in surgical bacteriology and did important work on the micro-organisms of tetanus and suppuration. His early bacteriological investigations were done in his own house as he had no laboratory. (*Deutsche med. Wchnschr.*, 1924, l, 184; Autobiog.: *Med. d. Gegenwart*, 1923, ii. 187–92 [portrait].)

ROUX, [PIERRE-PAUL-]ÉMILE (born 1853, died 1933). Eminent French bacteriologist and associate of Pasteur. Born at Confolens (Charente). Educated at Aurillac, medical school of Clermont-Ferrand and Faculty of Medicine in Paris. Préparateur in Pasteur's Laboratory from 1878 to 1883. Sub-director of Pasteur's Laboratory 1883. Chef de service in Pasteur Institute, Paris, 1888–95. Sub-director 1893–1904. Director 1904–33. Roux published conjointly with Pasteur and Chamberland classical work on anthrax and rabies. Later with Yersin he made important contributions to bacteriology of diphtheria and with Vaillard on tetanus. By himself he published many researches of the greatest importance. Successfully transmitted syphilis to monkeys. As an original worker, teacher, and inspirer Roux is regarded as the greatest French exponent of bacteriology after Pasteur. He received the Copley medal of the Royal Society in 1917. Foreign member Royal Society 1913. (*J. Path. and Bacteriol.*, 1934, xxxviii. 99–105 [portrait].)

ROWLAND, SYDNEY DOMVILLE (born 1872, died 1917). Educated at Cambridge and St. Bartholomew's Hospital, London. Was on the staff of the Lister Institute in London and successfully devised important mechanical apparatus for disintegrating bacteria at low temperatures. Studied plague in India, and in England. Went to France in the Great War in Oct. 1914 and after three years' service died of cerebro-spinal meningitis, aged 45. (*J. Path. and Bacteriol.*, 1916–17, xxi. 453; *Lancet*, 1917, i. 552.)

RUFFER, MARC ARMAND (born 1859, died 1916). English bacteriologist born of French parents. Studied in Paris and Oxford, and worked on chemotaxis and phagocytosis. Became Director of the British (Lister) Institute of Preventive Medicine and was the first to manufacture antitoxin in England, 1894. Appointed President of International Board of Quarantine in Egypt and published many reports on his work. Later he turned to palaeopathology. In the Great War he served on the Red Cross and during his duties his ship was torpedoed and he was drowned in the Mediterranean.

SACHS, HANS (born 1877). German immunologist and serologist. Born at Kattowitz (Upper Silesia). Was for years a close associate with Ehrlich in Frankfurt-a.-M. Later he became Director of the scientific department in the Inst. f. exp. Krebsforschung in Heidelberg. Sachs has been a voluminous writer and a very accurate worker in many branches of serology.

SALMON, DANIEL ELMER (born 1850, died 1914). American veterinary pathologist. Born in New Jersey. Studied at Cornell Univ. and at Alfort. Graduated as Bachelor of Veterinary Science at Cornell 1872. Practised as veterinary surgeon in Newark, N.J. In 1878 he investigated diseases of swine and Texas fever, and, 1883, established at Washington a veterinary division in the Department of Agriculture and shortly after became chief of the Bureau of Animal Industry. He held this appointment till 1905, when he became head of the Veterinary Department of the Univ. of Montevideo (Uruguay). After five years he returned to U.S.A. and took up work in the West, where he was in charge of a plant for making anti-hog cholera serum at Butte, Montana. He died there of pneumonia 30 Aug. 1914, aged 64. (*Amer. Vet. Review*, 1914–15, xlvi. 1, 93, 95.)

SALOMONSEN, CARL JULIUS (born 1847, died 1924). Danish bacteriologist and a pioneer in bacteriology. Born in Copenhagen. Studied under P. L. Panum. Published (1877) the first important bacteriological work in Denmark on the decomposition of the blood. He demonstrated bacteria in putrid blood and cultivated them in long capillary tubes. He early stained bacteria with fuchsin, and supported the views of Cohn in opposition to those of Billroth. In 1878 he made experiments with Cohnheim which established the specificity of tubercle. In 1877 founded bacteriological laboratory in Copenhagen and became Prof. of General Pathology. With Madsen he published important researches on the production of antitoxin. Salomonsen was an excellent teacher and had a profound knowledge of almost every aspect of pathology, epidemiology, and medical history, and did much to promote the interest in these subjects in Denmark. (*J. Path. and Bacteriol.*, 1925, xxviii. 702–8 [portrait].)

SANDERSON, JOHN SCOTT BURDON (born 1828, died 1905). English pathologist and physiologist. Born at North Jesmond (Northumberland). Graduated at Edinburgh Univ., where he was a pupil of Goodsir and Hughes Bennett. Studied in Paris. Was appointed medical officer of health for Paddington (London) and undertook work on infective processes and contagion for medical department of Privy Council. Later he became Prof. of Physiology in Univ. College, London 1874–83, and in Oxford 1883–95. Finally he was Regius Prof. of Medicine in Oxford 1895–1905. Burdon Sanderson was an early worker on the pathology of infection and supported the view that its cause was corpuscular rather than fluid. He later supported the 'germ theory'. (*Proc. Roy. Soc.*, Lond., 1907, B, lxxix, pp. iii–xviii [portrait]; *Brit. M. J.*, 1905, ii. 1481 [portrait]; *Lancet*, Lond., 1905, ii. 1652 [portrait]; Burdon Sanderson, Lady, *Sir John Burdon Sanderson, a memoir*, Oxf. 1911, 315 pp.

SCHATTENFROH, ARTUR (born 1869, died 1923). Austrian bacteriologist and hygienist. Born in Salzburg. Studied in Vienna with Gruber and Paltauf. Became assistant and successor to Gruber as Prof. of Hygiene in Vienna. Wrote extensively on anaerobic bacteria, esp. *B. butyricus* and its relations. (*Wiener med. Wchnschr.*, 1923, lxxiii. 1889.)

SCHICK, BELA (born 1877). Inventor of Schick test for diphtheria. Born in Boglar, Hungary, in 1877. Studied at Graz. From 1902 to 1923 was connected with the Children's Department of the Univ. of Vienna, as assistant, privat docent, and in 1918 as Prof. of Pediatrics. In 1923 was called to New York City as Director of the Children's Department of Mount Sinai Hospital. Schick wrote with von Pirquet important work on serum sickness and published his test for diphtheria in 1913.

SCHIMMELBUSCH, C[URT] (born 1860, died 1895). German pathologist and surgeon. Born in West Prussia. Studied medicine in Würzburg, Göttingen, Berlin, and Halle, where he was assistant to Eberth. Took up surgery and was assistant to von Bergmann for several years. He published several bacteriological papers and on the aseptic treatment of wounds. With Eberth he wrote a classical work *Die Thrombose*, 1888. Died of thrombosis following septic infection, aged 35. (*Deutsche med. Wchnschr.*, 1895, xxi. 524.)

SCHÖNLEIN, JOHANN LUCAS (born 1793, died 1864). German clinician and pathologist. Born in Bamberg. Taught in Würzburg,

Zürich, and finally in Berlin 1839–59. In 1839 he discovered the parasitic fungus Achorion Schönleinii in certain 'Impetigines' (Favus).

SCHOTTELIUS, MAX (born 1849, died 1919). German pathologist, bacteriologist, and hygienist. Born in Brunswick. Trained as morbid anatomist under Rindfleisch. Took to bacteriology and studied in Berlin, Munich, and Paris. Became Prof. of Hygiene in Freiburg-i.-Br. Wrote on cholera, bacillus prodigiosus, and other bacteria.

SCHOTTMÜLLER, HUGO (born 1867, died 1936). German physician and bacteriologist. Born at Trebbin (Brandenburg). Became Prof. and Director of the Medical Polyclinic in Hamburg-Eppendorf. Wrote extensively on bacteriology applied to clinical medicine. In 1900 he discovered paratyphoid fever and worked out its aetiology. (*München med. Wchnschr.*, 1936, lxxxii. 1097–9.)

SCHRÖDER, HEINRICH GEORG FRIEDRICH (born 1810, died 1885). Carried out, partly in conjunction with von Dusch (1854), classical researches on the protection of putrescible fluids in the presence of air filtered through cotton-wool. Was for some years School Director in Mannheim and died in Carlsruhe.

SCHROETER, JOSEPH (born 1835, died 1894). Was a military doctor and privat docent in Breslau, where he worked with Ferdinand Cohn. He was a distinguished mycologist and bacteriologist, and was one of the first to differentiate cultures of chromogenic bacteria by culture on solid media. (*Arch. f. path. Anat.*, 1895, cxxxix. 587.)

SCHULZE, FRANZ (born 1815, died 1873). Worked in Mitscherlich's laboratory in Berlin and carried out important experiments on the protection of putrescible fluids in the presence of purified air. Became Prof. of Chemistry in Rostock and wrote extensively on agricultural chemistry.

SCHÜTZ, [JOHANN] WILHELM (born 1839, died 1920). German veterinary pathologist. Born in Berlin, where for many years he was Prof. of Pathological Anatomy in the Kgl. Tierärztl. Hochschule. Was associated with Loeffler and with R. Koch in their researches on glanders and tuberculosis, and made many other important contributions to the pathology of diseases in animals. (*Berl. tierärztl. Wchnschr.*, 1920, xxxvi. 624.)

SCHWANN, THEODOR (born 1810, died 1882). Great German physiologist, pathologist, and experimenter. One of the founders of the cell doctrine and of the idea of the living nature of yeast. Born at Neuss, near Düsseldorff. A catholic, educated in the Jesuit gymnasium in Cologne. Intended for the church but took to medicine. He was a pupil of Johannes Müller and a colleague and lifelong friend of J. Henle, the anatomist. In Berlin Schwann was Johannes Müller's assistant for five years, and it was then that he discovered pepsin,

1836. He now took up the subject of alcoholic fermentation, and proved that it was due to the vital activity of the yeast plant, 1837. His results were ridiculed by Liebig but were confirmed and amplified by Pasteur twenty years later. His famous *Mikroskopische Untersuchungen über die Uebereinstimmung in der Struktur und dem Wachsthum der Thiere und Pflanzen* (1839) was rewarded with the Copley Medal (1845) of the Royal Society. From Berlin Schwann went (1839) to be Prof. of Anatomy in Louvain, and in 1848 in a similar capacity to Liège, where he spent the rest of his life. In his later years he published little. Schwann was in stature a minute man of genial appearance. He was a bachelor and a devout catholic. (*Arch. f. mikroskop. Anatomie*, 1882, xxi, pp. i–xlix [Henle].)

SÉDILLOT, CHARLES-EMMANUEL (born 1804, died 1883). French surgeon. Born in Paris. Was Prof. of Surgery in the Val-de-Grâce Military College. A prolific writer on surgical matters. Wrote on Pyaemia (1849) and blue pus, and early took an interest in the germ theory of wounds. By him the term 'microbes' was introduced (1878). (*Arch. de méd. et pharm. mil.*, 1883, i. 294.)

SELMI, FRANCESCO (born 1817, died 1881). Chemist and toxicologist. Prof. of Pharmaceutical Chemistry in Bologna. He discovered putrefactive alkaloids or ptomaines. A very successful investigator. (*Mem. Accad. d. sc. d. Ist. di Bologna*, 1881–2, 4th s. iii. 3–7.)

SHATTOCK, SAMUEL GEORGE (born 1852, died 1924). Eminent English pathologist. Born in Camden Town, London, and was originally called Betty but took the name of Shattock. Educated at Univ. College, London. In 1884 became Curator of the Museum at St. Thomas's Hospital and later Curator of the Pathological Section of the Museum of the Royal College of Surgeons. Shattock was an industrious and sincere man who wrote a great deal on almost every branch of pathology and bacteriology, and he was held in the highest repute as a pathologist and scholar for many years in England. (*Proc. Roy. Soc.*, Lond., 1924–5, B. xcvi, pp. xxx–xxxii.)

SHIGA, KIYOSHI (born 1870). Born in Sendai, Japan. Studied at Tokyo 1892–6. Assistant to Kitasato. Worked with Ehrlich 1900–3. Director of Department in Institute for Infectious Diseases in Tokyo 1904–20. Dean of medical faculty of Keijo Imperial University. Chosen, Japan. Discovered bacillus of dysentery 1898, and worked at chemotherapeutic and bacteriological subjects with great success.

SMITH, ERWIN FRANK (born 1854, died 1927). Was chief of Laboratory of Plant Pathology in Bureau of Plant Industry, Department of Agriculture, U.S.A., and over a series of years made important contributions to infective diseases of plants.

SMITH, THEOBALD (born 1859, died 1934). American pathologist and bacteriologist. Born in Albany. Educated at Cornell Univ. and

Albany Medical College. Director of Pathological Lab. Bureau of Animal Industry, Washington, 1884–95. Prof. of Comparative Pathology, Harvard, 1896–1915. Director of Department of Animal Pathology, Rockefeller Institute, 1915–29. America's foremost bacteriologist, who enriched many fields of bacteriological science. Noted especially for pioneer work on Texas fever, immunization by dead vaccines, differentiation of tubercle bacilli, anaphylaxis, and many other subjects. He died Dec. 1934. Foreign Member Roy. Soc. 1932. Copley medallist Roy. Soc. 1933. (*J. Path. and Bacteriol.*, 1935, xl. 621–35 [portrait].)

SNOW, JOHN (born 1813, died 1858). English anaesthetist and founder of water-borne doctrine of Asiatic cholera. Born at York. Practised in London and introduced the scientific use of ether in England. He chloroformed Queen Victoria in 1853 and again in 1857. Snow made elaborate studies on the dissemination of Asiatic cholera by water 1855. (*Asclepiad*, Lond., 1887, iv. 274–300 [portrait].)

SPALLANZANI, LAZZARO (born 1729, died 1799). Famous Italian naturalist and one of the world's greatest experimenters. A man of the highest intellect and indomitable perseverance. Born at Scandiano near Modena. Educated at Reggio and Modena. Became Prof. of Greek Logic and Mathematics at Reggio at the age of 26. Transferred to Modena, where he remained eight years. Here he did his work on the regeneration of lost parts, digestion, the circulation of the blood and his epoch-making experiments destroying the doctrine of spontaneous generation. From 1769 till 1799 he was conservator of the Natural History Cabinet in Pavia and became one of the most famous men of his time. His last work, on respiration, was of monumental proportions and included the record of over 12,000 experiments, a selection of which was published after his death.

STERNBERG, GEORGE MILLER (born 1838, died 1915). American bacteriologist and hygienist. Born at Hartwick Seminary, Otsego Co., New York. Entered American Army as Assistant Surgeon 1861, and rose to be Surgeon-General. He studied yellow fever in Brazil, Cuba, and Mexico. As a bacteriologist he investigated the microbe of sputum septicaemia and wrote a large manual of Bacteriology. (*Milit. Surg. Chicago*, 1915, xxxvii. 644 [portrait]; *J. Am. M. Ass.*, 1915, lxv. 1745; Sternberg, Martha L., *George Miller Sternberg*, Chicago, 1920, 331 pp. [portrait].)

STRAUS, ISIDORE (born 1845, died 1896). Born in Alsace but after the war of 1870–1 settled in Paris, and was associated with Claude Bernard. Later he worked with Roux and Chamberland. In 1888 he succeeded Vulpian as Prof. of Experimental and Comparative Pathology in Paris. He served in the French Cholera Expedition to Egypt in 1883. Straus was a man of great activity and learning and wrote much on bacteriology. He investigated tuberculosis, the early diagnosis of glanders by the 'Straus reaction', and demonstrated the passage of anthrax bacilli from mother to foetus. A large book, *La tuberculose et son bacille*, 1895, was written by him and he was one of the editors of the *Arch. de méd. expér. et d'anat. path.* (*Arch. de méd. expér. et d'anat. path.*, 1897, ix. 1–8.)

TALAMON, CHARLES (born 1850, died 1929). French physician and early worker in bacteriology. In 1882 he found the pneumococcus and gave the first accurate description of its characters. He published classical work on appendicitis and Bright's disease. He practised in Paris and was on the staff of the Hôpital Bichat. He led an isolated life and died in obscurity 1929. (*Le Nourisson*, 1929, 17ᵉ année, pp. 126–8.)

TARASSEVITCH, LYOV ALEKSANDROVICH (born 1868, died 1927). Russian bacteriologist and epidemiologist. Pupil of Metchnikoff. Taught bacteriology in Odessa and in Moscow. Director of State Institute for serum in Moscow. After the revolution he became President of the Medical and Scientific Advisory Council to the Soviet Government. T. published much on phagocytosis and phagocytic immunity, tuberculosis, and other subjects. After the War he went through the great typhus epidemic in Russia. Died in Germany. (*Seuchenbekämpfung*, 1927, iv. 192; *Presse méd.*, 1927, xxxv. 845.)

TAVEL, ERNEST (born 1858, died 1912). Swiss surgeon and bacteriologist. Born near Payerne (Switzerland). Studied in Lausanne, Strassburg, Berlin, Berne, and Paris. Worked under Koch 1885. Assistant to Kocher the surgeon. Taught bacteriology in Berne for many years and practised as a surgeon there. Published papers on bacteriology in relation to surgery. (*Corresp.-Bl. f. schweiz. Aerzte*, 1912, xlii. 1269.)

TEAGUE, OSCAR (born 1878, died 1924). American bacteriologist. Killed in automobile accident near Vineland, N.Y. (*J. Immunol.*, 1924, ix. 1–5 [portrait]; *J. Am. Med. Ass.*, 1923, lxxxi. 1300.)

THUILLIER, LOUIS-[FERDINAND] (born 1856, died 1883). French bacteriologist and assistant to Pasteur, with whom he worked on the cause of swine erysipelas 1882, and protection against it 1883. He went out to Egypt on the French Cholera Expedition and died in Alexandria from the disease in its most virulent form, aged 27. (*Compt. rend. Acad. d. sc.*, Paris, 1883, xcvii. 689.)

TIZZONI, GUIDO (born 1853, died 1932). Italian bacteriologist and pathologist. Born in Pisa. Studied there and in Naples and subsequently in Berlin and Turin. In 1878 was Prof. of Pathology in Catania and from 1880 in Bologna. Wrote much on various subjects but is specially known through his researches on tetanus and its antitoxin. (Pagel's *Biog. Lex.*, Berl.-Wien, 1901; *Biochem. e terap. sper.*, 1932, xix. 353–61.)

TOUSSAINT, H. (born 1847, died 1890). He studied at the Veterinary College at Lyons and was attached to the laboratory of Chauveau. Later he was head of the laboratory of anatomy and physiology, and in 1876 was Prof. of Anatomy and Physiology in the Veterinary School at Toulouse. Between 1877 and 1880 Toussaint published several important researches on anthrax and immunity. He died of a lingering nervous disease, aged 43. (*Lyon Méd.*, 1890, lxv. 55–65.)

TRAUBE, MORITZ (born 1826, died 1894). Born in Ratibor. Younger brother of Ludwig Traube the physiologist and clinician. Moritz Traube studied under Liebig and took degree of Doctor of Philosophy. He took over his father's wine business but continued his successful work on the chemistry of fermentation. (*Arch. f. path. Anat.*, 1895, cxxxix. 571.)

TWORT, FREDERICK WILLIAM (born 1877). English bacteriologist. Educated at St. Thomas's Hospital, London. Demonstrator of bacteriology, London Hospital, 1902–9. Superintendent of the Brown Institution, London, 1909– . F.R.S., 1929. He was the first to cultivate Johne's bacillus and was the original discoverer of bacteriophagic phenomena.

TYNDALL, JOHN (born 1820, died 1893). English physicist. Born at Leighlin Bridge, Co. Carlow (Ireland). Studied at Marburg and Berlin. Became Prof. at the Royal Institution in London 1853. Colleague of Faraday, whom he succeeded as Superintendent 1867–87. Wrote extensively on natural philosophy and was an admirable popular lecturer and experimenter. In 1870 he began to interest himself in atmospheric germs and dust, and he carried out numerous exact experiments on sterilization by heat which led him to the discovery (1877) of fractional sterilization, now called Tyndallization. His famous experiments were carried out in the laboratories of the Royal Institution and in the Jodrell laboratories in the Royal Gardens at Kew. Tyndall's *Essays on the floating-matter of the air in relation to putrefaction and infection* was published 1881. Tyndall, by his lectures and writings, did much to further the teaching of Pasteur, and he was an uncompromising opponent of Pouchet and Bastian. He was accidentally poisoned and died 1893. Buried (without gravestone) in Haslemere Churchyard.

UHLENHUTH, PAUL [THEODOR] (born 1870). German bacteriologist and serologist. Born in Hanover, privat docent in Greifswald. Director of Bacteriological Department in Reichsgesundheitsamt, 1906. Prof. in Strassburg 1911, Marburg 1921, and Freiburg-i.-B. Wrote extensively on bacteriological and serological subjects, such as biological differentiation of proteins, swine fever, experimental transmission of syphilis, spirochaetal jaundice, tuberculosis, typhus fever, and relapsing fever. He made several discoveries of great practical importance. With Kraus he edited a large *Handbuch der mikrobiologischen Technik*, 1923-4, 3 vols., 2745 pp.

UHLWORM, Oskar (born 1849, died 1929). Founder and editor of the *Centralblatt für Bakteriologie*. Born in Arnstadt, Thuringia. He was Librarian of the Univ. Library in Leipzig in 1880, and from 1881 to 1901 was in charge of the Murhard Library in Kassel, and from 1901 Librarian of the Stadtsbibliothek in Berlin, with title of Professor. In 1880 he founded *Botanisches Centralblatt* and in 1887 *Centralblatt für Bakteriologie*, which in 1895 appeared in two divisions. He died 13 March 1929, aged 80. He was a quiet, learned, and courteous man.

UNNA, P[aul] G[erson] (born 1850, died 1929). Dermatologist and histologist. Born in Hamburg and practised there. A most voluminous writer who introduced important staining methods into pathological anatomy and bacteriology.

VAILLARD, Louis (born 1850, died 1935). French military surgeon and bacteriologist. Born at Montauban. Prof. of Military Hygiene at the Val-de-Grâce Military School in Paris. Medical Inspector-General of French Army. Published accurate work on pathology and serotherapy of tetanus.

VALLISNERI, Antonio (born 1661, died 1730). Italian physician and naturalist. Born at Tresilico. Educated at Modena, Reggio, and Bologna, where he was pupil of Malpighi. Became Prof. of Medicine in Padua. His collected works were published in three magnificent folio volumes (1733).

VAUGHAN, Victor Clarence (born 1851, died 1929). American biological chemist and hygienist. Born at Mount Airy, Mo. He was Director of the hygienic laboratory and Prof. of Hygiene and Physiological Chemistry in Univ. of Michigan for many years. Published with Novy a work on ptomaines. Died at Richmond, Va., aged 78.

VIALA, Eugène (born (?), died 1926). Was a laboratory attendant for many years with Pasteur and after 1888, in the Pasteur Institute in Paris, he carried out all the inoculations for the preservation of the rabies virus.

VIGNAL, William (born 1852, died 1894). French histologist and bacteriologist. Son of a French officer, who was killed in the Crimea, and of an English mother. Vignal fought and was wounded at the siege of Metz and sent to Germany as a prisoner of war. He afterwards studied at Bonn with Max Schultze and at Edinburgh with Rutherford. Returned to Paris and worked in the histological laboratory of the Collège de France under Ranvier. Vignal wrote bacteriological papers on the flora of the mouth, tuberculosis and anaerobic methods. He developed consumption and travelled abroad in search of health but died 1894, aged 42. (*Compt. rend. Soc. de biol.*, 1894, 10th s., i. 845–55.)

VILLEMIN, JEAN ANTOINE (born 1827, died 1892). Born at Prey (Vosges). Studied at Strassburg. Entered medical department of French Army 1848, was subsequently Prof. agrégé in Val-de-Grâce Military Hospital, Paris. Retired from service as medical Inspector-General. He proved the inoculability of tuberculosis 1865. (*Rev. de la tuberculose*, Paris, 1893, pp. 52–9; *Presse méd.*, 1927, xxxv. 1273; *Bull. Acad. de méd.*, 1927, xcviii. 225.)

VINCENT, HYACINTHE JEAN (born 1862). Born at Bordeaux. M.D. Member of the Academy of Sciences. Prof. at the Val-de-Grâce Hospital, Paris. Inspector-General in the French army.

WARD, HARRY MARSHALL (born 1854, died 1906). English botanist and authority on diseases of plants. Born in Hereford. Studied under Huxley in London and at Cambridge, where he became Prof. of Botany. He devoted much time to the study of bacteriology, especially of water, and did much to disseminate knowledge of Schizomycetes by his admirable article in the *Encyclopædia Britannica*. Died at Cambridge.

WASHBOURN, JOHN WICHENFORD (born 1863, died 1902). English bacteriologist and physician. Born at Gloucester. Educated at Guy's Hospital, London, Vienna, and Königsberg. Lectured at Guy's Hospital and was physician on the staff. He worked especially at the bacteriology of pneumonia. Died of miliary tubercle 1902. (*Brit. M. J.*, 1902, i. 1627; *Med.-Chir. Tr.*, Lond., 1903, lxxxvi, pp. cxvii–cxx; *Trans. Epidemiol. Soc.*, Lond., 1901–2, N.S. xxi. 151.)

VON WASSERMANN, AUGUST (born 1866, died 1925). Born in Bamberg. Studied in Strassburg, Vienna, and Berlin. He early became associated with the Institute of Infectious Diseases in Berlin, where under Koch most of his work was done. He ultimately became Director of the Serum Department of Koch's Institute, and in 1913 Director of the large Institute of Experimental Therapy in Dahlem (Berlin). Ennobled in 1910. Wassermann's life work was immunology and he investigated it in many aspects. In 1906 he attained great fame by the syphilitic reaction which bears his name. With W. Kolle he edited the large *Handbuch der pathogenen Mikroorganismen* (1903–13).

WEICHSELBAUM, ANTON (born 1845, died 1920). Eminent Austrian bacteriologist and morbid anatomist. Born at Schiltern near Langenlois (Austria). Educated in Vienna. Was trained as military surgeon. In 1893 became ordinary Prof. of Pathological Anatomy in Vienna. Weichselbaum early took up Koch's methods and made many additions to bacteriology of the greatest importance. Discovered Meningococcus 1887, and enriched our knowledge of pneumonia, tuberculosis, and other subjects. As a man, a teacher, and a scientist, he was held in high repute not only in Vienna but in the outside world. He died in Weidling near Vienna, aged 75. (*Wiener med. Wchnschr.*, 1927, lxxvii. 738; *Wiener klin. Wchnschr.*, 1920, xxxiii. 979; *Lancet*, 1920, ii. 921.)

WEIGERT, CARL (born 1845, died 1904). Famous German pathologist and histologist. Born in Münsterberg (Silesia). Educated in Breslau and Berlin. Served in Franco-German War 1870–1. Assistant to Cohnheim in Breslau and Leipzig. From 1885 to 1904 Director of Senckenberg Institute in Frankfurt a. Main. Weigert's work covered a wide range and included pathological anatomy, histology, bacteriology, neurology, and technique. He introduced many of the best staining methods used to-day in histology and bacteriology. He was a cousin of Paul Ehrlich, and was a man of the highest integrity and intellect. (Rieder, R., 'Carl Weigert und seine Bedeutung für die medizinische Wissenschaft unserer Zeit', 1906, in Weigert's *Ges. Abhandl.*, Berlin, 1906, i. 1–132 [portrait]; *Berl. klin. Wchnschr.*, 1904, xli. 938 [Salomonsen]; *Deutsche med. Wchnschr.*, 1904, xxx. 1318 [Lubarsch]. Portrait in Salomonsen's *Smaa-Arbejder*, Kjøbenh., 1917, p. 353.)

WELCH, WILLIAM HENRY (born 1850, died 1934). American pathologist and bacteriologist. Born in Norfolk (Conn.). Educated at Yale, Strassburg, Breslau, Berlin, and Leipzig. Prof. of Pathology in Johns Hopkins Medical School from 1884 to 1918. Director of School of Hygiene and Public Health at Baltimore. He did much to develop pathology and bacteriology in U.S.A. He died 1 May 1934, aged 84.

WERNICKE, E[RICH ARTHUR EMMANUEL] (born 1859, died 1929). German bacteriologist and hygienist. Trained in the Kaiser Wilhelm Academy. Assistant in Koch's Institute, where he worked especially in connexion with the early preparation of diphtheria antitoxin. He became Prof. in Marburg, and for many years was Director of the Hygienic Institute in Posen. After the War he became Director of the Institute of Hygiene in Landsberg a. W.

WERTHEIM, ERNST (born 1864, died 1920). Austrian gynaecologist and bacteriologist. Studied in Graz and Vienna, and was assistant in gynaecological clinic in Prague. Became Director of the gynaecological and obstetrical clinic in Vienna, where he became famous for his operations for uterine cancer. In bacteriology Wertheim carried out exact researches on gonococcus and was the first to show the great importance of this microbe as a cause of diseases in women. His most important paper on the subject entitled *Die ascendierende Gonorrhoe beim Weibe* was published in 1891.

WESBROOK, FRANK FAIRCHILD (born 1868, died 1918). Bacteriologist and hygienist. Born in Ontario and educated at Manitoba,

Toronto, and Cambridge. In 1895 he became Director of Pathology, Bacteriology, and Hygiene at Univ. of Minnesota, and in 1913 President of the Univ. of British Columbia. Died in Vancouver, aged 50. Wesbrook studied especially bacteriology in reference to hygiene, and wrote on the diagnosis of diphtheria and the sterilization of drinking-water.

WIDAL, [GEORGES]-FERNAND-[ISIDORE] (born 1862, died 1929). French clinician and bacteriologist. Born at Dellys (Algiers). Educated in Paris. Prof. in the Faculty of Medicine, Paris. In 1896 proposed the sero-diagnostic test for enteric fever, since known by his name.

WILLIS, THOMAS (born 1621, died 1675). English physician and anatomist. Born in Wiltshire. Practised in Oxford and in London. Wrote many works; among them one on fermentation, 1659. Buried in Westminster Abbey.

WINOGRADSKY, SERGE (born 1856). Russian bacteriologist. Made classical researches on sulphur bacteria, iron bacteria and nitrification. After the revolution in Russia he lived in France.

WOLFFHÜGEL, GUSTAV (born 1854, died 1899). Was an early associate of R. Koch in Berlin and wrote important memoirs on bacteriology and hygiene. He became Prof. of Hygiene in Göttingen.

WOODHEAD, GERMAN SIMS (born 1855, died 1921). English pathologist and bacteriologist. Born in Huddersfield. Graduated in Edinburgh 1877, where he excelled as an athlete. Became assistant Prof. of Pathology in Edinburgh, and (1890) Director of the Laboratories of the Conjoint Board of the Royal College of Physicians and Royal College of Surgeons, London. In 1899 he became Prof. of Pathology in the Univ. of Cambridge. Woodhead wrote many reports on bacteriology, especially in relation to public health, and he also published several text-books which had a wide vogue in England. He founded and edited the *Journal of Pathology and Bacteriology*. He was knighted for military services in the Great War.

WRIGHT, ALMROTH EDWARD (born 1861). Educated at Dublin University, Leipzig, Strassburg, and Marburg. Demonstrator of physiology, Univ. of Sydney, 1889. Prof. of Pathology Army Medical School, Netley, 1892–1902. Director of Institute of Pathology, St. Mary's Hospital, Paddington. Member of Indian Plague Commission, 1898–1900. He has made very numerous contributions to various aspects of bacteriology, and was a pioneer in typhoid inoculation and developed the subject of therapeutic immunization generally. He was knighted in 1906.

WRIGHT, JAMES HOMER (born 1870, died 1928). American bacteriologist and pathologist. Born at Pittsburg. Studied at Johns

Hopkins and Univ. of Maryland. Was for many years Assistant Prof. of Pathology in Harvard Univ. and Director of Pathological Lab., Massachussets General Hospital. Homer Wright was a high-class worker and made many important and accurate contributions to bacteriology, protozoology, and pathological anatomy. His principal works were on blood platelets, neurocytoma, actinomycosis, aortic syphilis, cerebro-spinal meningitis, tropical ulcer. (*Arch. Pathology*, 1928, v. 494 [portrait].)

WURTZ, ROBERT-THÉODORE (born 1858, died 1919). Son of Adolphe Wurtz the famous chemist. He studied in Paris and worked in the laboratory of Isidore Straus. In 1898 he went to Abyssinia to vaccinate the natives against variola. Later he had a private hygienic and bacteriological laboratory in Paris. (*Arch. de méd. exp. et d'anat. path.*, 1918–19, xxviii. 669–73.)

WYMAN, JEFFRIES (born 1814, died 1874). American naturalist and archaeologist. Was Prof. of Anatomy in Harvard. In 1862 he published important observations on the sterilization of infusions in connexion with spontaneous generation. (*Boston M. and S. J.*, 1924, cxci. 429.)

WYSSOKOWITSCH, VLADIMIR CONSTANTINOVICH (born ?, died 1912). Russian pathologist and early worker on bacteria. Became Prof. of Morbid Anatomy in Kieff. In Flügge's laboratory he carried out important researches on the fate of bacteria injected into the vascular system, 1886; he also studied the bacteriology of endocarditis.

YERSIN, ALEXANDRE-[EMILE-JOHN] (born 1863). French bacteriologist. Born at La Vaux (Lausanne). Surgeon in the French Colonial army. Director of the Pasteur Institute at Nha-trang (Annam). With Roux carried out classical researches on diphtheria bacillus, including discovery of diphtheria toxin 1888–90. At Hong-Kong he independently discovered *Bacillus pestis*, 1894.

ZAMMIT, THEMISTOCLES (born 1864, died 1935). Maltese bacteriologist and chemist. M.D. Malta, 1889. He was a member of the Malta Fever Commission, and was largely instrumental in discovering that the goat carried Malta fever. He played an important part in Malta and was knighted in 1930. He was also famous as an anthropologist.

ZETTNOW, HUGH OSCAR EMIL (born 1842, died 1927). Honorary member of the R. Koch Institute in Berlin. Known chiefly as an expert technician and photographer of bacteria. He invented an excellent method of cilia staining. (*Seuchenbekämpfung*, Wien, 1927, iv. 273.)

ZIEHL, FRANZ [HEINRICH PAUL] (born 1857, died 1926). Born in Wismar. While assistant in the medical clinic in Heidelberg

(1882–6) wrote on staining of tubercle bacillus. Later, nerve specialist in Lübeck.

ZINGHER, ABRAHAM (born 1885, died 1927). Born in Rumania. Educated at Cornell Univ., N.Y. Assistant Director of Bureau of Laboratories, New York City, Department of Health, and took a prominent part in the work of immunizing children against diphtheria in U.S.A. Z. was asphyxiated with gas in the laboratory of the Willard Parker Hospital, N.Y. (*Am. J. Dis. Child.*, 1927, xxxiv. 105.)

ZINSSER, HANS (born 1878). American bacteriologist. Born in New York City. Instructor of Bacteriology and Hygiene, College of Physicians and Surgeons, N.Y. Prof. of Bacteriology in Columbia Univ. and in Harvard. Has made many important contributions to bacteriology and immunology.

ZOPF, WILHELM (born 1846, died 1909). Born at Rossleben. Was for a time a school teacher at Eisleben. Studied science in Berlin and was taught botany by Brefeld. Zopf became privat docent in Halle 1883, and succeeded Brefeld in Münster, where he was Director of the botanic garden. Zopf was a prolific writer, and in bacteriology is chiefly known as an upholder of the doctrine of pleomorphism. (*Ber. d. deutschen botan. Gesellsch.*, 1909, xxvii. 58 [portrait].)

INDEX OF PERSONAL NAMES

INDEX OF SUBJECTS